U0903329

周叔子论文选

湖南大学数学学院　编

科 学 出 版 社
北　京

内 容 简 介

本书汇编了周叔子教授的部分重要学术成果，包括他在微分方程、变分不等式及最优化等领域中的理论和数值算法方面的主要成果。内容基本反映了周叔子教授在计算数学方面研究成果的概貌。此外，还选取了他指导过的学生的几篇论文，最后附上周叔子教授发表文章的目录。

本书适合从事微分方程数值解和最优化等领域的研究生和科研人员参阅。

图书在版编目（CIP）数据

周叔子论文选/湖南大学数学学院编. —北京：科学出版社, 2019.6
ISBN 978-7-03-061624-1

Ⅰ. ①周… Ⅱ. ①湖… Ⅲ. ①数学-文集 Ⅳ. ①O1-53

中国版本图书馆 CIP 数据核字 (2019) 第 114440 号

责任编辑：郭勇斌 彭婧煜／责任校对：邹慧卿
责任印制：张 伟／封面设计：无极书装

科学出版社 出版
北京东黄城根北街 16 号
邮政编码：100717
http://www.sciencep.com

北京虎彩文化传播有限公司 印刷
科学出版社发行 各地新华书店经销
*

2019 年 6 月第 一 版 开本：787×1092 1/16
2019 年 6 月第一次印刷 印张：21 1/4 插页：2
字数：496 000

定价：128.00 元

(如有印装质量问题，我社负责调换)

周叔子教授简介

周叔子(1940—2009 年)，教授，湖南长沙人，1962 年毕业于湖南大学数学力学系数学专业并留校任教，1978 年评为湖南大学数学系讲师；1980—1982 年威斯康星大学访问学者，1985 年破格晋升为湖南大学数学系教授；1992 年被国务院评为有突出贡献的专家，享受国务院政府特殊津贴；1993 年经国务院学位委员会批准，获博士生导师资格。长期从事计算数学研究工作，在微分方程数值解、变分不等式理论与算法、优化算法等方面取得了一系列重要成果，在国内外重要学术刊物上发表学术论文 100 余篇，曾获得机械工业部科技进步奖二等奖和湖南省科技进步奖二等奖；指导博士、硕士研究生 70 余人。历任湖南大学应用数学系主任、湖南大学图书馆馆长、湖南省计算数学暨应用软件学会理事长、中国数学会理事、中国数学会计算数学分会理事，是湖南大学计算数学学科的主要创始人。

序

周叔子教授是我国从事计算数学工作的著名学者，长期从事计算数学的教学和研究工作；他在微分方程数值解、变分不等式理论与算法及最优化算法等许多领域都取得了突出的成就，得到了国内外学者们的广泛认可。

周叔子教授是中国计算数学领域的重要代表人物之一，是湖南大学计算数学学科的主要创始人。我与周教授相识多年，从早期的书信、电话联系，到后来各种学术交流，以及成为短期同事，我们既是同行，也是朋友，我很享受与他交谈和探讨各种学术问题的时光。在与周教授的交流中，他给我留下了深刻的印象——周教授雅量高致、彬彬有礼，是计算数学界有名的谦谦君子；他学识渊博、思维灵活，得到同行、同事们的高度评价；他治学严谨，因材施教，深得学生们的喜爱和崇敬。

我很高兴看到《周叔子论文选》的出版。该文选涵盖了周叔子教授在不同时期、不同研究方向所做出的部分重要成果，并选择了他指导过的部分学生的代表性论文；比较全面地反映了周教授在计算数学的研究工作概貌。当然，周教授的在计算数学方面的成果远不止这些。

我相信，无论是对数学专业的本科生、研究生，还是计算数学工作者，该书都具有很好的参考价值。

石钟慈

2019 年 5 月

目　录

第三篇　指导的部分学生论文

第一篇
微分方程数值计算理论与方法

The Linear Finite Element Method for a Two-Dimensional Singular Boundary Value Problem*

S.Z. Zhou (周叔子)†

Abstract The following model problem is studied:

$$\Omega : -\left[\frac{1}{r}\frac{\partial}{\partial r}\left(r\beta\frac{\partial u}{\partial r}\right)+\frac{\partial}{\partial z}\left(\beta\frac{\partial u}{\partial z}\right)\right]=f,$$

$$\Gamma_1 : u=0$$

where Ω is a bounded open domain with r >0 in the (r,z) plane, $\Gamma_1 = \partial\Omega/\Gamma_0, \Gamma_0 = \partial\Omega\bigcap\{(r,z):r=0\}$. We introduce weighted Sobolev spaces $V^k(k=1,2)$, and prove:

(1) The problem has a unique solution u, and $u\epsilon V_0^1(\Omega)\bigcap V^2(\Omega)$.

(2) The linear finite element solution u_h exists and is unique.

(3) The error $u-u_h$ in "energy norm" is of $O(h^2)$. Particularly, if is a polygon, then

$$\|u-u_h\|_{1,\Omega}=O(h)$$

$$\|u-u_h\|_{0,\Omega}=O(h^2)$$

where$\|\cdot\|_{k,\Omega}$ $(k=1,2)$are the V^k norms.

Keywords Finite element method, two-dimensional singular boundary value problem, weighted Sobolev spaces, order of convergence.

1 Introduction

The numerical solution of singular boundary value problems have been studied by several authors. The finite difference methods and its theory for a type of two-dimensional singular boundary value problems are given in [1, 2]. The finite element method for axisymmetric elastic solids is proposed in [3]. [4 − 7], give a proof of the convergence of the finite element

* 发表于: SIAM Journal on Numerical Analysis, Vol. 20, 1983, pp. 976-984.

† Department of Applied Mathematics, Hunan University, Changsha 410082, China.

methods for one-dimensional singular problems. [8] proves the "optimal" order of convergence for the method of [3] provided the loads are axisymmetric and the solution is in $C^{k+1}(\bar{\Omega})$. The convergence of the linear finite element method for tow-dimensional singular Dirichlet problem is proved in [9]. Recently, [10] has proved that the error is of order $O(h)$ for a variant linear element including a logarithmic term. In this paper we will prove the so-called "optimal" order of convergence of the linear finite element method for the following model problem:

$$\begin{cases} \Omega : -\left[\dfrac{1}{r}\dfrac{\partial}{\partial r}\left\{r\beta\dfrac{\partial u}{\partial r}\right\} + \dfrac{\partial}{\partial z}\left\{\beta\dfrac{\partial u}{\partial z}\right\}\right] = f, \\ \Gamma_1 : u = 0 \end{cases} \tag{1.1}$$

where Ω is bounded open domain with $r > 0$ in (r, z)-plane,

$$\Gamma_1 = \partial\Omega/\Gamma_0, \qquad \Gamma_0 = \partial\Omega\bigcap\{(r, z) : r = 0\}.$$

We assume:

(i) The function β is uniformly Lipschitz continuous in Ω.

(ii) $\beta \geqslant \beta_0 > 0, \beta_0$ is a constant.

(iii) $r^{1/2} f \epsilon L^2(\Omega)$.

2 Transformation of Coordinates

We define domains Ω_1, Ω^* as follows:

$$\Omega_1 \equiv \{(r, \theta, z) : (r, z) \in \Omega, 0 \leqslant \theta < 2\pi\},$$

$$\Omega^* \equiv \{(x, y, z) : x = r\cos\theta,\ y = r\sin\theta,\ r = 0 \text{ and } (r, z) \in \bar{\Omega},\ \text{or } (r, \theta, z) \in \Omega_1\}.$$

Ω^* is the axisymmetric domain in (x, y, z)-space, corresponding to the domain Ω. In cylindric coordinates the domain Ω^* becomes Ω_1. For any real function v^* defined in Ω^* we denote by v the corresponding function defined in Ω_1. If $v^* \in C^1(\Omega^*)$, then we have in Ω_1

$$\frac{\partial v}{\partial r} = \frac{\partial v^*}{\partial x}\cos\theta + \frac{\partial v^*}{\partial y}\sin\theta, \tag{2.1}$$

$$\frac{\partial v}{\partial \theta} = -\frac{\partial v^*}{\partial x} r\sin\theta + \frac{\partial v^*}{\partial y} r\cos\theta, \tag{2.2}$$

$$\frac{\partial v}{\partial z} = \frac{\partial v^*}{\partial z}. \tag{2.3}$$

Now we generalize these formulae for the functions in $H^1(\Omega^*)$ by using the method used in [11, p. 64].

Lemma 2.1. *If $V^* \in H^1(\Omega)$, then the generalized derivatives $\partial v/\partial r$, $\partial v/\partial\theta$ defined in Ω_1 exist and the formulae (2.1)–(2.3) are still valid.*

Proof. We prove only (2.1). By virtue of the denseness of $C^\infty(\Omega^*)$ in $H^1(\Omega^*)$, there exists a sequence $v_n^* \in C^\infty(\Omega^*)$ such that v_n^* converges to v^* in $H^1(\Omega^*)$. The formula (2.1) is valid for v_n^*:

$$\frac{\partial v_n}{\partial r} = \frac{\partial v_n^*}{\partial x}\cos\theta + \frac{\partial v_n^*}{\partial y}\sin\theta.$$

For any $\phi \in C_0^\infty(\Omega_1)$ we obtain from integration by parts that

$$-\int_{\Omega_1} v_n \frac{\partial \phi}{\partial r}\mathrm{d}r\,\mathrm{d}\theta\,\mathrm{d}z = \int_{\Omega_1}\left(\frac{\partial v_n^*}{\partial x}\cos\theta + \frac{\partial v_n^*}{\partial y}\sin\theta\right)\phi\mathrm{d}r\,\mathrm{d}\theta\,\mathrm{d}z. \tag{2.4}$$

In the (x, y, z) coordinate system we have

$$-\int_{\Omega^*} v_n^* \frac{\partial \phi^*}{\partial r} r^{-1}\mathrm{d}x\,\mathrm{d}y\,\mathrm{d}z = \int_{\Omega^*}\left(\frac{\partial v_n^*}{\partial x}\cos\theta + \frac{\partial v_n^*}{\partial y}\sin\theta\right)\phi^* r^{-1}\mathrm{d}x\,\mathrm{d}y\,\mathrm{d}z. \tag{2.5}$$

Since $r^{-1}\phi^*$ and $r^{-1}\partial\phi^*/\partial r$ are bounded in Ω^*, $v^* \in H^1(\Omega^*)$, we may take the limit through (2.5) as $n \to \infty$ and hence we obtain (2.5) as well as (2.4) with v, v^* replacing v_n , v_n^* respectively. (2.1) is proved. □

Simple calculation derives the following results.

Corollary 2.2. *If* $v^* \in H^1(\Omega^*)$, *then*

$$\frac{\partial v^*}{\partial x} = \frac{\partial v}{\partial r}\cos\theta - \frac{\partial v}{\partial \theta}\frac{\sin\theta}{r},$$

$$\frac{\partial v^*}{\partial y} = \frac{\partial v}{\partial r}\sin\theta + \frac{\partial v}{\partial \theta}\frac{\cos\theta}{r} \quad in\ \Omega^*.$$

Corollary 2.3. *Assume v is independent of θ. Then we have for* $v^* \in H^1(\Omega^*)$:

$$\frac{\partial v^*}{\partial x} = \frac{\partial v}{\partial r}\cos\theta, \quad \frac{\partial v^*}{\partial y} = \frac{\partial v}{\partial r}\sin\theta;$$

for $v^* \in H^2(\Omega^*)$

$$\frac{\partial^2 v^*}{\partial x^2} = \frac{\partial^2 v}{\partial r^2}\cos^2\theta + \frac{\partial v}{\partial r}\frac{\sin^2\theta}{r}$$

$$\frac{\partial^2 v^*}{\partial y^2} = \frac{\partial^2 v}{\partial r^2}\sin^2\theta + \frac{\partial v}{\partial r}\frac{\cos^2\theta}{r}$$

$$\frac{\partial^2 v^*}{\partial x\partial y} = \left(\frac{\partial^2 v}{\partial r^2} - \frac{1}{r}\frac{\partial v}{\partial r}\right)\sin\theta\cos\theta,$$

$$\frac{\partial^2 v^*}{\partial z^2} = \frac{\partial^2 v}{\partial z^2}, \quad \frac{\partial^2 v^*}{\partial x\partial z} = \frac{\partial^2 v}{\partial r\partial z}\cos\theta, \quad \frac{\partial^2 v^*}{\partial y\partial z} = \frac{\partial^2 v}{\partial r\partial z}\sin\theta.$$

3 Spaces V^1, V^2 ([12, 13])

We define functionals $\|\cdot\|_{k,\Omega}, k=0,1,2$, as follows:

$$\|v\|_{0,\Omega}=\left(\int_{\Omega} v^2 r\mathrm{d}r\mathrm{d}z\right)^{1/2},$$

$$\|v\|_{1,\Omega}=\left(\sum_{|\alpha|\leqslant 1}\|\partial^\alpha v\|_{0,\Omega}^2\right)^{1/2},$$

$$\|v\|_{2,\Omega}=\left(\sum_{|\alpha|\leqslant 2}\|\partial^\alpha v\|_{0,\Omega}^2\left\|\frac{1}{r}\frac{\partial v}{\partial r}\right\|_{0,\Omega}^2\right)^{1/2}.$$

Definition 3.1. *Assume that D is an open or closed set in the (r,z)-plane, D^* the correspondent axisymmetric set in (x,y,z)-space, $A(D)$ the set of real functions defined in D, and*

$$A^*(D^*)=\{v^*:\ v^* \text{ real function defined in } D^*, \text{ and there exists } v\in A(D) \text{ such that } v^*(x,y,z)=v((x^2+y^2)^{1/2},z)\}.$$

We define a mapping T: $A^*(D^*)\to A(D)$ as follows:

$$Tv^*(x,y,z)=v(r,z).$$

Obviously, the mapping T is one-to-one.

Definition 3.2.

$$U^k(\Omega^*)=H^k(\Omega^*)\cap A^*(\Omega^*),\quad k=0,1,2.$$

It is easy to see that $U^k(\Omega^*)$ is a closed subspace in $H^k(\Omega^*)$. Now establish the relations between the norms $\|\cdot\|_{H^k(\Omega^*)}$ and the functionals $\|\cdot\|_{k,\Omega}$, a for the elements of $U^k(\Omega^*)$.

Lemma 3.3. *Assume $v^*\epsilon U^K(\Omega^*), v=Tv^*$. Then $\|v\|_{k,\Omega}<\infty$,and*

$$\|v^*\|_{H^k(\Omega^*)}^2=2\pi\|v\|_{k,\Omega}^2\quad \forall u^*\epsilon U^K(\Omega^*), k=0,1, \tag{3.1}$$

$$\frac{3\pi}{2}\|v\|_{2,(\Omega^*)}^2\leqslant\|v^*\|_{H^2(\Omega^*)}^2\leqslant 2\pi\|v\|_{2,\Omega}^2\quad \forall u^*\in U^2(\Omega^*). \tag{3.2}$$

Proof. By direct computation and Corollary 2.3. □

Definition 3.4.

$$V^k(\Omega)=\{v:\ v=Tv^*, v^*\in U^k(\Omega^*)\},\quad k=0,1,2.$$

It follows from Lemma 3.3 and the closedness of $U^k(\Omega^*)$ in $H^k(\Omega^*)$ that $V^k(\Omega)$, $k=0,1,2$, are Banach spaces. We need the following subspace $V_0^1(\Omega)$ of $V^1(\Omega)$:

$$V_0^1(\Omega)=\{v:\ v=Tv^*, v^*\in U^1(\Omega^*)\cap H_0^1(\Omega^*)\}.$$

Let $v \in V_0^1(\Omega)$, $v = Tv^*$, tr v^* be the trace of v^* on $\partial\Omega^*$. We define $T(\text{tr } v^*)$ as the trace of v on Γ_1. Obviously, it is zero.

By Lemma 3.3 and the embedding theorems of $H^k(\Omega^*)$, we obtain the correspondent theorems of $V^k(\Omega)$. Particularly, we have the following result.

Lemma 3.5. *There exists a constant C' such that*

$$\|v\|_{1,\Omega}^2 \leqslant C' \int_\Omega \left[\left(\frac{\partial v}{\partial r}\right)^2 + \left(\frac{\partial v}{\partial z}\right)^2\right] r\mathrm{d}r\,\mathrm{d}z \quad \forall v \in V_0^1(\Omega). \tag{3.3}$$

Finally, the following statement on denseness may be proved. (See [13] for V^1. The proof is similar for V^2.)

Lemma 3.6. *Assume that the domain Ω has a locally Lipschitz boundary. Then $C^\infty(\bar{\Omega})$ is dense in $V^k(\Omega)$, $k = 1, 2$.*

Remark 3.1. Lemma 3.6 is not a direct corollary of the denseness theorem of $H^k(\Omega^*)$. If $v^* \in H^k(\Omega^*)$, then there exists a sequence $v_n^* \in C^\infty(\bar{\Omega}^*)$ converging to $v^* \in H^k(\Omega^*)$. But we can not claim that $v_n^* \in A^*(\Omega^*)$.

Remark 3.2. The facts $v \in C^\infty(\bar{\Omega})$ and $v = Tv^*$ do not imply that $v^* \in C^\infty(\bar{\Omega}^*)$. Counterexample: $v = r$. But $v \in C^0(\bar{\Omega}) \Leftrightarrow v^* \in C^\infty(\bar{\Omega}^*)$.

4 Solution of Problem (1.1)

We define a bilinear form $B(\cdot,\cdot)$ on $V^1(\Omega) \times V^1(\Omega)$ and a linear functional $F(\cdot)$ on $V^1(\Omega)$ as follows:

$$B(u,v) = \int_\Omega \beta \left(\frac{\partial u}{\partial r}\frac{\partial v}{\partial r} + \frac{\partial u}{\partial z}\frac{\partial v}{\partial z}\right) r\mathrm{d}r\mathrm{d}z,$$

$$F(v) = \int_\Omega f v r\mathrm{d}r\mathrm{d}z$$

Then we have the variational formulation of problem (1.1): Find $u \in V_0^1(\Omega)$ such that

$$B(u,v) = F(v), \qquad \forall v \in V_0^1(\Omega). \tag{4.1}$$

From now on we assume that Ω has a locally Lipschitz boundary.

Theorem 4.1. *Problem (4.1) has a unique solution.*

Proof. It follows from Lemma 3.5 and assumptions (i)–(ii) that the bilinear form $B(u,v)$ is coercive and continuous on $V_0^1(\Omega) \times V_0^1(\Omega)$. And the linear functional $F(v)$ is continuous on $V_0^1(\Omega)$ by virtue of assumption (iii). Hence the conclusion of the theorem is a result of the Lax-Milgram theorem. □

Remark 4.1. Let u be the solution of problem (4.1). Since $B(u,v)$ is symmetric, u is also the solution of the following problem: Find $u \in V_0^1(\Omega)$ such that

$$J(u) = \min_{v \in V_0^1(\Omega)} J(v)$$

where $J(v) = B(v,v) - 2F(v)$.

Consider the boundary value problem in Ω^* corresponding to problem (1.1):

$$\begin{cases} \Omega^* : \ -\left[\dfrac{\partial}{\partial x}\left(\beta^* \dfrac{\partial w^*}{\partial x}\right) + \dfrac{\partial}{\partial y}\left(\beta^* \dfrac{\partial w^*}{\partial y}\right) + \dfrac{\partial}{\partial z}\left(\beta^* \dfrac{\partial w^*}{\partial z}\right)\right] = f^*, \\ \partial\Omega^* : \ w^* = 0, \end{cases} \tag{4.2}$$

where $\beta^* = T^{-1}\beta, f^* = T^{-1}f$. The correspondent variational problem is: Find $w^* \in H_0^1(\Omega^*)$ such that

$$B_1(w^*, v^*) = F_1(v^*) \qquad \forall v^* \in H_0^1(\Omega^*), \tag{4.3}$$

where

$$B_1(w^*, v^*) = \int_{\Omega^*} \beta^* \left(\frac{\partial w^*}{\partial x}\frac{\partial v}{\partial x} + \frac{\partial w^*}{\partial y}\frac{\partial v^*}{\partial y} + \frac{\partial w^*}{\partial z}\frac{\partial v^*}{\partial z}\right) \mathrm{d}x\mathrm{d}y\mathrm{d}z,$$

$$F_1(v^*) = \int_{\Omega^*} f^* v^* \mathrm{d}x\mathrm{d}y\mathrm{d}z.$$

From now on we assum:

(iv). The boundary $\partial\Omega^*$ is smooth enough to ensure that problem (4.3) has a unique solution w^* and $w^* \in H_0^1(\Omega^*) \cup H^2(\Omega^*)$. For example, we may assume that $\partial\Omega^*$ is of class C^2 (see, for instance, [14, p.176]) or that the domain Ω^* is convex.

Theorem 4.2. *Let u be the solution of problem (4.1). Then*

$$u \in V_0^1(\Omega) \cap V^2(\Omega).$$

Proof. Let $u^* = T^{-1}u$. We define for $v^* \in H_0^1(\Omega^*)$ that

$$v(r,\theta,z) = v^*(x,y,z), \qquad \bar{v}(r,z) = \int_0^{2\pi} v(r,\theta,z)\mathrm{d}\theta$$

It is easily proved that

$$\bar{v} \in V_0^1(\Omega), \tag{4.4}$$

$$\frac{\partial \bar{v}}{\partial r} = \int_0^{2\pi} \frac{\partial v}{\partial r}\mathrm{d}\theta, \quad \frac{\partial \bar{v}}{\partial z} = \int_0^{2\pi} \frac{\partial v}{\partial z}\mathrm{d}\theta. \tag{4.5}$$

Now we prove that u^* is the solution of problem (4.3). It follows from Lemma 2.1, (4.4), (4.5) and (4.1) that for $v^* \in H_0^1(\Omega^*)$

$$\begin{aligned} B_1(u^*, v^*) - F_1(v^*) &= \int_\Omega \left[\int_0^{2\pi} \left(\frac{\partial u}{\partial r}\frac{\partial v}{\partial r} + \frac{\partial u}{\partial z}\frac{\partial v}{\partial z} - fu\right)\mathrm{d}\theta\right] r\mathrm{d}r\mathrm{d}z \\ &= \int_\Omega \left[\left(\frac{\partial u}{\partial r}\frac{\partial \bar{v}}{\partial r} + \frac{\partial u}{\partial z}\frac{\partial \bar{v}}{\partial z}\right) - f\bar{v}\right] r\mathrm{d}r\mathrm{d}z = B(u,\bar{v}) - F(\bar{v}) = 0. \end{aligned}$$

Hence u^* is the solution of problem (4.3), and $u^* \in H_0^1(\Omega^*) \cap H^2(\Omega^*)$ by assumption (iv). According to Definition 3.4, we obtain the conclusion of the theorem. □

5 Linear Finite Element Solution and Its Order of Convergence Order

Assume that the domain Ω is convex. Let $T_h = \{C_1, \cdots, C_m\}$ be a triangulation of Ω, h_i the maximum edge of the triangle C_i, θ_i the minimum angle of C_i, $h = \max_i h_i$, $\theta = \min_i \theta_i$, $\Omega = \sum_{k=1}^{m} C_i$, and assume:

(a). $\theta \geqslant \theta_0 > 0$, θ_0 is independent of h [14], [15].

Define a linear finite element space V^h as follows:

$$V^h = \{v_h \in C^0(\bar{\Omega}) : \ v_h \text{ is linear function in } C_i,\ i = 1, \cdots, m;\ v_h = 0 \text{ on } (\Omega - \Omega_h) \cup \Gamma_1\}.$$

Then it is easy to prove that $V^h \subset V_0^1(\Omega)$. We have the correspondent discrete problem for problem (4.1): Find $u_h \in V^h$ such that

$$B(u_h, v_h) = F(v_h) \quad \forall v_h \in V^h. \tag{5.1}$$

Theorem 5.1. *Problem (5.1) has a unique solution.*

The proof is similar to that of Theorem 4.1.

Remark 5.1. The solution u_h of (5.1) is also the solution of the minimization problem: Find $u_h \in V^h$ such that $J(u_h) = \min_{v_h \in V^h} J(v_h)$.

Assume that u is the solution of problem (4.1), u_I the piecewise linear interpolation corresponding to the triangulation T_h. For any triangle $C \in T_h$, we now estimate $\|u - u_I\|_{1,C}$. Let $P_j = (r_j, z_j)$, $j = 1, 2, 3$ be the vertexes of C,$\lambda_j = (r, z)$, $j = 1, 2, 3$ the so-called barycentric coordinates [16, p.45], i.e. the basis functions for the linear interpolation on C:

$$\lambda_j(P_i) = \delta_{ij} \quad (i, j = 1, 2, 3).$$

Then we have for any function v defined on C and its linear interpolation v_I:

$$\sum_j \lambda_j(P) v(P_j) = v_I(P) \quad \forall P \in C, \tag{5.2}$$

in particular,

$$\sum_j \lambda_j = 1, \quad \sum_j \lambda_j r_j = r, \quad \sum_j \lambda_j z_j = z, \qquad \forall (r, z) \in C. \tag{5.3}$$

It follows from (5.3) that

$$\sum_j \lambda_j (r_j - r) = \sum_j \lambda_j (z_j - z) = 0, \qquad \forall (r, z) \in C. \tag{5.4}$$

The proof of the following lemma belongs to [17].

Lemma 5.2. *Assume that $v \in V^2(C)$, and the condition (a) is true. Then*

$$\|v - v_I\|_{1,c}^2 \leqslant M h^2 \|v\|_{2,c}^2, \tag{5.5}$$

where the constant M is independent of C and V.

Proof. Assume $v \in C^\infty(C)$ temporarily. Expand v at the point $P=(r,z)$ by using the Taylor's formula with integral remainder (see, for instance [18, p.36]):

$$v(P_j)-v(P)=d_jv(P)+\int_0^1(1-t)d_j^2v(M_j)\mathrm{d}t, \qquad j=1,2,3, \tag{5.6}$$

where

$$\begin{aligned}d_j&=(r_j-r)\frac{\partial}{\partial r}+(Z_j-z)\frac{\partial}{\partial z}, \quad d_j^2=d_jd_j,\\ M_j&=P_jt+P(1-t).\end{aligned}$$

It follows from (5.2), (5.3) and (5.6) that

$$\begin{aligned}v_I(P)-v(P)&=\sum_j\lambda_j(P)[v(P_j)-v(P)]\\ &=\sum_j\left[\lambda_j(P)(r_j-r)\frac{\partial v(P)}{\partial z}+\lambda_j(P)(z_j-z)\frac{\partial v(P)}{\partial z}\right]\\ &\quad+\sum_j\int_0^1(1-t)\lambda_j(P)d_j^2v(M_j)\mathrm{d}t.\end{aligned}$$

By virtue of (5.4) the first sum vanishes, and we have

$$v_I(P)-v(P)=\sum_j\int_0^1(1-t)\lambda_j(P)d_j^2v(M_j)\mathrm{d}t. \tag{5.7}$$

Differentiating (5.7) we obtain

$$\begin{aligned}\frac{\partial v_I}{\partial r}-\frac{\partial v}{\partial r}&=\sum_j\int_0^1(1-t)\left(\frac{\partial\lambda_j}{\partial r}d_j^2-2\lambda_jd_j\frac{\partial}{\partial r}\right)v(M_j)\mathrm{d}t\\ &\quad+\sum_j\int_0^1(1-t)\lambda_jd_j^2\left[\frac{\partial v(M_j)}{\partial r}(1-t)\right]\mathrm{d}t.\end{aligned} \tag{5.8}$$

Integrating by parts the integrals in the second sum, noting (5.4) and that $d/\mathrm{d}t[d_jv(M_j)]=d_j^2v(M_j)$, we derive from (5.8) that

$$\frac{\partial v_I}{\partial r}-\frac{\partial v}{\partial r}=\sum_j\int_0^1(1-t)\frac{\partial\lambda_i}{\partial r}d_j^2v(M_j)\mathrm{d}t. \tag{5.9}$$

It follows from the uniform basis condition (a) that (see, for instance, [19] or [20, p. 137])

$$\left|\frac{\partial\lambda_j}{\partial r}\right|,\ \left|\frac{\partial\lambda_j}{\partial z}\right|\leqslant M_1h^{-1}, \tag{5.10}$$

where h is the maximum edge of C, $M_1=4/\sin\theta_0$. Hence we have

$$\left|\frac{\partial v_I}{\partial r}-\frac{\partial v}{\partial r}\right|\leqslant M_1h^{-1}\sum_j\int_0^1(1-t)\mid d_i^2v(M_j)\mid\mathrm{d}t,$$

and then

$$
\begin{aligned}
&\int_c \left|\frac{\partial v_I}{\partial r}-\frac{\partial v}{\partial r}\right|^2 r\mathrm{d}r\mathrm{d}z\\
\leqslant& M_1^2h^{-2}\int_c\left(\sum_j\int_0^1(1-t)^{-1/4}(1-t)^{5/4}\mid d_j^2v(M_j)\mid \mathrm{d}t\right)^2 r\mathrm{d}r\mathrm{d}z\\
\leqslant& 3M_1^2h^{-2}\sum_j\int_c\left(\int_0^1(1-t)^{5/2}\mid d_j^2v(M_j)\mid^2\mathrm{d}t\int_0^1(1-t)^{-1/2}\mathrm{d}t\right)r\mathrm{d}r\mathrm{d}z\\
=& 6M_1^2h^{-2}\sum_j\int_0^1\mathrm{d}t\int_c(1-t)^{5/2}\left|\left[(r_j-r)\frac{\partial}{\partial r}+(z_j-z)\frac{\partial}{\partial z}\right]^2 v(M_j)\right|^2 r\mathrm{d}r\mathrm{d}z\\
\leqslant& 6M_1^2h^{-2}\sum_j\int_0^1\mathrm{d}t\int_c(1-t)^{5/2}h^4\left(\left|\frac{\partial^2v(M_j)}{\partial r\partial z}\right|+2\left|\frac{\partial^2v(M_i)}{\partial r\partial z}\right|+\left|\frac{\partial^2v(M_j)}{\partial z^2}\right|\right)^2 r\mathrm{d}r\mathrm{d}z \quad (5.11)
\end{aligned}
$$

where $M_2=72M_1^2$. Make variable transformations in the integrals as follows:

$$\zeta=r_jt+r(1-t),\quad \eta=z_j+z(1-t).$$

Then $M_i=(\zeta,\eta)$,and the triangle C reduces to a similar triangle $C_{j,t}$ with the similarity transformation center P_j. Hence the right side of (5.11) becomes

$$
\begin{aligned}
&M_2h^2\sum_j\int_0^1\mathrm{d}t\int_{c_{j,t}}(1-t)^{-1/2}(\zeta-r_jt)\left(\left|\frac{\partial^2v(\zeta,\eta)}{\partial\zeta^2}\right|^2+\left|\frac{\partial^2v(\zeta,\eta)}{\partial\zeta\partial\eta}\right|^2+\left|\frac{\partial^2v(\zeta,\eta)}{\partial\eta^2}\right|^2\right)\mathrm{d}\zeta\mathrm{d}\eta\\
\leqslant& M_2h^2\sum_j\int_0^1\mathrm{d}t\int_{c_{j,t}}(1-t)^{-1/2}(---)\zeta\mathrm{d}\zeta\mathrm{d}\eta\quad(\text{ since }\zeta-r_jt\leqslant\zeta)\\
\leqslant& M_2h^2\sum_j\int_0^1\mathrm{d}t\int_c(1-t)^{-1/2}(---)\zeta\mathrm{d}\zeta\mathrm{d}\eta\quad(\text{ since }C_{j,t}\in C)\\
=& 3M_2h^2\int_c(---)\zeta\mathrm{d}\zeta\mathrm{d}\eta\int_0^1(1-t)^{-1/2}\mathrm{d}t\leqslant M_3h^2\|v\|_{2,c}^2.
\end{aligned}
$$

Hence we obtain by (5.11) that

$$\int_c\left(\frac{\partial v_I}{\partial r}-\frac{\partial v}{\partial r}\right)^2 r\mathrm{d}r\mathrm{d}z\leqslant M_3h^2\|v\|_{2,c}^2.$$

Similarly we obtain

$$\int_c\left(\frac{\partial v_I}{\partial z}-\frac{\partial v}{\partial z}\right)^2 r\mathrm{d}r\mathrm{d}z\leqslant M_4h^2\|v\|_{2,c}^2,$$

$$\int_c(v_I-v)^2r\mathrm{d}r\mathrm{d}z\leqslant M_5h^2\|v\|_{2,c}^2.$$

Therefore,

$$\|v-v_I\|_{1,c}^2\leqslant Mh^2\|v\|_{2,c}^2\forall v\in C^\infty(C). \quad (5.12)$$

Finally (5.5) is deduced from (5.12) and Lemma 3.6. □

Define the "energy norm" $B_h(u,u)$ on Ω_h as follows:

$$B_h(u,u)=\int_{\Omega_h}\beta\left[\left(\frac{\partial u}{\partial r}\right)^2+\left(\frac{\partial u}{\partial z}\right)^2\right]rdrdz.$$

Theorem 5.3. *Assume that the domain Ω is convex, u_h is the solution of (5.1), u the solution of (4.1). Then*

$$B_h(u-u_h,u-u_h)=O(h^2). \tag{5.13}$$

Proof. u_h minimizes the error $u-u_h$ in the "energy norm" on $\Omega - B(v,v)$, i.e. (see [20, p.39])

$$B(u-u_h,u-u_h)=\min_{u_h\in V^h}B(u-v_h,u-v_h).$$

Since $u_h=v_h=0$ on $\Omega-\Omega_h$, we have

$$B_h(u-u_h,u-u_h)=\min_{u_h\in V^h}B_h(u-v_h,u-v_h).$$

Let u_I be the piecewise linear interpolant of u on Ω_h corresponding to T_h. Define $u_I=0$ on $\Omega-\Omega_h$. Then $u_I\in V_h$. So

$$B_h(u-u_h,u-u_h)\leqslant B_h(u-u_I,u-u_I)\leqslant \max_{\Omega}\beta\cdot\|u-u_I\|_{1,\Omega_K}^2. \tag{5.14}$$

By virtue of Lemma 5.2 we have

$$\|u-u_I\|_{1,\Omega_h}^2=\sum_{i=1}^m\|u-u_I\|_{1,c_i}^2\leqslant Mh^2\sum_{i=1}^m\|u\|_{2,c_i}^2\leqslant Mh^2\|u\|_{2,\Omega}^2. \tag{5.15}$$

Relations (5.14) and (5.15) prove that (5.13) is valid. □

If Ω is a convex polygon, then $\Omega_k=\Omega, B_h(v,v)=B(v,v)$. Since $B(u,v)$ is coercive on $V_0^1(\Omega)$, we have the following.

Corollary 5.4. *If Ω is a convex polygon, then*

$$\|u-u_h\|_{1,\Omega}=O(h),\quad \|u-u_h\|_{0,\Omega}=O(h^2).$$

Acknowledgments The author wishes to express his sincere appreciation to Professor Parter for his very valuable suggestions. The author also appreciates the helpful comments of the referee.

References

[1] P. Jamet and S.V. Parter, Numerical methods for elliptic differential equations whose coefficients are singular on a portion of the boundary, *SIAM Journal on Numerical Analysis*, 4 (1967), 131-146.

[2] S.V. Parter, Numerical methods for generalized axially symmetric potentials, *SIAM Journal on Numerical Analysis*, 2 (1965), 500-516.

[3] E. Wilson, Structural analysis of axisymmetric solids, *AIAA Journal*, 3 (1965), 2269-2274.

[4] M. Crouzeix and J.M. Thomas, Elements finis et problémes elliptiques dégénérés, *RAIRO Anal-Numér.*, 7 (1973), 77-104.

[5] D. Jesperson, Ritz-Galerkin methods for singular boundary value problems, *SIAM Journal on Numerical Analysis*, 15 (1978), 813-834.

[6] R.D. Russell and L.F. Shampine, Numerical methods for singular boundary value problems, *SIAM Journal on Numerical Analysis*, 12 (1975), 13-36.

[7] R. Shreiber and S.C. Eesenstat, Finite element methods for spherically symmetric elliptic equations, *SIAM Journal on Numerical Analysis*, 18 (1981), 546-558.

[8] E.X. Jiang, The error bound of the finite element method for the axisymmetric solid in elastic mechanics, *Fudan Journal (Natur. Sci.)*, 19 (1980), 87-96.

[9] S. Z. Zhou and B.B. Tang, The convergence of the semi-analytical finite element method, *Journal of Hunan University*, 1979, No.2, 1981, No. 1.

[10] A. Bendali, Approximation of a degenerated elliptic boundary value problem by a finite element method, *RAIRO Anal-Numér.*, 15 (1981), 87-99.

[11] R. A. Adams, *Sobolev Space*, Academic Press, New York, 1975.

[12] K.C. Chang and L,S, Jiang, *The free boundary problem of the stationary water cone*, Acta Scicentiarum Naturalum Universitis Pekinesis, 1978, No.1, 1-25.

[13] S.Z. Zhou, Functional spaces $W^m_{p,2}$, *Journal of Hunan University*, 1980, No. 4.

[14] D. Gilberg and S. Trudinger, *Elliptic Partial Differential Equations of Second Order*, Springer-Verlag, New York, 1977.

[15] K. Feng, Differencing scheme based on variational principle, *Applied and Computational Mathematics*, 4 (1965), 238-262. (In Chinese.)

[16] P.G. CIARLET, *The Finite Element Method for Elliptic Problems*, North-Holland, New York, 1978.

[17] Q.M. Chen, Personal communication, 1981.

[18] T. Dupont and R. Scott, Constructive polynomial approximation in Sobolev spaces, in: *Recent Advances in Numerical Analysis*, C. de Boor and G. H. Golub, eds., Academic Press, New York, 1978.

[19] K. Feng, Finite element method (III), *Practice and Theory of Math.* 1975: 2. (In Chinese.)

[20] G. Strang and G.J. Fix, *An Analysis of the Finite Element Method, Prentice-Hall*, Englewood Cliffs, NJ, 1973.

The Solution of Free Boundary Problem for an Axisymmetric Partially Penetrating Well*

C.W. Cryer† S.Z.Zhou (周叔子)‡

Abstract The weak form of the free boundary problem for an axisymmetric partially penetrating well may be formulated as follows: find $\varphi(r) \in C^0([r_0, r_1])$ and $u \in C^0(\overline{\Omega}) \bigcap V^1(\Omega)$ such that:

$$\int_{\Omega} r\nabla u \cdot \nabla v \mathrm{d}r\mathrm{d}z = 0, \quad \forall v \in K_1$$

and u satisfies appropriate boundary conditions. Here, u is related to the hydraulic head, $\varphi(r)$ is the unknown water-air interface, Ω is the region of saturated flow:

$$\Omega = \{(r, z) \mid 0 < r \leqslant r_0, 0 < z < h\} \cup \{(r, z) \mid r_0 < r < r_1, 0 < z < \varphi(r)\}$$

K_1 is a convex set in the weighted Sobolev space $V^1(\Omega)$.

We reduce the problem to three families of variational inequalities by using a type of ≪ Baiocchi transform ≫, study equivalence of the three families and regularity of the solutions of the variational inequalities. Finally, we prove the existence of the solution for the well problem.

Keywords Free boundary problem, variational inequalities, equivalence.

Introduction

The free boundary problem for a fully penetrating well in a layer of soil of permeability $K(x, y) = \exp[f(x) + g(y)]$ has been solved by Cryer and Fetter [1979] using variational inequalities. In this paper we consider the problem for a partially penetrating well. A type of ≪ Baiocchi transform ≫ (Baiocchi [1974]) is used to derive a corresponding family of variational inequalities. Existence of the solution is proved. To this end we use the theory of weighted Sobolev spaces and some results in Chang and Jiang [1978].

We have omitted some technical details which the interested reader will find in a report

* 发表于: Annali Di Matematica Pura Ed Applicata, Vol. 135, 1983, pp. 219-235.

† Computer Sciences Department and Mathematics Research Center, University of Wisconsin Madison. Current Address: Dept. Appl. Math. Theor. Physics, University of Cambridge.

‡ Hunan University, Changsha, China, and Mathematics Research Center, University of Wisconsin-Madison.

with the same title (Technical Summary Report No. 2245, Mathematics Research Center, University of Wisconsin, Madison, Wisconsin).

1 Weighted Sobolev Spaces

Our problem is governed by a degenerate elliptic equation. Degenerate elliptic equations can often be associated with a weighted Sobolev space (e.g., Murty and Stampacchia [1968], Trudinger [1973]). Various kinds of Sobolev spaces have been studied (e.g., Jakovlev [1966], Cryer [1980], Chang and Jiang [1978], Leventhal [1975], and Zhou [1980]). We recall some results.

Let A be a bounded domain in the (r,z)−plane with a locally Lipschitz boundary Γ, and with $r>0$; $C_0^\infty(A)$-the space of functions infinitely differentiable and with support compact in A; $C_0^\infty(A;\Gamma_i)$-the space of functions infinitely differentiable in $\bar{A}$ and vanishing in some neighborhood of Γ_i, where $\Gamma_i\subset\Gamma$. $L^p(A;r)$-the space of measurable functions satisfying:

$$\|v\|_{L^p(A;r)}=\int_A r|v|^p\mathrm{d}r\ \mathrm{d}z<\infty. \tag{1.1}$$

We define weighted Sobolev spaces as follows:

$$\begin{cases} V^0(A)=L^2(A;r),\\ V^1(A)=\{v\mid \partial^\alpha v\in L^2(A;r)\ \ |\alpha|\leqslant 1\},\\ V^2(A)=\left\{v\ \middle|\ \dfrac{1}{r}\dfrac{\partial v}{\partial r},\ \partial^\alpha v\in L^2(A;r),\ |\alpha|\leqslant 2\right\}\end{cases} \tag{1.2}$$

with norms, respectively,

$$\begin{cases} \|v\|_{V^0(A)}=\|v\|_{L^p(A;r)},\\ \|v\|_{V^1(A)}=\displaystyle\sum_{|\alpha|\leqslant 1}\|\partial^\alpha v\|_{V^0(A)},\\ \|v\|_{V^2(A)}=\displaystyle\sum_{|\alpha|\leqslant 2}\|\partial^\alpha v\|_{V^0(A)}+\left|\frac{1}{r}\frac{\partial v}{\partial r}\right|_{V^0(A)}.\end{cases} \tag{1.3}$$

Denote by $V_0^1(A)$, $V_0^1(A;\Gamma_i)$ respectively the closure of $C_0^\infty(A)$, $C_0^\infty(A;\Gamma_i)$ in $V^1(A)$.

Lemma 1.1. *$V^0(A)$, $V^1(A)$ and $V^2(A)$ are Banach spaces.*

Lemma 1.2 (Green's Formula). *If $u,v\in V^1(A)$, then*

$$\begin{aligned}\int_A ru\frac{\partial v}{\partial r}\mathrm{d}r\ \mathrm{d}z&=-\int_A v\frac{\partial(ru)}{\partial r}\mathrm{d}r\ \mathrm{d}z+\int_\Gamma ruv\ \cos(n,r)\mathrm{d}s\\ &=-\int_A v\frac{\partial(ru)}{\partial z}\mathrm{d}r\ \mathrm{d}z+\int_\Gamma ruv\ \cos(n,z)\mathrm{d}s,\end{aligned}$$

where n is the outer normal of Γ.

Lemma 1.3. *If A_ε, is a closed subdomain of A and $\partial A_\varepsilon\cap\{r=0\}=\varnothing$, then*

$$V^1(A_\varepsilon) = H^1(A_\varepsilon).$$

Now let $\bar{A}^*$ be the three dimensional domain formed by rotating $\bar{A}$ about z-axis, and let S_i be the surface formed by rotating Γ_i about the z-axis.

Lemma 1.4. *If $v(r,z) \in V^k(A)$, $k = 0, 1, 2$ and*

$$f(x,y,z) = v\left(\sqrt{x^2+y^2}, z\right) \tag{1.4}$$

then $f \in H^k(A^*)$, where $H^k(A^*)$ is the usual Sobolev space, and A^* is the interior of $\bar{A}^*$.

Lemma 1.5. *If $v \in V^1(A)$, and $\Gamma_i \bigcap \{r=0\} = \varnothing$, then*

$$\|f\|_{H^1(A^*)} = 2\pi \|v\|_{V^1(A)},$$
$$\int_{S_i} f^2 \mathrm{d}S = 2\pi \int_{\Gamma_i} rv^2 \mathrm{d}s.$$

By using Lemma 1.5 and results in Sobolev [§10, 1950] we obtain:

Lemma 1.6. *If $v \in V_0^1(A;\Gamma_i)$, and*

$$\mathrm{mes}\,[\Gamma_i \setminus (\Gamma_i \cap \{r=0\})] > 0,$$

then

$$\|v\|^2_{V^1(A)} \leqslant C \int_A \left[\left(\frac{\partial v}{\partial r}\right)^2 + \left(\frac{\partial v}{\partial Z}\right)^2\right] r \,\mathrm{d}r\,\mathrm{d}z.$$

where C does not depend on v.

2 Description of the Problem

The problem to be considcred is shown in Figure 1.

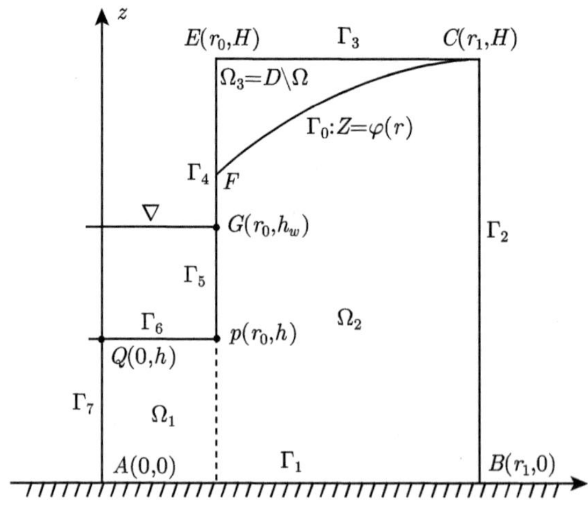

Figure 1

A cylindrical well of radius r_0 partially penetrates a layer of soil of depth H and radius r_1. Take the axis of symmetry as the z-axis. The bottom of the soil layer is impermeable. The distance of the well bottom from the bottom of the soil layer is h. We assume that the soil layer is homogeneous and isotropic; that the water is incompressible; that the flow is irrotational and steady (in particular the height of water on the outer boundary of soil and in the well is respectively H and h_w); that the permeability $k(r,z)\equiv 1$.

The cross section of the soil layer is

$$D=\Omega_1\cup\Omega_2\cup\Omega_3. \tag{2.1}$$

where

$$\begin{aligned}\Omega_1&=\{(r,z)\mid 0\leqslant r<r_0, 0<z<h\}\\ \Omega_2&=\{(r,z)\mid r_0<r<r_1, 0<z<\varphi(r)\}\\ \Omega_3&=\{(r,z)\mid r_0<r<r_1, \varphi(r)<z<H\}\end{aligned}$$

and $z=\varphi(r)$ is the boundary between the wet region $\Omega=\Omega_1\cup\Omega_2$ and dry region Ω_2. It is called the free boundary as it is the unknown part of $\partial\Omega$.

Denoted by $p(r,z)$ and $u(r,z)$ respectively the pressure at point (r,z) of D (the atmospheric pressure being measured by zero) and the hydraulic head, then we have

$$u(r,z)=p(r,z)+z \quad \text{in} \quad \Omega. \tag{2.2}$$

It follows from Darcy's law and the equation of continuity that (see Hantush[1964], or Cryer [1976, p. 86])

$$Lu=\frac{\partial^2 u}{\partial r^2}+\frac{1}{r}\frac{\partial u}{\partial r}+\frac{\partial^2 u}{\partial z^2}=0 \quad \text{in} \quad \Omega. \tag{2.3}$$

We introduce the notation

$$\begin{aligned}\Gamma_1&=\{(r,z)|0<r<r_1,\ z=0\},\\ \Gamma_2&=\{(r,z)|r=r_1,\ 0<z<H\},\\ \Gamma_3&=\{(r,z)|r_0<r<r_1,\ z=H\},\\ \Gamma_4&=\{(r,z)|r=r_0,\ h_w<z<H\},\\ \Gamma_5&=\{(r,z)|r=r_0,\ h<z<h_w\},\\ \Gamma_6&=\{(r,z)|0<r<r_0,\ z=h\},\\ \Gamma_7&=\{(r,z)|r=0,\ 0<z<h\},\\ \Gamma_0&=\{(r,z)|r_0<r<r_1,\ z=\varphi(r).\}\end{aligned}$$

Then $u(r,z)$ satisfies the following boundary conditions:

$$\begin{cases} u=H & \text{on } \Gamma_2 & \text{(constant hydraulic head)},\\ u=z & \text{on } \Gamma_0\cup(\Gamma_4\cap\partial\Omega) & \text{(interface with air)},\\ u=h_w & \text{on } \Gamma_5\cup\Gamma_6 & \text{(interface with water at rest)}\end{cases} \tag{2.4}$$

and

$$\begin{cases} \dfrac{\partial u}{\partial n}=0 \quad \text{on } \Gamma_0 \cup \Gamma_1 \quad \text{(streamline)}, \\ \dfrac{\partial u}{\partial n}=0 \quad \text{on } \Gamma_7 \quad \text{(symmetry)}. \end{cases} \tag{2.5}$$

Now we can state our problem in weak form.

Problem (PPW)

Given the domain as in (2.1), and a real number h_w such that $h < h_w < H$, find functions $\varphi(r)$ and $u(r,z)$ such that u satisfies (2.4) and

$$\varphi \in C^0([r_0,r_1]), \varphi(r_1) = H, \varphi(r_0) > h_w, \tag{2.6}$$

$$\varphi \text{ is strictly increasing}, \tag{2.7}$$

$$u \in V^1(\Omega) \cap C^0(\bar{\Omega}), \tag{2.8}$$

$$\int_\Omega r\nabla u \cdot \nabla v \mathrm{d}r \, \mathrm{d}z = 0, \quad \forall\, v \in K_1, \tag{2.9}$$

where

$$\Omega = \Omega_1 \cup \{(r,z) \in D | r > r_0, 0 < z < \varphi(r)\},$$

$$K_1 = \{v \in V^1(\Omega) | v = 0 \text{ on } \Gamma_2 \cup (\Gamma_4 \cap \partial\Omega) \cup \Gamma_5 \cup \Gamma_6\}.$$

Remark 2.1. This problem can be regarded as a plane problem with permeability $K = \exp(\ln\ r)$, but it is not covered by the work of Benci [1974] because, $\ln\ r \notin H^{1,2+\mu}([0,r_1])$. Also, it can not be included in Alt [1979] as a two dimensional problem. Rama and Das [1976] solved this problem by numerical methods.

Remark 2.2. Chang and Jiang [1978] has solved a similar problem by using so-called ≪ Sequence of set-valued mappings ≫ instead of the method of variational inequalities. Further results about obstacle problems have been obtained recently by Chang [1980]. We use some results on linear equations given by Chang and Jiang [1978]. But we solve our problem by using the method of variational inequalities because the corresponding numerical method is more convenient, and because in our case the boundary conditions and right term of the nonlinear equation for the Baiocchi function w are different from those in Chang and Jiang [1978].

3 The Baiocchi Function w and Its Properties

Assuming a priori the existence of the solution u of (PPW), set (see Baiocchi [1974], [1976], [1978])

$$\bar{u}(r,z) = \begin{cases} u(r,z), & \text{in } \bar{\Omega} \\ z, & \text{in } \bar{D}\backslash\bar{\Omega}. \end{cases} \tag{3.1}$$

$$w(r,z) = \int_0^z [\bar{u}(r,t) - t]\mathrm{d}t. \tag{3.2}$$

It is easy to show the following properties of u:

Lemma 3.1.

$$Lu = 0 \quad \text{in} \quad \Omega \tag{3.3}$$

$$u \ \text{is analytic in} \ \bar{\Omega} \setminus \{\Gamma_0, A, B, C, F, G, P, Q\} \tag{3.4}$$

$$\frac{\partial u}{\partial n} = 0 \ \text{ on } \Gamma_1 \cup \Gamma_7 \tag{3.5}$$

$$\frac{\partial u}{\partial n} = 0 \text{ in the weak sense on } \Gamma_0 \tag{3.6}$$

$$u(r, z) > z, \quad \text{in} \quad \Omega \tag{3.7}$$

$$\bar{u} \in V^1(D) \cap C^0(\bar{D}) \tag{3.8}$$

$$L\bar{u} = -\frac{\partial \Phi_\Omega}{\partial z} \text{ in the sense of distributions,} \tag{3.9}$$

where Φ_Ω *is the characteristic function of* Ω *in* D.

Based on Lemma 3.1 we can derive some properties of w.

Proposition 3.2. *Let* w *be defined by (3.2). Then*

$$Lw = -\Phi_\Omega \text{ in the sense of distributions} \tag{3.10}$$

$$r\frac{\partial w(r, H)}{\partial r} = \text{constant } = q \text{ for } r \in]r_0, r_1[\tag{3.11}$$

$$w(r, z) = g_q \text{ on } \Gamma_D \tag{3.12}$$

$$\frac{\partial w}{\partial n} = g_N \text{ on } \Gamma_N \tag{3.13}$$

$$w(r, z) = g_q(r, H) \ \text{ in } D \setminus \Omega \tag{3.14}$$

$$w(r, z) < g_q(r, H) \ \text{ in } \Omega \setminus \Omega_1 \tag{3.15}$$

where

$$\Gamma_D = \bigcup_{i=1}^{5} \Gamma_i, \ \Gamma_N = \Gamma_6 \cup \Gamma_7. \tag{3.16}$$

$$g_q = \begin{cases} 0, & \text{on } \Gamma_1 \\ Hz - \dfrac{z^2}{2}, & \text{on } \Gamma_2 \\ \dfrac{H^2}{2} + q\ln\dfrac{r}{r_1}, & \text{on } \Gamma_3 \cup \Gamma_4 \\ \dfrac{H^2}{2} + q\ln\dfrac{r_0}{r_1} - \dfrac{(h_w - z)^2}{2}, & \text{on } \Gamma_5 \end{cases} \tag{3.17}$$

$$g_N = \begin{cases} h_w - h, & \text{on } \Gamma_6 \\ 0, & \text{on } \Gamma_7 \end{cases} \tag{3.18}$$

Lemma 3.3. *There exists a function* v_q *such that*

$$\begin{cases} v_q|_{\Gamma_D} = g_q \\ \dfrac{\partial v_q}{\partial n}|_{\Gamma_N} = g_N \\ v_q \in V^2(D) \cap C^1(\bar{D}), \ Lv_q \in L^\infty(D; r). \end{cases} \tag{3.19}$$

By using Lemma 3.3 and a result in Chang and Jiang [1978] we can prove

Theorem 3.4.

$$w \in V^2(D) \cap C^1(\bar{D}). \tag{3.20}$$

4 Variational Inequalities (VI) Satisfied by w: Regularity of the Solution of VI

Define functions for every $v \in V^1(D)$:

$$v' = \begin{cases} v, & \text{in} \quad \Omega_1 \cup \{v \leqslant g_{qH}\}, \\ g_q(r,H), & \text{in} \quad \{v > g_{qH}\}. \end{cases} \tag{4.1}$$

$$v'' = \begin{cases} 0, & \text{in} \quad \Omega_1 \cup \{v \leqslant g_{qH}\}, \\ v - g_q(r,H), & \text{in} \quad \{v > g_{qH}\}. \end{cases} \tag{4.2}$$

where

$$\{v \leqslant g_{qH}\} = \{(r,z) \in D | r \leqslant r_0, v(r,z) \leqslant g_q(r,H)\},$$

$$\{v > g_{qH}\} = \{(r,z) \in D | r > r_0, v(r,z) > g_q(r,H)\}.$$

Then, clearly, we have

$$v = v' + v'',\ v'' \geqslant 0,\ v' \leqslant g_q(r,H) \ \text{ for } \ r \in]r_0, r_1[,\ v',\ v'' \in V^0(D). \tag{4.3}$$

Let

$$K_q = \{v \in V^1(D) | v = g_q \ \text{ on } \ \Gamma_D\}. \tag{4.4}$$

We have

Theorem 4.1. *If u is a solution of (PPW), then w defined by (3.2) is a solution of the VI:*

$$\begin{cases} w \in K_q, \\ \displaystyle\int_D r\nabla w \cdot \nabla(v-w)\mathrm{d}r\mathrm{d}z - (h_w - h)\int_0^{r_0} r(v-w)|_{z=h}\mathrm{d}r - \int_D (v'-w')r\mathrm{d}r\mathrm{d}z \geqslant 0, \ \text{ for } \ v \in K_q. \end{cases} \tag{4.5}$$

Proof. By (3.20) we have $w \in V^2(D)$. Apply to w and to any $v \in K_q$ the following Green's formula:

$$\begin{aligned}
&\int_D r\nabla w \cdot \nabla(v-w)\mathrm{d}t\mathrm{d}z \\
&= -\int_D rLw \cdot (v-w)\mathrm{d}r\mathrm{d}z + \int_{\Gamma_D} r(v-w)\frac{\partial w}{\partial n}\mathrm{d}s + \int_{\Gamma_N} r(v-w)\frac{\partial w}{\partial n}\mathrm{d}s \\
&= \int_D r\Phi_\Omega \cdot (v-w)\mathrm{d}r\mathrm{d}z + (h_w - h)\int_0^{r_0} (v-w)|_{z=h} r\mathrm{d}r \\
&= \int_\Omega r(v-w)\mathrm{d}r\mathrm{d}z + (h_w - h)\int_0^{r_0} (v-w)|_{z=h} r\mathrm{d}r \\
&\geqslant \int_\Omega r(v'-w')\mathrm{d}r\mathrm{d}z + (h_w - h)\int_0^{r_0} (v-w)|_{z=h} r\mathrm{d}r \quad (v'' \geqslant 0 \ \textit{and} \ w'' = 0).
\end{aligned}$$

But $-\int_{D\setminus\Omega}(v'-w')r\mathrm{d}r\mathrm{d}z = -\int_{D\setminus\Omega}[v'-g_q(r,H)]r\mathrm{d}r\mathrm{d}z \geqslant 0$ (by (3.17)). Hence (4.5) is valid. □

Remark 4.1. It is easily seen (by (3.14) and (3.15)) that w is also a solution of the VI

$$\begin{cases} w \in K_q^*, \\ \displaystyle\int_D r\nabla w\cdot\nabla(v-w)\mathrm{d}r\mathrm{d}z-(h_w-h)\int_0^{r_0}(v-w)|_{z=h}r\mathrm{d}r-\int_D r(v-w)\mathrm{d}r\mathrm{d}z \geqslant 0, \ \text{for} \ v\in K_q^*, \end{cases} \tag{4.6}$$

where

$$K_q^* = \{v\in V^1(D)|v=g_q \ \text{ on } \ \Gamma_D, v\leqslant g_q(r,H) \ \text{ in } \ D\setminus\Omega_1\}. \tag{4.7}$$

Remark 4.2. Noting (3.7) we have that w is also a solution of VI

$$\begin{cases} w \in K_q^{**}, \\ \displaystyle\int_D r\nabla w\cdot\nabla(v-w)\mathrm{d}r\mathrm{d}z-(h_w-h)\int_0^{r_0}(v-w)|_{z=h}r\mathrm{d}r-\int_D r(v-w)\mathrm{d}r\mathrm{d}z \geqslant 0, \ \text{for} \ v\in K_q^{**}, \end{cases} \tag{4.8}$$

where

$$K_q^{**} = \{v\in K_q^*|v\geqslant 0 \ \text{ in } \ D\}. \tag{4.9}$$

Remark 4.3. For numerical solutions (4.8) is the most convenient VI. By the well-known result (Lions [1971, p.9]), problem (4.8) creates a minimization problem on a convex set as follows: find $w\in K_q^{**}$ such that

$$J(w) = \min_{v\in K_q^{**}} J(v), \tag{4.10}$$

where

$$J(v) = \int_D r|\nabla v|^2\mathrm{d}r\mathrm{d}z - 2(h_w-h)\int_0^{r_0} v_{z=h}r\mathrm{d}r - 2\int_D vr\mathrm{d}r\mathrm{d}z. \tag{4.11}$$

(4.11) is the basis of numerical solutions to (PPW) by using VI's.

For $q\in\mathcal{R}$, (4.5) is a family of VI's. So are (4.6) and (4.8). Now study these families.

Proposition 4.2. *$\forall q\in\mathcal{R}$, (4.5) has unique solution W_q.*

Proof. Set

$$V = \{v\in V^1(D)|v=0 \ on \ \Gamma_1\} \equiv V_0^1(D;\Gamma_1)$$

$$a(u,v) = \int_D r\nabla u\cdot\nabla v\mathrm{d}r\mathrm{d}z$$

$$f(v) = \int_D v'r\mathrm{d}r\mathrm{d}z + (h_w-h)\int_0^{r_0} rv|_{z=h}\mathrm{d}r$$

Then V is a Hilbert space with inner product $(u,v)_V = \int_D r(uv+\nabla u\cdot\nabla v)\mathrm{d}r\mathrm{d}z$; K_q is a closed, convex, non-empty (e.g., $v_q\in K_q$; see (3.19)) subset of V; $a(u,v)$ is a bilinear, continuous and coercive form on $V\times V$ (by Lemma 1.6); and it is easy to show that $f(v)$ is a convex, continuous functional on V with $f(v)\neq-\infty$ and $f(v)\not\equiv+\infty$. By the well known theorem (Lions and Stampacchia [1967, Theorem 2.2]), we obtain the conclusion of our proposition. □

Proposition 4.3. *$\forall q \leqslant q_0$, where*

$$q_0 = \frac{H^2 - (h_w - h)^2}{2\ln(r_1/r_0)} \tag{4.12}$$

(4.6) has a unique solution. (4.8) also has a unique solution.

The proof is similar to that of Proposition 4.2. The condition $q \leqslant q_0$ ensures that both K_q^* and K_q^{**} are non-empty.

We will prove later that the problems (4.5), (4.6) and (4.8) are equivalent for $q < q_0$ (Theorem 4.15). Now we study (4.5) in detail.

Proposition 4.4. *$\forall q \leqslant \mathcal{R}$, the solution w_q of (4.5) satisfies, in the sense of distributions, that*

$$-1 \leqslant Lw_q \leqslant 0 \tag{4.13}$$

$$Lw_q \in L^\infty(D; r). \tag{4.14}$$

Proposition 4.5. *If w_q is a solution of (4.5), then*

$$\left.\frac{\partial w_q}{\partial n}\right|_{\Gamma_N} = g_N. \tag{4.15}$$

The proof of the proposition is based on the generalized Green's formula (see, for instance, Baiocchi and Capelo [1978], Appendix 4 of V.I).

Now we need the following results (see Chang and Jiang [1978]).

Lemma 4.6. *Let $f \in L^p(D; r), p \geqslant 2$. Then the problem*

$$\begin{cases} Lv = f, & in\ D, \\ v|_{\Gamma_D} = \left.\dfrac{\partial v}{\partial n}\right|_{\Gamma_N} = 0. \end{cases}$$

has unique weak solution v in $V^1(D)$, and $v \in C^0(\bar{D})$. Moreover, the linear operator $K: f \to v$, mapping $L^p(D; r)$ $(p \geqslant 2)$ into $C^0(\bar{D})$, is compact.

Lemma 4.7. *Let*

$$U(D) = \left\{ v \in V^2(D) | v = 0 \ on\ \Gamma_D,\ \frac{\partial v}{\partial n} = 0 \ on\ \Gamma_N \right\}.$$

Denote by $R(L)$ the range of the operator L as a map from $U(D)$ into $V^0(D)$. Denote by $R(L)^\perp$ the orthocomplement of $R(L)$ in $V^0(D)$. Then

$$\dim R(L)^\perp = 1$$

i.e. there exits $v^0 \in R(L)^\perp$ such that

$$R(L)^\perp = \{v \in V^0(D) | v = \mu v_0, \mu \in R\}.$$

Remark 4.4. It is easy to show that K is also a compact operator mapping $L^p(D; r)$ into $V^1(D)$.

The following result follows from Lemma 4.6.

Proposition 4.8. *If w_q is a solution of (4.5), then*

$$w_q \in C^0(\bar{D}). \tag{4.16}$$

We state some results without proofs.

Proposition 4.9. *Assume w_q is a solution of (4.5). Let*

$$W_q = \{(r,z) \in D | r > r_0, w_q < g_q(r,H)\} \cup \Omega_1$$

$$\Omega_q^* = \{(r,z) \in D | r > r_0, w_q < g_q(r,H)\}.$$

Then, in the sense of distributions,

$$Lw_q = \begin{cases} -1, & in \ \ \Omega_q, \\ 0, & in \ \ \Omega_q^*. \end{cases} \tag{4.17}$$

Lemma 4.10. *Let $f_2 = Lv_2$, where v_2 is defined by (3.19). If $v \in R(L)^{\perp}$, then*

$$\beta = \int_D v f_2 r \mathrm{d}r \mathrm{d}z \neq 0. \tag{4.18}$$

Theorem 4.11. *Assume v_q is defined by (3.19), $f_q = Lv_q; w_q$ is the solution of (4.5), $F_q = Lw_q$. Let $v \in R(L)^{\perp}$,*

$$G(q) = \int_D (F_q - f_q) v r \mathrm{d}r \mathrm{d}z. \tag{4.19}$$

Then the following two assertions are equivalent:

(1) $\bar{q}$ is a root of the equation

$$G(q) = 0; \tag{4.20}$$

(2)

$$w_{\bar{q}} \in V^2(D) \cap C^1(\bar{D}). \tag{4.21}$$

Lemma 4.12. *(4.20) has at least one real root.*

We call the solution of (4.5) regular if $w_q \in C^1(\bar{D}) \cap V^2(D)$. It follows immediately from Lemma 4.12 and Theorem 4.11 that the following theorem is valid.

Theorem 4.13. *There exits at least one $\bar{q} \in R$ such that the solution $w_{\bar{q}}$ of (4.5) is regular.*

Proposition 4.14. *If $w_{\bar{q}}$ is a regular solution of (4.5), then*

$$\bar{q} < q_0. \tag{4.22}$$

where q_0 is defined by (4.12).

To complete this section we display the relation between (4.5), (4.6) and (4.8).

Theorem 4.15. *If $q \leqslant q_0$, and w_q is the solution of (4.5), then*

$$\begin{cases} w_q \geqslant 0, & in \ \ \bar{D}, \\ w_q \leqslant g_q(r,H), & in \ \ D \setminus \Omega_1. \end{cases} \tag{4.23}$$

Moreover, (4.5), (4.6) and (4.8) are equivalent for $q \leqslant q_0$.

Proof. By Propositions 4.2 and 4.3, it is sufficient to prove (4.23). Let $q \leqslant q_0$ and w_q be the solution of (4.5). It follows from (4.13), (4.17), (3.15) and (3.16) that

$$Lw_q \leqslant 0, w_q|_{\Gamma_D} \geqslant 0, \frac{\partial w_q}{\partial n}|_{\Gamma_N} \geqslant 0. \tag{4.24}$$

It is easily shown by the maximum principle that $w_q \geqslant 0$ in $\bar{D}$. We now prove the second part of (4.23).

Let Ω_q^* be defined as in Proposition 4.9. Noting that

$$w_q = \begin{cases} w_q, & \text{in } \Omega_1 \\ \max(w_q, g_q(r,H)), & \text{in } D \setminus \Omega_1. \end{cases}$$

and $w_q \in C^0(\bar{D})$ we may prove that $w_q' \in V^1(D)$ (see for instance Kinderlehrer and Stampacchia [1980, p.50]). Hence $w_q'' \in V^1(D) \cap C^0(\bar{D})$. We have

$$Lw_q'' = Lw_q - Lg_q(r,H) = 0, \quad \text{in } \Omega_q,$$

$$w_q''|_{\partial\Omega_q^*} = 0, \quad (\text{by that } w_q'' \in C^0(\bar{D}) \text{ and } w_q''|_{\partial D} = 0)$$

It follows from the maximum principle that $w_q'' \equiv 0$ in Ω_q^*. Hence $w_q'' \equiv 0$ in D and $w_q = w_q'$. □

It follows from the theorem that

Corollary 4.16. *Under the same assumption as in Theorem 4.15 we have*

$$\left.\frac{\partial w}{\partial z}\right|_{\Gamma_D} \geqslant 0. \tag{4.25}$$

5 The Existence of the Solution of (PPW)

In this section we prove that a regular solution of (4.5) corresponds to a solution of (PPW). Following the framework of Baiocchi et al.[1973], we can easily prove the following lemmas and theorem.

Throughout this section let w_q be a regular solution of (4.5) and let Ω_q be defined as in Proposition 4.9.

Lemma 5.1.

$$\frac{\partial w_q}{\partial z} \geqslant 0 \quad in \ \bar{D}. \tag{5.1}$$

Lemma 5.2. *If $q > 0$, then*

$$0 \leqslant \frac{\partial w_q}{\partial r} \leqslant \frac{q}{r} \quad in \ D. \tag{5.2}$$

Remark 5.1. Similarly we have that if $q < 0$ then

$$\frac{q}{r} \leqslant \frac{\partial w_q}{\partial r} \leqslant 0 \quad \text{in} \quad D. \tag{5.3}$$

Remark 5.2. If $q = 0$, then $\partial w_0/\partial r = 0$. Hence

$$Hz - \frac{z^2}{2} = \frac{H^2}{2} - \frac{(h_w - z)^2}{2} \qquad \text{for} \quad h \leqslant z \leqslant h_w.$$

It requires that $H = h_w$. We have assumed that $h_w < H$. So w_0 is not regular solution.

Remark 5.3. By using (5.1), (5.2) and (5.3) we may easily show that if $(r, z) \in D \setminus \Omega_q$ then

$$r \geqslant r_0, \quad z \geqslant h_w.$$

Lemma 5.3.

$$\frac{\partial w_q}{\partial z} = 0 \quad on \ \ \Gamma_3. \tag{5.4}$$

Lemma 5.4. *$\partial\Omega_q \cap D$ does not contain any vertical or horizontal line segment, and $\partial\Omega_q \cap \Gamma_3 = \emptyset$.*

Theorem 5.5. *If $q > 0$, and*

$$\Omega_q = \Omega_1 \cup \{(r, z) \in D | r > r_0, w_q < g_q(r, H)\}, \tag{5.5}$$

$$\varphi_q(r) = \sup\{z | (r, z) \in \Omega_q\} \quad for \ \ r \in]r_0, r_1[, \tag{5.6}$$

$$\varphi_q(r_0) = \lim_{r \to r_0+0} \varphi_q(r), \ \ \varphi_q(r_1) = \lim_{r \to r_1-0} \varphi_q(r), \tag{5.7}$$

$$\bar{u}_q = \frac{\partial w_q}{\partial z} + z \quad in \ \ \bar{D}, \ \ u_q = \bar{u}_q|_{\Omega_q}. \tag{5.8}$$

then $\{u_q, \varphi_q(r)\}$ is the solution of (PPW).

If $q < 0$, then by using similar arguments we obtain that $\varphi_q(r)$ is strictly decreasing, continuous for $r \in]r_0, r_1[$, and that $\varphi_q(r_1) = H$. This is absurd. Hence we obtain (recall (4.22) and Remark 5.2):

Proposition 5.6. *If w_q is a regular solution, then*

$$0 < q < q_0. \tag{5.9}$$

Proposition 5.7. *Let*

$$Q^* = \{q | w_q \text{ is a regular solution of (4.5)}\}. \tag{5.10}$$

Then,

$$Q^* \subset]0, q_0[, \tag{5.11}$$

$$Q^* \text{ is a closed set}, \tag{5.12}$$

$$w_q \text{ is nonincreasing on } \ Q^*. \tag{5.13}$$

Proof. (5.11)is clear by virtue of (5.9). (5.12) follows immediately from Theorem 4.11 and the continuity of $G(q)$ (see the proof of Lemma 4.12). Now we prove (5.13). Let $q_1, q_2 \in Q^*, q_1 < q_2$, and

$$E = \{(r, z) \in D | w = w_{q_1} - w_{q_2} < 0\}.$$

Then $w_{q_1} < w_{q_2} < g_{q_2}(r, H) < g_{q_1}(r, H)$ in $E \cap \{r > r_0\}$. Hence $E \subset \Omega_{q_1}$, and $Lw \leqslant 0$ in E (by Propositions 4.4 and 4.9). w has strictly negative minimum on $\bar{E}$ which lies on ∂E; but not on $\partial E \cap D$ where $w = 0$; nor on Γ_D where $w \geqslant 0$; nor on Γ_6 where $\partial w / \partial n = 0$; nor on Γ_7 (by Remark 3.2). This is absurd. Hence $E = \emptyset$. □

Let $q_m = \inf_{Q^*}\{q\}$, $q_M = \sup_{Q^*}\{q\}$. By (5.11) and (5.12) we have

$$q_m, q_M \in Q^*, \quad q_m > 0, \ q_M < q_0.$$

From (5.13) follows immediately the theorem

Theorem 5.8. *For any $q \in Q^*$ we have*

$$w_{q_m} \geqslant w_q \geqslant w_{q_M} \quad in\ \bar{D}. \tag{5.14}$$

References

[1] H. W. Alt, Strömungen druch inhomogene poröse Medien mit freiem Rand, *Journal Für Die Reine Und Angewandte Mathematik*, 305 (1979), 89-115.

[2] C. Baiocchi, Problém á frontiére libre en hydraulique, *Comptes Rendus de l'Académie des Sciences-Series A*, 278 (1974), 1201-1204.

[3] C. Baiocchi, Inequations quasi-variationnelles dans les problemes a frontiere libre en hydraulique, *Lecture Notes in Mathematics*, Springer, Berlin, 503 (1976), 1-7.

[4] C. Baiocchi and A. Capelo, *Disequazioni variazionali e quasi-variazionali; applicazioni a problemi di frontiera libera*, Quaderni U.M.I., Pitagora, 1978.

[5] C. Baiocchi, V. Comincioli, E. Magenes and G.A. Pozzi, Free boundary problems in the theory of fluid flow through porous media: existence and uniqueness theorems, *Annali Di Matematica Pura Ed Applicata*, 97 (1973), 1-82.

[6] V. Benci, On a filtration problem through a porous medium, *Annali Di Matematica Pura Ed Applicata*, 100 (1974), 191-209.

[7] K.C. Chang, The obstacle problem and partial differential equations with discontinuous nonlinearities, *Communications on Pure & Applied Mathematics*, 33 (1980), 117-146.

[8] K.C. Chang and L.S. Jiang, The tree boundary problem of the stationary water cone, *Acta Scicentiarum Naturalum Universitis Pekinesis*, 1 (1978), 1-15.

[9] C. W. Cryer, *A survey of steady-state porous flow free boundary problems*, Technical Summary Report No.1657, Wisconsin Univ. Madison. Mathematics Research Center, 1976.

[10] C. W. Cryer, The solution of the axisymmetric elastic-plastic torsion of a shaft using variational inequalities, *Journal of Mathematical Analysis & Applications*, 76 (1980), 535-570.

[11] C. W. Cryer and H. Fetter, The numerical solution of axisymmetric free boundary porous flow well problems using variational inequalities, *Constructive Methods for Nonlinear Oscillations*, pp. 177-191, Birkhäuser Verlag, Basel, 1979.

[12] M. S. Hantush, Hydraulics of wells, *Advances of Hydroscience*, 1 (1964), 281-432.

[13] G. N. Jakovlev, On the density of finite functions in weight spaces, *Doklady Akademii Nauk SSSR*, 170 (1966), 1300-1302.

[14] D. Kinderlehrer and G. Stampacchia, *An instroduction to variational inequalities and their applications*, Academic Press, New York, 1980.

[15] S.H. Leventhal, Method of moments for singular problems, *Computer Methods in Applied Mech. and Engineering*, 6 (1975), 79-100.

[16] J.L. Lions, *Optimal control of systems governed by partial differential equations*, Springer, Berlin, 1971.

[17] J.L. Lions and E. Magenes, *Non-homogeneous boundary value problems and applications, I*, Springer-Verlag, New York, 1972.

[18] J.L. Lions and G. Stampacchia, Variational inequalities, *Communications on Pure & Applied Mathematics*, 20 (1967), 493-519.

[19] M.K.V. Murthy and G. Stampacchia, Boundary value problems for some degenerate elliptic operators, *Annali Di Matematica Pura Ed Applicata*, 80 (1968), 1-222.

[20] B.S. Rama Rao and R.N. Das, Free surface flow to a partial penetrating well, *Second Int. Symp. on FEM in Flow Problems, Santa Margherita, Italy, Preprint*, (1976), 473-484.

[21] L. Schwarz, *Théorie des distributions, (Nouvelle edition)*, Hermann, Paris, 1973.

[22] S.L. Sobolev, *Applications of functional analysis in mathematical physics*, Izdat. Leningrad, Gos. Univ., Leningrad, 1950.

[23] N.S. Trudinger, Linear elliptic operators with measurable coefficients, *Annali Scuola Norm. Sup. di Pisa*, 17 (1973), 265-308.

[24] S.Z. Zhou, Functional spaces $W_{p,1}^{m}$, *Journal of Hunan University*, 1980, No. 1, 1-9.

Perturbation for Elliptic Variational Inequalities*

S.Z. Zhou (周叔子)†

Abstract In this paper we consider the perturbation problems for linear elliptic variational in equalities in Hilbert spaces.A more general sufficient condition for the convergence of perturbed solution to the original solution is derived and applied to deal with the boundary value perturbation problem and two practical variational inequalities.

Keywords Variational inequality, perturbation, Hausdorff distance.

Perturbation problems of elliptic boundary value problems have been attracted people's attention for a long time because the mathematical models and data are a kind of approximation to the object of study. In the classical theory of partial differential equations, the stability of classical solutions with respect to boundary data and the right-side term of equations was discussed. The author of [1] studied the stability of weak solutions with respect to the perturbation of the bilinear forms and linear functionals corresponding to the differential equations. After that, [2] studied the stability of weak solution with respect to the differential operator, right-side term, boundary values as well as the domain of definition of solution. The study of the perturbation for variational inequality appeared some time later. Recently, many authors [3] discussed the perturbation problems for variational inequalities in R^n. As for perturbation of the variational inequalities corresponding to the differential operator, as far as the author knows, there have been only the simpler results in [4] so far.

The main difference between perturbation problems for variational inequalities and the perturbation problems for boundary value problems of elliptic differential equations is that the perturbation of the closed, convex subset K defined by constraint conditions must be studied, which is the main difficulty of the research work. In order to overcome this difficulty, Dafermos [3] has proposed a more general and efficient method for the variational inequalities in R^n. This paper extends the method to an infinite dimensional space. A more general sufficient condition for the convergence of perturbed solution to the original solution is derived and applied to deal with the boundary value perturbation problem and two practical variational inequalities.

* 发表于: Science in China, Series A, Vol. 36, 1991, 650-659.

† Department of Applied Mathematics, Hunan University, Changsha 410082, China.

1 Basic Theorem

Let V be a Hilbert space, $a(u,v)$ be a bilinear form on $V \times V$, f be a linear continuous functional on V, and K be a nonempty closed and convex subset of V. The correspondent variational inequality is: find $u \in K$ such that

$$a(u, v-u) \geqslant f(v-u), \quad \forall\, v \in K. \tag{1.1}$$

The corresponding perturbed problem is: find $u_\varepsilon \in K_\varepsilon$ such that

$$a_\varepsilon(u_\varepsilon, v-u_\varepsilon) \geqslant f_\varepsilon(v-u_\varepsilon), \quad \forall\, v \in K_\varepsilon, \tag{1.2}$$

where a_ε, f_ε and K_ε denote the perturbed bilinear form, linear functional and nonempty, closed, convex subset respectively. For convenience, let

$$a_0 = a,\ f_0 = f,\ K_0 = K,\ u_0 = u.$$

One of the basic problems in perturbation theory is to prove that u_ε is continuous at $\varepsilon = 0$ with respect to ε under some conditions, or more precisely, is to prove that $\| u_\varepsilon - u\| \to 0$ when $\varepsilon \to 0$ under some conditions.

From now on we suppose that there exist positive numbers ε_0, α and M such that there uniformly hold

$$a_\varepsilon(v,v) \geqslant \alpha\|v\|^2, \tag{1.3}$$

$$|a_\varepsilon(v,w)| \leqslant M\|v\| \cdot \|w\|, \tag{1.4}$$

$$|f_\varepsilon(v)| \leqslant M\|v\|, \tag{1.5}$$

for $|\varepsilon| \leqslant \varepsilon_0$ and $v,w \in V$. Then it follows from a well-known theorem (e.g., see [5, 6]) that both problems (1.1) and (1.2) have a unique solution, and there exists linear continuous operator A_ε mapping V into V such that

$$(A_\varepsilon v, w) = a_\varepsilon(v-w), \quad \forall\, v, w \in V, \tag{1.6}$$

which and (1.4) imply that we have for $|\varepsilon| \leqslant \varepsilon_0$

$$\|A_\varepsilon\| \leqslant M. \tag{1.7}$$

Let P_ε be the projector upon K_ε. Denote still by f_ε the element in V corresponding to the functional $v \to f_\varepsilon(v)$ according to the Riesz representation theorem. For any positive number ρ, write

$$G(v,s) = P_\varepsilon[v - \rho(A_\varepsilon v - f_\varepsilon)].$$

Then it is easy to prove (see [6, p.49]) that the solution u_ε of (1.2) satisfies

$$u_\varepsilon = G(u_\varepsilon, \varepsilon).$$

The proof of the following lemma may be found also on [6, p.49].

Lemma 1.1. *Suppose* $0 < \rho < 2\alpha M^{-2}$. *Then the operator* $v \to G(v,\varepsilon)$ *mapping* V *onto* K_ε *is a contraction operator with contraction number*

$$\beta = (1 - 2\rho\alpha + \rho^2 M^2)^{1/2}.$$

In order to ensure the convergence of the perturbed solution to the original solution we suppose that the following conditions hold (see [1]): there exist non-negative functions $b(\varepsilon)$ and $c(\varepsilon)$ such that

$$b(\varepsilon), c(\varepsilon) \to 0, \quad (\varepsilon \to 0) \tag{1.8}$$

$$|a_\varepsilon(v,w) - a(v,w)| \leqslant b(\varepsilon)\|v\| \cdot \|w\|, \tag{1.9}$$

$$|f_\varepsilon(v) - f(v)| \leqslant c(\varepsilon)\|v\|. \tag{1.10}$$

Theorem 1.2. *Suppose that the conditions (1.3)–(1.5) and (1.8)–(1.10) hold, and that for fixed* $v \in V$ *the mapping* $\varepsilon \to P_\varepsilon v$ *is continuous at* $\varepsilon = 0$. *Then* u_ε *is continuous at* $\varepsilon = 0$ *with respect to* ε, *i.e.* u_ε *converges to* u *in* V *when* $\varepsilon \to 0$.

Proof. By Lemma 1.1, $v \to G(v,\varepsilon)$ is a contraction operator with contraction number β for sufficiently small ρ. Then we have

$$\|u_\varepsilon - u\| \leqslant (1-\beta)^{-1}\|G(u,s) - G(u,0)\|. \tag{1.11}$$

Since the projector is nonexpansive, we have

$$\begin{aligned}\|G(u,\varepsilon) - G(u,0)\| &= \|P_\varepsilon(u - \rho(A_\varepsilon u - f_\varepsilon)) - P_0(u - \rho(A_0 u - f_0))\| \\ &\leqslant \rho\|A_\varepsilon u - A_0 u - (f_\varepsilon - f_0)\| + \|P_\varepsilon(u - \rho(A_0 u - f_0)) - P_0(u - \rho(A_0 u - f_0))\|.\end{aligned}$$

For $\varepsilon \to 0$, at the right side of the above inequality, the second term tends to zero due to the continuity of mapping $\varepsilon \to P_\varepsilon v$ at $\varepsilon = 0$, and the first term tends to zero due to (1.8)–(1.10) and (1.6). This fact in combination with (1.11) implies the conclusion of the theorem. □

For the continuity of the mapping $\varepsilon \to P_\varepsilon v$ at $\varepsilon = 0$, Theorem 1.4 gives a sufficient condition, which is also a necessary condition in some cases, just as Theorem 1.6 says.

Lemma 1.3. *Let* K *be a nonempty, closed and convex set,* $v \in V$, $w \in K$. *Then*

$$\|w - v\|^2 \geqslant \|v - P_k v\|^2 + \|w - P_k v\|^2,$$

where P_k *is projector upon* K.

The lemma can be easily proved by using the properties of projector.

Theorem 1.4. *Suppose the mapping* $\varepsilon \to K_\varepsilon$ *is continuous under the Hausdorff distance at* $\varepsilon = 0$. *Then the mapping* $\varepsilon \to P_\varepsilon v$ *is continuous at* $\varepsilon = 0$ *with respect to* ε *for any* $v \in V$.

Proof. We need only discuss the following two cases.

(i) If $v \in K$, then $P_0 v = v$. Thus we have

$$\|P_\varepsilon v - P_0 v\| = \text{dist}(v, K_\varepsilon) \leqslant d^*(K, K_\varepsilon), \tag{1.12}$$

where d^* is the Hausdorff distance. By (1.12) and the conditions of our theorem, we know that the mapping $\varepsilon \to P_\varepsilon v$ is continuous at $\varepsilon = 0$.

(ii) If $v \notin K$, then $l = \|v - P_0 v\| \neq 0$. If the conclusion of our theorem is wrong, then there exist $v \in V$, $\rho_0 > 0$, and an arbitrarily small number $\varepsilon \neq 0$ such that

$$\|P_\varepsilon v - P_0 v\| \geqslant \rho_0. \tag{1.13}$$

Suppose $w \in K$. By Lemma 1.3 we have

$$\|w - v\| \geqslant \left(l^2 + \|w - P_0 v\|^2\right)^{1/2}.$$

If $\|w - P_0 v\| \geqslant \rho/2$, then for ρ_0 small enough we have

$$\|w - v\| \geqslant l + l^{-1}\rho_0^2/9. \tag{1.14}$$

By the assumptions of our theorem, there exists $\varepsilon^* > 0$ such that for $|\varepsilon| < \varepsilon^*$ the following holds

$$d^*(K_\varepsilon, K) < \min\{l^{-1}\rho_0^2/18,\ \rho_0/2\}. \tag{1.15}$$

Hence there exists $w \in K$ such that

$$\|P_\varepsilon v - w\| \leqslant \min\{l^{-1}\rho_0^2/18,\ \rho_0/2\}. \tag{1.16}$$

(1.13), (1.16) and (1.14) imply that

$$\|v - P_\varepsilon v\| \geqslant l + l^{-1}\rho_0^2/18. \tag{1.17}$$

On the other hand, by (1.15) we know that there exists $z \in K_\varepsilon$ such that

$$\|z - P_0 v\| < l^{-1}\rho_0^2/18,$$

which together with the properties of projector yields

$$\|v - Pv\| \leqslant \|v - z\| < l + l^{-1}\rho_0^2/18.$$

It contradicts (1.17). The theorem has been proved. □

Corollary 1.5. *Suppose that (1.3)–(1.5) and (1.8)–(1.10) hold, and the mapping $\varepsilon \to K_\varepsilon$ is continuous under the Hausdorff distance at $\varepsilon = 0$. Then u_ε converges to u in V for $\varepsilon \to 0$.*

Theorem 1.6. *Suppose that K and K_ε are contained in some compact set. Then the converse proposition of Theorem 1.4 is also true.*

Proof. Suppose the mapping $\varepsilon \to K_\varepsilon$ is not continuous under the Hausdorff distance at $\varepsilon = 0$. Then there exist $\rho_0 > 0$, sequence $\{\varepsilon_n\}$ and corresponding nonempty closed convex sets $\{K_n\}$ such that

$$\varepsilon_n \to 0, \quad (n \to \infty),$$

and

$$\max\{\sup_{v\in K} \text{dist}(v, K_n), \sup_{w\in K} \text{dist}(K, w)\} \geqslant \rho_0.$$

This means that we need only consider the following two cases:

(i) There exists a subsequence of $\{K_n\}$, denoted still by $\{K_n\}$, and the corresponding $\{v_n\} \subset K$ such that

$$\text{dist}(v_n, K_n) \geqslant \rho_0/2, \quad n = 1, 2, \cdots.$$

(ii) There exists a subsequence of $\{K_n\}$, denoted still by $\{K_n\}$, and $w_n \in K_n$ such that

$$\text{dist}(K, w_n) \geqslant \rho_0/2, \quad n = 1, 2, \cdots. \tag{1.18}$$

Suppose (i) holds. Since $\{v_n\}$ is contained in a compact set, there exists a subsequence (denoted still by $\{v_n\}$) convergent strongly to $v_0 \in K$. So when n is large enough we have

$$\|P_0 v_0 - P_n v_0\| = \|v_0 - P_n v_0\| \geqslant \rho_0/4,$$

where P_n is a projector upon K_n. The above expression means that $P_n v_0$ does not converge to $P_0 v_0$ when $n \to \infty$, i.e., the mapping $\varepsilon \to P_\varepsilon v_0$ is not continuous at $\varepsilon = 0$.

Suppose (ii) holds. Since $\{w_n\}$ is contained in a compact set, there exists a subsequence, denoted still by $\{w_n\}$, convergent to $w_0 \in V$. We have by (1.18)

$$\|w_0 - P_0 w_0\| \geqslant \rho_0/2. \tag{1.19}$$

On the other hand, we know that

$$\|w_n - P_n w_0\| \leqslant \|w_n - w_0\| \to 0, \quad (n \to \infty). \tag{1.20}$$

If the mapping $\varepsilon \to P_\varepsilon w_0$ is continuous at $\varepsilon = 0$, then

$$\|w_n - P_n w_0\| \to \|w_0 - P_0 w_0\|, \quad (n \to \infty),$$

which together with (1.20) contradicts (1.19). So the mapping $\varepsilon \to P_\varepsilon w_0$ is not continuous at $\varepsilon = 0$. The proof is completed. □

2 Two Lemmas on Hausdorff Distance

In the argument of Sections 3 and 4, we need the following properties of the Hausdorff distance, for which we have the following two lemmas.

Lemma 2.1. *Let V be a linear normed space,K_1 and K_2 nonempty, closed subsets, w_1^*,$w_1^* \in V$ and*

$$K_i^* = K_i - w_i^* := \{v^* : v^* = v - w_i^*, v \in K_i\}, i = 1, 2.$$

Then

$$d^*(K_1^*, K_2^*) \leqslant d^*(K_1, K_2) + \|w_1^* - w_2^*\|.$$

Proof. For any $v_i^* \in K_i^*$ there exists a unique $v_i \in K_i$ such that

$$v_i^* = v_i - w_i^*, \quad i = 1, 2.$$

Thus,

$$\|v_1^* - v_2^*\| \leqslant \|v_1 - v_2\| + \|w_1^* - w_2^*\|, \forall v_i^* \in K_i^*, i = 1, 2.$$

Hence it is obvious that

$$\sup_{v_1^* \in K_1^*} \text{dist}(v_1^*, K_2^*) \leqslant \sup_{v_1 \in K_1} \text{dist}(v_1, K_2) + \|w_1^* - w_2^*\|,$$
$$\sup_{v_2^* \in K_2^*} \text{dist}(K_1^*, v_2^*) \leqslant \sup_{v_2 \in K_2} \text{dist}(K_1, v_2) + \|w_1^* - w_2^*\|.$$

By using the two inequalities and the definition of the Hausdorff distance we may easily derive the conclusion of our lemma. □

Lemma 2.2. *Let V be a metric space,and suppose A, B_1, B_2 and C are subsets of V with $A \subset B_i \subset C$,$i = 1, 2$. Then*

$$d^*(B_1, B_2) \leqslant d^*(A, C).$$

Proof. Since $A \subset C$, we have

$$\sup_{v \in A} \text{dist }(v, C) = 0$$

and

$$d^*(A, C) = \sup_{w \in C} \text{dist }(A, w).$$

However there exists $A \subset B_i$; it follows that

$$\text{dist}(A, w) \geqslant \text{dist }(B_i, w), \quad i = 1, 2.$$

Noting $B_j \subset C$, we obtain

$$\sup_{w \in C} \text{dist }(B_i, w) \geqslant \sup_{w \in B_j} \text{dist }(B_i, w), \quad i, j = 1, 2$$

and

$$d^*(A, C) \geqslant \sup_{w \in B_j} \text{dist }(B_i, w), \quad i, j = 1, 2.$$

The conclusion of the lemma may easily be derived now. □

3 Perturbation of Boundary Value

In this section we study perturbation problems of boundary value for $V = H^m(\Omega)$. Consider the variational inequality corresponding to elliptic differential operator of order $2m$: find $u \in K$ such that

$$a(u, v-u) \geqslant f(v-u), \quad \forall v \in K, \tag{3.1}$$

where $a(u,v)$ is a bilinear form on $H^m(\Omega) \times H^m(\Omega)$, $f(v)$ stands for a linear, continuous functional on $H^m(\Omega)$, Ω refers to a bounded,open subset of R^n, K indicates a nonempty, closed and convex subsets of $H^m(\Omega)$, which may be denoted by $K = C \bigcap B$, C being a subset of $H^m(\Omega)$ defined by the given inequality constraints, and B means a subset of $H^m(\Omega)$ defined by the given boundary conditions, i.e.,

$$B = \{v \in H^m(\Omega) : \partial^j v/\partial \gamma^j = g_j \text{ on } \partial\Omega, j = 0, 1, \cdots, m-1\},$$

γ being the outward unit normal of $\partial\Omega$. For example,for the obstacle problem about elliptic operator of the second order with homogeneous boundary condition, we have

$$\begin{aligned} K &= \{v \in H^1(\Omega) : v \geqslant \varphi \text{ in } \Omega, v = 0 \text{ on } \partial\Omega\}, \\ C &= \{v \in H^1(\Omega) : v \geqslant \varphi \text{ on } \Omega\}, \\ B &= \{v \in H^1(\Omega) : v = 0 \text{ on } \partial\Omega\}. \end{aligned}$$

The perturbed problem relevant to the problem (3.1) is to find $u_\varepsilon \in K_\varepsilon$ such that

$$a_\varepsilon(u_\varepsilon, v-u_\varepsilon) \geqslant f_\varepsilon(v-u_\varepsilon), \quad \forall v \in K_\varepsilon, \tag{3.2}$$

where the implication of a_ε and f_ε is self-evident, $K_\varepsilon = C_\varepsilon \bigcap B_\varepsilon$, C_ε is obtained by perturbing the inequality constraints defined to C, B_ε is found by perturbing B, i.e.

$$B_\varepsilon = \{v \in H^m(\Omega) : \partial^i v/\partial r^i = g_{i\varepsilon} \text{ on } \partial\Omega,\ j - 0, 1, \cdots, m-1\}.$$

As the foregoing, let $a_0 = a$, $f_0 = f$, $C_0 = C$, $\cdots$.

In this section we always suppose that (1.3)–(1.5) and (1.8)–(1.10) are true, but among them we replace $\|\cdot\|$ by $\|\cdot\|_m$. We want to prove that under more general conditions the solution of (3.2) converges to the solution of (3.1) for $\varepsilon \to 0$. The method of argument is classical, i.e. reducing nonhomogeneous boundary value problem to homogeneous bound value problem and turning the perturbation of boundary value into that of "right-side term",then dealing with the problem by the basic theorem in Section 1.

Suppose $\partial\Omega \in C^m$ (suppose for $m = 1$ only that $\partial\Omega$ is locally Lipschitz continuous). Then by the trace theorem (see [7]) there exist w^* , $w_\varepsilon^* \in H^m(\Omega)$ such that

$$\partial^j w^*/\partial r^j = g_i, \partial^j w_\varepsilon^*/\partial r^j = g_{j\varepsilon} \text{ on } \partial\Omega, \quad j = 0, 1, \cdots, m-1,$$

and there exists a positive number N such that

$$\|w^* - w_\varepsilon^*\| \leqslant N \sum_{j=0}^{m-1} \|g_i - g_{i\varepsilon}\|_{m-i-1/2,\partial\Omega}. \tag{3.3}$$

Let

$$K^* = K - w^*, K_\varepsilon^* = K_\varepsilon - w_\varepsilon^*, C^* = C - w^*, C_\varepsilon^* = C_\varepsilon - w_\varepsilon^*.$$

Then it is easy to see

$$K^* = H_0^m(\Omega) \cap C^*, \quad K_\varepsilon^* = H_0^m(\Omega) \cap C_\varepsilon^*$$

and K^*,K_ε^* are nonempty, closed and convex subset of $H^m(\Omega)$.

The following lemma can be easily proved.

Lemma 3.1. *Suppose u_ε is the solution of (3.2) and $u_\varepsilon^* = u_\varepsilon - w_\varepsilon^*$. Then u_ε^* is the unique solution of the following variational inequality: find $u_\varepsilon^* \in K_\varepsilon^*$ such that*

$$a_\varepsilon(u_\varepsilon^*, v_\varepsilon^* - u_\varepsilon^*) \geqslant f_\varepsilon(v_\varepsilon^* - u_\varepsilon^*), \quad \forall v_\varepsilon^* \in K_\varepsilon^*, \tag{3.4}$$

where

$$f_\varepsilon^*(v) = f_\varepsilon(v) - a_\varepsilon(w_\varepsilon^*, v). \tag{3.5}$$

Conversely, if u_ε^* is the solution of (3.4), then $u_\varepsilon = u_\varepsilon^* + w_\varepsilon^*$ is the solution of (3.2).

Now we can prove the main result of this section.

Theorem 3.2. *Suppose*

(i) a_ε, f_ε satisfy (1.3)–(1.5) and (1.8)–(1.10), where $\|\cdot\|$ is replaced by $\|\cdot\|_m$;

(ii) the mapping $\varepsilon \to C_\varepsilon$ is continuous under the Hausdorff distance at $\varepsilon = 0$;

(iii) for $j = 0, 1, \cdots, m-1$, $g_{i\varepsilon}$ converges to g_i in $H^{m-i-1/2}(\Omega)$ for $\varepsilon \to 0$;

(iv) u, u_ε are the solution of (3.1), (3.2) respectively.

Then u_ε converges to u in $H^m(\Omega)$ when $\varepsilon \to 0$.

Proof. By Lemma 3.1 and (3.3) we know that

$$\|u_\varepsilon - u\|_{m'\Omega} \leqslant \|u_\varepsilon^* - u^*\|_{m,\Omega} + N \sum_{j=0}^{m-1} \|g_i - g_{i\varepsilon}\|_{m-i-1/2,\vartheta\Omega}. \tag{3.6}$$

In this case it suffices to prove

$$|u_\varepsilon^* - u^*\|_{m,\Omega} \to 0, \quad (\varepsilon \to 0) \tag{3.7}$$

According to Corollary 1.5 we need only verify the following two points.

(a) f_ε^* satisfies (1.5) and (1.10) except that the constant M and function $c(\varepsilon)$ may be varied.

(b) The mapping $\varepsilon \to K_\varepsilon^*$ is continuous under the Hausdorff distance at $\varepsilon = 0$.

Now we proceed to verify (a) first. (1.8)–(1.10), (3.3) and (3.5) imply that

$$|f_\varepsilon^*(v) - f^*(v)| \leqslant c^*(\varepsilon)\|v\|_{m,\Omega}, \tag{3.8}$$

where

$$c^*(\varepsilon) = c(\varepsilon) + MN \sum_{j=0}^{m-1} \|g_{i\varepsilon} - g_i\|_{m-i-1/2,\vartheta\Omega} + b(\varepsilon)\|w^*\|_{m,\Omega}.$$

Then it is easy to prove that (a) holds.

Next we verify (b). By Lemma 2.1, we have

$$d^*(C_\varepsilon^*, C^*) < d^*(C_\varepsilon, C) + \|w^* - w\|_{m,\Omega},$$

by which in combination with (3.3), (ii) and (iii) we can derive

$$d^*(C_\varepsilon^*, C^*) \to 0, \ (\varepsilon \to 0).$$

However, $K^* = H_0^m(\Omega)\bigcap C^*$, $K_\varepsilon^* = H_0^m(\Omega)\bigcap C_\varepsilon^*$. Hence we conclude that

$$d^*(K_\varepsilon^*, K^*) \to 0, \ (\varepsilon \to 0).$$

(b) is verified and Theorem 3.2 has been proved. □

Remark 3.1. The above arguments are valid and Theorem 3.2 holds if $H^m(\Omega)$ is replaced by its subspace. Then the main result in [4] is a corollary of Theorem 3.2 for $K = K_\varepsilon$, $a_\varepsilon = a$, $f_\varepsilon = f$.

4 Two Examples

In this section, by using the basic theorem we deal with the perturbation of two well-known variational inequalities; namely, the problem of elastic-plastic torsion of bar and the bending problem of thin plate with "mean curvatrue" constraints. Since the perturbation conditions are given by (1.8)–(1.10), we need only discuss the continuity of mapping $\varepsilon \to K_\varepsilon$ at $\varepsilon = 0$ under the Hausdorff distance provided some more general conditions on K_ε hold.

Example 1. The problem of elastic-plastic torsion (see [6]). In this case,$V = H_0^1(\Omega)$, $K = \{v \in H_0^1(\Omega) : \ |\nabla v| \leqslant 1 \ \text{ in } \ \Omega\}$. We can solve the problem in three steps.

(i) *Positive Perturbation.* The perturbed set is

$$K_\varepsilon^+ = \{v \in H_0^1(\Omega) : \ |\nabla v| \leqslant 1 + \alpha(\varepsilon) \ \text{ in } \ \Omega\},$$

where $\alpha(\varepsilon)$ defined for $|\varepsilon| \leqslant \varepsilon_0$,$0 \leqslant \alpha(\varepsilon) \leqslant 1$, and $\alpha(\varepsilon) \to 0$ if $\varepsilon \to 0$. Obviously, $K \subset K_\varepsilon^+$. As a result, we have

$$d^*(K, K_\varepsilon^+) = \sup_{w \in K_\varepsilon^+} \min_{v \in K} \| \ v - w \ \|_{1,\Omega} \, .$$

Since semi-norm $|w|_{1,\Omega}$ on $H_0^1(\Omega)$ is indeed a norm equivalent to $\|w\|_{1,\Omega}$, Ω is bounded, and $|\nabla w_\varepsilon| \leqslant 2$ for $|\varepsilon| \leqslant \varepsilon_0$, we know that K_ε is uniformly bounded in $H_0^1(\Omega)$ for $|\varepsilon| \leqslant \varepsilon_0$. Denote by M the bound. Then we see

$$d^*(K, K_\varepsilon^+) \leqslant \alpha(\varepsilon) M. \tag{4.1}$$

(ii) *Negative Perturbation.* The perturbed set is

$$K_\varepsilon^- = \{v \in H_0^1(\Omega) : \ |\nabla v| \leqslant 1 - \alpha(\varepsilon) \ \text{ in } \ \Omega\},$$

where $\alpha(\varepsilon)$ is the same as in (i) of this example. Similarly to (4.1), we deive that

$$d^*(K, K_\varepsilon^-) \leqslant \alpha(\varepsilon)M. \tag{4.2}$$

(iii) *General Perturbation.* The perturbed set is

$$K_\varepsilon = \{v \in H_0^1(\Omega) : \ |\nabla v| \leqslant \alpha(\varepsilon) \ \text{ in } \ \Omega\},$$

where $|\varphi(\varepsilon) - 1| \leqslant \alpha(\varepsilon)$, $\alpha(\varepsilon)$ is te same as in (i). Since $1 - \alpha(\varepsilon) \leqslant \varphi(\varepsilon) \leqslant 1 + \alpha(\varepsilon)$, we have $K_\varepsilon^- \subset K_\varepsilon \subset K_\varepsilon^+$. It follows from Lemma 2.2, (4.1) and (4.2) that

$$d^*(K, K_\varepsilon) \leqslant 2\alpha(\varepsilon)M, \tag{4.3}$$

which implies that the mapping $\varepsilon \to K_\varepsilon$ is continuous under the Hausdorff distance. Moreover, if $\alpha(\varepsilon) \leqslant L\varepsilon$ for $|\varepsilon| \leqslant \varepsilon_0$, then

$$d^*(K, K_\varepsilon) \leqslant 2LM\varepsilon. \tag{4.4}$$

Example 2. Bending problem of thin plate with "mean curvature" constraint. (see [8]). In the case,we have $V = H_0^2(\Omega)$, $K = \{v \in H_0^2(\Omega) : \ |\nabla v| \leqslant \beta \ \text{ in } \ \Omega\}$, $\alpha < 0 < \beta$.

Consider the case of positive perturbation first. The perturbed set is

$$K_\varepsilon^+ = \{v \in H_0^2(\Omega) : \alpha - \theta(\varepsilon) \leqslant \Delta v \leqslant \beta + \theta(\varepsilon) \ \text{ in } \ \Omega\},$$

where $\theta(\varepsilon)$ is defined for $|\varepsilon| \leqslant \varepsilon_0$, $0 \leqslant \theta(\varepsilon) \leqslant \min\{-\alpha, \beta\}$, and $\theta(\varepsilon) \to 0$ if $\varepsilon \to 0$. Obviously $K \subset K_\varepsilon^+$. Thus

$$d^*(K, K_\varepsilon^+) = \sup_{w \in K_\varepsilon^+} \text{dist } (K, w).$$

For $w_\varepsilon \in K_\varepsilon^+$, let $v_\varepsilon = (1 + \lambda(\varepsilon))^{-1} w_\varepsilon$, where

$$\lambda(\varepsilon) = \max\{-\alpha^{-1}\theta(\varepsilon), \beta^{-1}\theta(\varepsilon)\}.$$

Then $\lambda(\varepsilon) \to 0$ for $\varepsilon \to 0$, and $v_\varepsilon \in K$. For $H_0^2(\Omega)$, $\|\Delta v\|_{0,\Omega}$ is a norm equivalent to $\|v\|_{2,\Omega}$ (see [8]). Hence we may take

$$\text{dist } (v_1, v_2) = \|\Delta(v - v)\|_{0,\Omega}.$$

Then

$$[\text{dist } (K, w\varepsilon)]^2 \leqslant \int_\Omega |\Delta[v_\varepsilon - w_\varepsilon)|^2 \mathrm{d}x \leqslant 2\lambda^2(\varepsilon) \max\{-\alpha, \beta\} \text{meas}(\Omega).$$

Let $M^2 = 2\max\{-\alpha, \beta\}\text{meas}(\Omega)$. We have

$$d^*(K, K_\varepsilon^*) \leqslant M\lambda(\varepsilon).$$

Similarly to Example 1, we study negative and general perturbations, and then obtain

$$d^*(K, K_\varepsilon) \leqslant 2M\lambda(\varepsilon).$$

which means that the mapping $\varepsilon \to K_\varepsilon$ is continuous under the Hausdorff distance at $\varepsilon = 0$ since $\lambda(\varepsilon) \to 0$ for $\varepsilon \to 0$.

References

[1] 冯康, 基于变分原理的差分格式, 应用数学与计算数学, 2 (1965), 238-262.

[2] F. Stummel, Perturbation of domains in elliptic boundary value problems, *Lecture Notes in Mathematics*, 503 (1976), 110-136.

[3] S. Dafermos, Sensitivity analysis in variational inequalities, *Mathematics of Operations Research*, 13 (1988), 421-434.

[4] 许嘉谟, 陆君安, 变分不等式的边值摄动问题, 数学杂志, (1985), 265-271.

[5] D. Kinderlehrer and G. Stampacchin, *An Introduction to Variational Inequelities and Their Applications*, SIAM Philadelphia, 1980.

[6] 周叔子, 变分不等式及其有限元法, 湖南大学出版社, 1988.

[7] J.L. Lions and E. Magenes, *Non-Homogeneous Boundary Value Prolems and Applications, Vol.I*, Spring-Verlag, New York, 1972.

[8] R. Glowinski, L.D, Marini and M. Vidrascu, Finite-element approximations and iterative solutions of a fourth-order elliptic variational inequality, *IMA Journal of Numerical Analysis*, 4 (1984), 127-167.

On Multigrid Methods for Parabolic Problems*

S. Larsson† V. Thomée† S.Z. Zhou (周叔子)‡

Abstract Multigrid methods with nested subspaces and inherited forms are analyzed in an abstract framework that permits application to linear systems of the type that have to be solved at each time level in time-stepping methods for finite element approximations of parabolic problems. Convergence rates that are independent of the space and time steps are obtained in an appropriate time step dependent norm.

Keywords Multigrid method, finite element method, parabolic problem.

1 Introduction

In this article we discuss the solution of linear systems of equations

$$Au = f \tag{1.1}$$

by iterative methods of multigrid type. We are particularly interested in equations of the kind that arise when a parabolic problem, such as

$$U_t(x,t) - \Delta U(x,t) = g(x,t), \quad x \in \Omega,\ t > 0 \tag{1.2}$$

together with initial and boundary conditions, is discretized with respect to the time variable by a time-stepping method, and with respect to the spatial variable by a finite element method, The operator A is then typically of the form $A = zI - k\Delta_h$, where z is a complex number with Re $z > 0$, k is a small positive parameter (the local time step) and Δ_h is a discrete version of the Laplacian generated by a finite element method with spatial mesh size h. For instance, if the backward Euler method with time step k is used, then (1.2) is first replaced by

$$(U_n - U_{n-1})/k - \Delta U_n = g_n, \quad U_n \approx U(nk),$$

or

$$(I - k\Delta)U_n = U_{n-1} + kg_n,$$

* 发表于: Journal of Computational Mathematics, Vol. 13, 1995, pp. 193-205.

† Department of Mathematical Sciences, Chalmers University of Technology, Goeteborg, Sweden.

‡ Department of Applied Mathematics, Hunan University, Changsha 410082, China.

and the finite element discretization of this elliptic problem has the form (1.1) with $A = I - k\Delta_h$. Analogous equations are obtained in connection with other time-stepping methods such as A-stable onestep or multistep methods, see Section 3 below.

We first formulate iterative methods of multigrid type for solving (1.1) and demonstrate convergence results within an abstract framework that permits application to the situation described above, where the special feature is the presence of the small parameter k. The framework is essentially that used in [1], where applications to elliptic problems are analyzed under weak assumptions, and our results and their proofs are close to those of earlier work, e.g., [2–5]. We restrict our discussion here to the case of nested subspaces and inherited forms, and we make regularity assumptions that are satisfied for convex polygonal domains Ω. This makes it possible to organize the theory in straightforward and compact manner, basing it on three simple assumptions, and to make our paper selfcontained.

Our convergence results for parabolic problems are expressed in a certain k-dependent energy norm and show rates of convergence that are uniform with respect to h and k. They are of the form required in the analysis of incomplete iterations in [6] and [7]. By combining our results with those of [6] and [7] one may obtain estimates of the total error caused both by the discretization and the iterative solution of the algebraic equations.

Various issues concerning multigrid methods for parabolic problems have been addressed in earlier work, for example, in [2,8–13], but in most cases (except [2] and [13]) the convergence analysis is restricted to model problems with a uniform mesh, where Fourier methods can be applied.

2 Abstract Multigrid Analysis

In the first subsection we define the multigrid algorithm and prove some convergence results in the context of symmetric equations in an abstract framework. In the second subsection we extend the analysis to a non-symmetric equation with a special structure.

2.1 Symmetric equations

Let M be a finite dimensional Hilbert space with inner product $(\cdot,\ \cdot)$ and norm $\|\cdot\| = (\cdot,\ \cdot)^{1/2}$ and let $A(\cdot,\ \cdot)$ be a symmetric, positive definite bilinear form on M. With the linear operator $A : M \to M$ defined by

$$(Au, v) = A(u, v), \quad \forall u, v \in M, \tag{2.1}$$

our concern is to solve the equation

$$Au = f, \qquad \text{for } f \in M. \tag{2.2}$$

Our multigrid method for (2.2) is then the iterative method

$$u^l = u^{l-1} - B(Au^{l-1} - f), \quad l = 1, 2, \cdots, \tag{2.3}$$

where $B : M \to M$ is defined as follows. W assume that we are given a nested sequence of subspaces $M_1 \subset \cdots \subset M_j \subset M_{j+1} \subset \cdots \subset M_J = M$, and define the local version $A_j = M_j \to M_j$ of A by

$$(A_j u, v) = A(u, v), \quad \forall u, v \in M_j. \tag{2.4}$$

The algorithm then defines $B_j : M_j \to M_j$ recursively for $j = 1, \cdots, J$, starting with $B_1 = A_1^{-1}$, and finally sets $B = B_J$.

Assuming that B_{j-1} is at our disposal and $g \in M_j$, we define $B_j g \in M_j$ in three steps, referred to as pre-smoothing, correction, and post-smoothing, The basic ingredient in the first and third steps is a so-called smoothing operator $R_j : M_j \to M_j$, which has the property that the corresponding operator $I - R_j A_j$ reduces particularly well the error components with high frequencies. In the middle step the lower frequencies are reduced by projecting the residual onto M_{j-1} and iterating with the aid of B_{j-1}. With $R_j^{(l)} = R_j$ if l is odd, $R_j^{(l)} = R_j^t$, if l is even, where R_j^t is the adjoint of R_j, and denoting by $Q_j : M \to M_j$ the orthogonal projection, the algorithm is defined more precisely as follows.

The Multigrid Algorithm. Let p, r, s be integers with $p \geqslant 1$ and $r + s \geqslant 1$. Set $B_1 = A_1^{-1}$ and let $2 \leqslant j \leqslant J$. Assume that $B_j : M_{j-1} \to M_{j-1}$ has been defined, and let $g \in M_j$. Then $B_j g \in M_j$ is defined by

(i) r pre-smoothing iterations: set $x^0 = 0$ and

$$x^l = x^{l-1} - R_j^{(l+r)}(A_j x^{l-1} - g), \quad \text{for } l = 1, \cdots, r,$$

(ii) p correction iterations: set $q^0 = 0$ and

$$q^i = q^{i-1} - B_{j-1}\Big(A_{j-1} q^{i-1} - Q_{j-1}(A_j x^r - g)\Big), \quad \text{for } i = 1, \cdots, p,$$

(iii) s post-smoothing iterations: set $y^0 = x^r - q^p$ and

$$y^l = y^{l-1} - R_j^{(l)}(A_j y^{l-1} - g), \quad \text{for } \ l = 1, \cdots, s.$$

Finally, set $B_j g = y^s$.

If instead of $x^0 = 0$ we set $x^0 = w$ in step (i) of the algorithm, then we obtain $y^s = w - B_j(A_j w - g)$, which is of the form required in step (ii) and in (2.3).

Note thus that in the second step we first project onto M_{j-1} the residual $\rho = A_j x^r - g$ of the error remaining after the pre-smoothing in M_j, and then apply the analog of the iteration (2.3) in M_{j-1} to the equation $A_{j-1} q = Q_{j-1}\rho$ to solve for a correction q^p which in the final step is interpreted as an element of M_j.

The reduction of error in each step of the iteration (2.3) is determined by the operator $D = I - BA$, and the purpose of the convergence analysis is thus to estimate some norm of D. In order to do so we introduce some more notation. Since $A(\cdot, \cdot)$ is symmetric, positive definite we may define an inner product $[\cdot, \cdot] = A(\cdot, \cdot)$ and a corresponding norm $|\cdot| = [\cdot, \cdot]^{1/2}$.

We use the same notation for the induced operator norm, and let T^*denote the adjoint of a linear operator T with respect to $[\cdot,\cdot]$ Moreover, we let $P_j : M \to M_j$ denote the orthogonal projection with respect to $[\cdot,\cdot]$ and $\lambda_j = \lambda_{\max}(A_j)$ the largest eigenvalue of A_j.

Noting that the error reduction operators of the smoothing iterations alternate between $K_j = I - R_jA_j$ and $K_j^* = I - R_j^tA_j$, we now show that $D_j = I - B_jA_j$ satisfies the recursion

$$D_j = K_{j,s}^*(I - P_{j-1} + D_{j-1}^p P_{j-1})K_{j,r}, \quad j = 2,\cdots,J, \tag{2.5}$$

where

$$K_{j,m} = \begin{cases} (K_j^*K_j)^n, & \text{if } m = 2n, \\ K_j^*(K_jK_j^*)^n, & \text{if } m = 2n+1. \end{cases} \tag{2.6}$$

In fact, with $g = A_jw$ in the definition of B_j, we have by (iii)

$$D_jw = w - B_jA_jw = w - y^s = (I - R_j^{(s)}A_j)(w - y^{s-1}) = \cdots = K_{j,s}^*(w - y^0),$$

where $w - y^0 = q^p + w - x^r$. Using the fact that $Q_{j-1}A_j = A_{j-1}P_{j-1}$, we get from (ii)

$$q^p = D_{j-1}q^{p-1} + B_{j-1}Q_{j-1}A_j(x^r - w) = D_{j-1}q^{p-1} - B_{j-1}A_{j-1}P_{j-1}(w - x^r),$$

and hence, since $q^0 = 0$,

$$q^p + P_{j-1}(w - x^r) = D_{j-1}(q^{p-1} + P_{j-1}(w - x^r)) = \cdots = D_{j-1}^p P_{j-1}(w - x^r),$$

so that $w - y^0 = (I - P_{j-1} + D_{j-1}^p P_{j-1})(w - x^r)$. Finally, by (i) $w - x^r = K_{j,r}(w - x^0) = K_{j,r}w$, which completes the proof of (2.5).

Our analysis is based on the following three assumptions: there are positive constants C_1, C_2, C_3 such that

(H1) $$\|(I - P_jv)\| \leqslant C_1\lambda_j^{-1/2}|(I - P_j)v|, \ v \in M, \quad j = 1,\cdots,J,$$

(H2) $$\lambda_j/\lambda_{j-1} \leqslant C_2, \quad j = 2,\cdots,J,$$

(H3) $$|K_jv|^2 \leqslant |v|^2 - C_3\lambda_j^{-1}\|A_jv\|^2, \ v \in M_j, \quad j = 2,\cdots,J.$$

The assumption (H1) is an error estimate. In typical finite element applications it is proved by means of the Aubin-Nitsche duality argument and expresses the fact that the error in the elliptic projection P_j is smaller in the L_2 norm than in the energy norm. Assumption (H2) means that the change from M_{j-1} to M_j is not too rapid. Assumption (H3) expresses the smoothing action of K_j: if v is an eigenvector of A_j with eigenvalue λ, then (H3) implies $|K_jv|^2 \leqslant (1 - C_3\lambda/\lambda_j)|v|^2$. High frequency eigenmodes are thus attenuated more by K_j than low frequency modes. Note that $C_3 \leqslant 1$ with $C_3 = 1$ for the "perfect smoother" $R_j = A_j^{-1}$ and for $R_j = \lambda_j^{-1}I$, cf. Lemma 2.5 below. Assumption (H3) is satisfied in finite element applications by other smoothers of practical interest, for example, the point and block Jacobi and Gauss-Seidel iterations, cf. [1].

By a spectral argument it follows from (H3) that $|K_j| \leqslant (1 - C_3/\kappa(A_j))^{1/2}$, where in typical applications the condition number $\kappa(A_j) = \lambda_{\max}(A_j)/\lambda_{\min}(A_j) \to \infty$ as $j \to \infty$. The convergence rate of the smoothing iteration may thus deteriorate as j grows large. By contrast we shall show that $|D_j| \leqslant \delta < 1$ independently of j. Hence the multigrid iteration (2.3) has a uniform rate of convergence.

The assumptions (H1), (H2) and (H3) enter our analysis combined into the inequality

$$|(I - P_{j-1})v|^2 \leqslant C\Big(|v|^2 - |K_j v|^2\Big), \quad v \in M_j,\ C = C_1^2 C_2/C_3. \tag{2.7}$$

To prove this inequality we first use (H1) to get

$$\begin{aligned}|(I - P_{j-1})v|^2 &= [(I - P_{j-1})v, v] = ((I - P_{j-1})v, A_j v)\\ &\leqslant \|(I - P_{j-1})v\|\,\|A_j v\| \leqslant C_1 \lambda_{j-1}^{-1/2}\,|(I - P_{j-1})v|\,\|A_j v\|,\end{aligned}$$

and hence

$$|(I - P_{j-1})v|^2 \leqslant C_1^2 \lambda_{j-1}^{-1} \|A_j v\|^2,$$

form which (2.7) follows in view of (H2) and (H3) .

We begin by stating and proving a convergence result for the so-called V-cycle algorithm, i.e., for $p = 1$ with one pre- and/or one post-smoothing iteration. Its proof may be considered as an adaptation to our simple situation of an argument in [4].

Theorem 2.1. *Assume that (H1), (H2) and (H3) hold. If $r = s = p = 1$ then*

$$|I - BA| \leqslant \delta_0 = 1 - 1/C, \quad \textit{where } C = C_1^2 C_2/C_3.$$

If $p = 1$ and $r + s = 1$ then $|I - BA| \leqslant \sqrt{\delta_0}$.

Proof. If $r = s = p = 1$ then the recursion (2.5) becomes

$$D_j = K_j(I - P_{j-1} + D_{j-1}P_{j-1})K_j^*. \tag{2.8}$$

This is an identity in M_j for $j = 2, \cdots, J$. We extend its scope to all of M by setting

$$\begin{aligned}\tilde{D}_j &= I - P_j + D_j P_j = I - B_j A_j P_j,\\ \tilde{K}_j &= I - P_j + K_j P_j = I - R_j A_j P_j,\\ \tilde{K}_j^* &= I - P_j + K_j^* P_j = I - R_j^t A_j P_j,\end{aligned}$$

and find that

$$\tilde{D}_j = \tilde{K}_j \tilde{D}_{j-1} \tilde{K}_j^*.$$

In fact, restricted to M_j this is the same as (2.8) and restricted to the orthogonal complement of M_j with respect to $[\cdot,\cdot]$ its left and right sides reduce to the identity operator. Since $\tilde{D}_1 = I - P_1$ we hence have

$$\tilde{D}_J = \tilde{K}_J \cdots \tilde{K}_2 (I - P_1)^2 \tilde{K}_2^* \cdots \tilde{K}_J^*,$$

and with $E_1 = I - P_1$, and $E_j = \tilde{K}_j E_{j-1}$ for $j = 2, \cdots, J$, this yields

$$I - BA = \tilde{D}_J = E_J E_J^*. \tag{2.9}$$

Note that E_J^* and E_J are the error reduction operators of the non-symmetric algorithms with $r = 1, s = 0$ and $r = 0, s = 1$, respectively. Since $|I - BA| = |E_J E_J^*| = |E_J^*|^2$ and $|E_J^*| = |E_J|$ it therefore suffices, in all the cases considered, to estimate the latter norm. In order to do so we take $v \in M$ and consider the expressions

$$|E_{j-1}v|^2 - |E_j v|^2 = |E_{j-1}v|^2 - |\tilde{K}_j E_{j-1} v|^2, \quad \text{for } j = 2, \cdots, J.$$

From the definition of $\tilde{K}_j$ and (2.7) it follows that

$$\begin{aligned} |\tilde{K}_j w|^2 &= |(I - P_j)w|^2 + |K_j P_j w|^2 \\ &\leqslant |(I - P_j)w|^2 + |P_j w|^2 - C^{-1}|(I - P_{j-1})P_j w|^2 \\ &= |w|^2 - C^{-1}|(P_j - P_{j-1})w|^2, \end{aligned}$$

for $w \in M$, $j = 2, \cdots, J$. Hence, setting $w = E_{j-1}v$,

$$C\Big(|E_{j-1}v|^2 - |E_j v|^2\Big) \geqslant |(P_j - P_{j-1})E_{j-1}v|^2 = |(P_j - P_{j-1})v|^2,$$

where in the last step we used the fact that $(I - E_{j-1})v \in M_{j-1}$ so that $(P_j - P_{j-1})(I - E_{j-1})v = 0$. This follows by induction and the recursion relation $I - E_1 = P_1$, $I - E_j = I - E_{j-1} + R_j A_j P_j E_{j-1}$ for $j = 2, \cdots, J$. Hence, by summation,

$$C\Big(|E_1 v|^2 - |E_J v|^2\Big) \geqslant \sum_{j=2}^{J} \Big(|P_j v|^2 - |P_{j-1}v|^2\Big) = |v|^2 - |P_1 v|^2 = |(I - P_1)v|^2,$$

which, since $E_1 v = (I - P_1)v$, yields

$$|E_J v|^2 \leqslant \delta_0 |(I - P_1)v|^2 \leqslant \delta_0 |v|^2$$

This implies the desired result. □

We now consider a more general case of the multigrid algorithm with an arbitrary but equal number of pre- and post-smoothings, and an arbitrary number of corrections. The result is a generalization of a result of [3] and may be seen as a special case of a more general result in [5].

Theorem 2.2. *Let $r = s = m \geqslant 1, p \geqslant 1$. Assume that (H1), (H2) and (H3) hold. Then*

$$|I - BA| \leqslant \delta_m = \frac{C}{C + m}, \quad \textit{where} \quad C = C_1^2 C_2 / C_3.$$

Note that the result of Theorem 2.1 is slightly better than that of Theorem 2.2 in the case $r = s = p = 1$ because $\delta_0 = 1 - 1/C < C/(1 + C) = \delta_1$. Note also that the convergence is faster if $m > 1$, but the present argument does not show that the performance is better for $p > 1$.

Proof. We shall prove, by induction on j, that $|D_j| \leqslant \delta_m$, for $j = 1, \cdots, J$. Since $D_1 = 0$ this is clearly true for $j = 1$. For $j \geqslant 2$ we use the recursion (2.5) which now takes the form

$$D_j = K_{j,m}^*(I - P_{j-1} + D_{j-1}^p P_{j-1})K_{j,m}.$$

Taking $v \in M_j$, setting temporarily $u = K_{j,m}v$, we have

$$[D_j v, v] = |(I - P_{j-1})u|^2 + [D_{j-1}^p P_{j-1}u, P_{j-1}u].$$

It follows by induction that D_j is selfadjoint and positive semidefinite with respect to $[\cdot,\cdot]$ so that $|D_j| = \sup_{v\in M_j}[D_j v, v]/|v|^2$. Moreover, assuming that $|D_{j-1}| \leqslant \delta_m < 1$ we have

$$[D_j v, v] \leqslant |(I - P_{j-1})u|^2 + \delta_m|P_{j-1}u|^2 = (1 - \delta_m)|(I - P_{j-1})u|^2 + \delta_m|u|^2$$

The desired result then follows from the following lemma. □

Lemma 2.3. *Assume that K_j satisfies (2.7). Let $m \geqslant 1$ and $\delta_m = C/(C+m)$. Then*

$$(1 - \delta_m)|(I - P_{j-1})K_{j,m}v|^2 + \delta_m|K_{j,m}v|^2 \leqslant \delta_m|v|^2, \quad v \in M_j.$$

Proof. Let $\delta \in [0, 1]$. Setting $u = K_{j,m}v$ we first use (2.7) to obtain

$$(1 - \delta)|(I - P_{j-1})u|^2 + \delta|u|^2 \leqslant C(1 - \delta)\Big(|u|^2 - |K_j u|^2\Big) + \delta|u|^2$$

We now want to express the norms on the right in terms of v. Recalling the de nition of $K_{j,m}$ in (2.6) we find that

$$K_{j,m}^* K_{j,m} = \begin{cases} (K_j^* K_j)^{2n} = (K_j^* K_j)^m, & \text{if } m = 2n, \\ (K_j K_j^*)^n K_j K_j^* (K_j K_j^*)^n = (K_j K_j^*)^m, & \text{if } m = 2n+1, \end{cases}$$

and, similarly,

$$(K_j K_{j,m})^*(K_j K_{j,m}) = \begin{cases} (K_j^* K_j)^{m+1}, & \text{if } m \text{ is even}, \\ (K_j K_j^*)^{m+1}, & \text{if } m \text{ is odd}, \end{cases}$$

so that with

$$\hat{K}_j = \begin{cases} K_j^* K_j, & \text{if } m \text{ is even}, \\ K_j K_j^*, & \text{if } m \text{ is odd}, \end{cases}$$

we have $|u|^2 = [\hat{K}_j^m v, v]$, and $|K_j u|^2 = [\hat{K}_j^{m+1} v, v]$. Hence

$$(1 - \delta)|(I - P_{j-1})u|^2 + \delta|u|^2 \leqslant \Big|C(1 - \delta)(I - \hat{K}_j)\hat{K}_j^m + \delta\hat{K}_j^m\Big|\,|v|^2.$$

The operator $\hat{K}_j$ is selfadjoint and positive semidefinite. Moreover, (2.7) implies that $|K_j| \leqslant 1$ so that $|\hat{K}_j| \leqslant |K_j^*|\,|K_j| = |K_j|^2 \leqslant 1$ and the largest eigenvalue of $\hat{K}_j$ is therefore $\leqslant 1$. Hence, by the spectral theorem we have

$$|C(1 - \delta)(I - \hat{K}_j)\hat{K}_j^m + \delta\hat{K}_j^m| \leqslant \max_{x\in[0,1]}\Big(C(1 - \delta)(1 - x)x^m + \delta x^m\Big).$$

Since $(1-x)x^m \leqslant m^{-1}(1-x^m)$ for $x \in [0,1]$, we have

$$0 \leqslant C(1-\delta)(1-x)x^m + \delta x^m \leqslant \frac{C}{m}(1-\delta)(1-x^m) + \delta x^m = \delta,$$

provided that $C(1-\delta)/m = \delta$, that is, $\delta = \delta_m = C/(C+m)$. This proves the lemma. □

This choice of δ is optimal, In fact, the above maximum is

$$f(\delta) = C(1-\delta)(1-x(\delta))x(\delta)^m + \delta x(\delta)^m, \quad \text{where } x(\delta) = \frac{m}{m+1}\frac{C(1-\delta)+\delta}{C(1-\delta)},$$

and it can be shown that $\delta_m = C/(C+m)$ is the smallest fixed point of f.

Next we discuss the non-symmetric multigrid algorithms without pre-smoothing or without post-smoothing.

Theorem 2.4. *Let $p \geqslant 1, r+s=m \geqslant 1$ and either $s=0$ or $r=0$ and assume that (H1), (H2) and (H3) hold. Then*

$$|I-BA| \leqslant \sqrt{\delta_m}, \quad \textit{where } \delta_m = C/(C+m) \textit{ with } C = C_1^2 C_2/C_3.$$

Proof. We first consider the case when $r=m \geqslant 1, s=0$. The recurrence relation (2.5) then becomes

$$D_j = (I - P_{j-1} + D_{j-1}^p P_{j-1})K_{j,m},$$

and D_j is no longer selfadjoint. But, if $|D_{j-1}| \leqslant \sqrt{\delta_m} < 1$, then we have for $u = K_{j,m}v$

$$\begin{aligned}|D_j v|^2 &= |(I-P_{j-1})u|^2 + |D_{j-1}^p P_{j-1}u|^2 \leqslant |(I-P_{j-1})u|^2 + \delta_m |P_{j-1}u|^2 \\ &= (1-\delta_m)|(I-P_{j-1})u|^2 + \delta_m |u|^2.\end{aligned}$$

Hence, Lemma 2.3 implies $|D_j| \leqslant \sqrt{\delta_m}$, which is the desired result in this case.

Finally when $r=0, s=m \geqslant 1$ the recursion (2.5) is

$$D_j = K_{j,m}^*(I - P_{j-1} + D_{j-1}^p P_{j-1}).$$

Taking adjoints we obtain

$$D_j^* = (I - P_{j-1} + (D_{j-1}^*)^p P_{j-1})K_{j,m}.$$

and since $|D_j| = |D_j^*|$ the desired result follows from the above. □

Finally we remark that when $p=1$ we have $I - B^S A = (I - B^N A)^*(I - B^N A)$, where B^S and B^N respectively, are the operators associated with the symmetric algorithm considered in Theorem 2.2 and the non-symmetric algorithm considered in the first part of Theorem 2.2, see (2.9). Hence $|I - B^S A| = |I - B^N A|^2$ when $p=1$. Moreover, the convergence rate for $p>1$ is always bounded by the convergence rate for $p=1$, see [14]. Thus it actually suffices to prove either Theorem 2.2 or Theorem 2.4 for $p=1$ in order to obtain the results of this subsection.

2.2　A non-symmetric equation

In this subsection we assume that M is a complex Hilbert space of finite dimension and consider a special non-symmetric bilinear form

$$A(u,v)=S(u,v)+i\beta(u,v),\quad u,v\in M,$$

where $i^2=-1$, $\beta\in\mathbf{R}$ and $S(\cdot,\cdot)$ is Hermitian, positive definite with

$$S(u,u)\geqslant\alpha\|u\|^2,\quad \alpha>0. \tag{2.10}$$

Defining again the operators A, A_j by (2.1), (2.4) we want to solve equation (2.2) by the iteration (2.3) defined by the multigrid algorithm described in the previous subsection.

In order to analyze this algorithm we now use the inner product $[\cdot,\cdot]=S(\cdot,\cdot)$ and corresponding norm $|\cdot|=[\cdot,\cdot]^{1/2}$. We also define operators $S_j:M_j\to M_j$ and $P_j:M\to M_j$ by the equations

$$\begin{aligned}(S_ju,v)&=S(u,v),\quad \forall u,v\in M_j,\\ A(P_ju,v)&=A(u,v),\quad \forall u\in M,\ v\in M_j,\end{aligned} \tag{2.11}$$

and let λ_j denote the largest eigenvalue of S_j. Note that A_j is invertible so that P_j exists with $P_j=A_j^{-1}Q_jA$. Note also that $A_j=S_j+i\beta I$, where S_j is Hermitian and thus A_j is normal with respect to $[\cdot,\cdot]$ and that λ_j is the largest of the real parts of the eigenvalues of A_j.

We still want to use the assumptions (H1), (H2) and (H3) for P_j,λ_j and K_j. Concerning K_j we note that it is no longer true in general that $I-R_j^tA_j$ is equal to K_j^*, so that alternating products of factors K_j and K_j^* would not occur as in (2.5). We therefore restrict our discussion now to operators of the form $K_j=I-\mu A_j$, corresponding to smoothing operators $R_j=\mu I$, where μ is a positive parameter. In this case K_j is a normal operator with respect to $[\cdot,\cdot]$ and, as we shall see in the convergence proof below the spectral argument in Lemma 2.3 can still be applied. In the following lemma we first show that (H3) and (2.7) are satisfied with the proper choice of μ.

Lemma 2.5. *Let $R_j=\mu I$ and $\gamma=(1+\beta^2/\alpha^2)^{-1}$. If $|\mu\lambda_j-\gamma|\leqslant\epsilon<\gamma$, then (H3) holds with $C_3=\gamma^2-\epsilon^2$. If , in addition, (H1) and (H2) hold, then (2.7) follows.*

The optimal choice $\epsilon=0$ gives $\mu=\gamma\lambda_j^{-1}, C_3=\gamma^2$ and, in particular, $\mu=\lambda_j^{-1}, C_3=1$ if $A(\cdot,\cdot)$ is symmetric. This is difficult to achieve exactly because λ_j is unknown, but the lemma permits λ_j to be replaced by an estimate.

Proof. We have

$$|v|^2-|K_jv|^2=2\mu\ \mathrm{Re}[A_jv,v]-\mu^2|A_jv|^2.$$

Here $|A_jv|^2\leqslant\lambda_j\|A_jv\|^2$ and $\mathrm{Re}[A_jv,v]=\|S_jv\|^2$. Since

$$\|A_jv\|^2=\|S_jv\|^2+\beta^2\|v\|^2\leqslant(1+\beta^2/\alpha^2)\|S_jv\|^2,$$

we conclude

$$|v|^2 - |K_j v|^2 \geqslant (2\gamma\mu - \mu^2\lambda_j)\|A_j v\|^2 = (2\gamma\mu\lambda_j - (\mu\lambda_j)^2\lambda_j^{-1})\|A_j v\|^2.$$

Here $2\gamma\mu\lambda_j - (\mu\lambda_j)^2 = \gamma^2 - (\mu\lambda_j - \gamma)^2 \geqslant \gamma^2 - \epsilon^2$ from which the first part of the lemma follows. In order to prove (2.7) we first use the definitions of P_j, A_j in (2.11), (2.4) together with (H1) to get

$$\begin{aligned}|(I-P_{j-1})v|^2 =& \mathrm{Re}\left(A((I-P_{j-1})v,(I-P_{j-1})v) - i\beta((I-P_{j-1})v,(I-P_{j-1})v)\right)\\ =& \mathrm{Re}\ A((I-P_{j-1})v,v) = \mathrm{Re}((I-P_{j-1})v, A_j^t v)\\ \leqslant& \|(I-P_{j-1})v\|\,\|A_j^t v\| \leqslant C_1\lambda_{j-1}^{-1/2}|(I-P_{j-1})v|\,\|A_j v\|,\end{aligned}$$

since $A_j^t = S_j - i\beta I$, so that $\|A_j^t v\| = \|A_j v\|$. Using also (H2) and (H3) we obtain (2.7). □

We now show using an argument of [2], that if the number of correction iterations p is greater than 1, and the number of smoothing iterations is sufficiently large, then we have a uniform rate of convergence.

Theorem 2.6. *Assume that (H1), (H2) hold and that $R_j = \mu I$ with μ chosen according to the condition in Lemma 2.5. Let $p \geqslant 2, r = m \geqslant 1, s = 0, C = C_1^2 C_2/C_3$, and $\sigma = 1 + |\beta|/\alpha$. Then*

$$|I - BA| \leqslant 2\sqrt{\frac{C}{m}},\ \ if\ m \geqslant 4C(2\sigma)^{2/(p-1)}.$$

Proof. We argue by induction as in the proof of Theorem 2.4. The recurrence relation for D_j is now

$$D_j = (I - P_{j-1} + D_{j-1}^p P_{j-1})K_j^m,$$

which is proved in the same way as (2.5). If $|D_{j-1}| \leqslant \delta < 1$, then since, as is easily seen, $|P_{j-1}| \leqslant 1 + |\beta|/\alpha = \sigma$ and, by (H3), $|K_j| \leqslant 1$, we have

$$|D_j v| \leqslant |(I - P_{j-1})K_j^m v| + \sigma\delta^p|v|.$$

Using (2.7) and the fact that K_j is normal we get with $\hat{K}_j = K_j^* K_j$

$$|(I-P_{j-1})K_j^m v|^2 \leqslant C\left(|K_j^m v|^2 - |K_j^{m+1}v|^2\right) = C[(I-\hat{K}_j)\hat{K}_j^m v, v].$$

Since $\hat{K}_j$ is selfadjoint, and since by (H3) its spectrum lies in the interval $[0,1]$, we conclude as in Lemma 2.3, using the inequality $(1-x)x^m \leqslant 1/m$ for $x \in [0,1]$, that

$$|(I-P_{j-1})K_j^m v|^2 \leqslant \frac{C}{m}|v|^2.$$

Hence

$$|D_j| \leqslant \sqrt{\frac{C}{m}} + \sigma\delta^p \leqslant \delta,$$

if, for example, we choose $\delta = 2\sqrt{C/m}$ and m so large that $\sigma\delta^p \leqslant \frac{1}{2}\delta$, which is the desired result. □

3 Application to Parabolic Problems

In this section we illustrate how the abstract results of the previous section can be applied to parabolic problems by considering the heat equation (1.2) with homogeneous Dirichlet boundary conditions written in weak form: find $U(t) \in H_0^1(\Omega)$ for $t \geqslant 0$ such that $U(0) = U_0$ and

$$(U_t, v) + (\nabla U, \nabla v) = (g, v), \quad \forall v \in H_0^1(\Omega),\ t > 0, \tag{3.1}$$

where $(\cdot,\cdot)$ denotes the standard inner product in $L_2(\Omega)$. We assume that $\Omega \subset \mathbf{R}^2$ is a bounded convex polygonal domain. For the approximation of (3.1) with respect to the spatial variable we introduce a nested sequence $M_1 \subset \cdots \subset M_{j-1} \subset M_j \subset \cdots \subset H_0^1(\Omega)$ of piecewise polynomial finite element spaces. Defining first M_1 by means of a coarse triangulation of Ω, we assume that the triangulation defining M_j with mesh-size h_j is obtained by subdividing each triangle corresponding to M_{j-1} into a fixed finite number N^2 of congruent triangles so that $h_{j-1} = Nh_j$.

We now consider the discretization of the time variable. Let $\Delta_j : M_j \to M_j$ denote the discrete Laplacian defined by

$$-(\Delta_j u, v) = (\nabla u, \nabla v), \quad \forall u, v \in M_j.$$

In an A-stable linear multistep method with time step k one has to solve an equation in M_J of the form

$$((\alpha + i\beta)I - k\Delta_J)u = F, \quad \text{with } \alpha > 0,\ \beta \in \mathbf{R}, \tag{3.2}$$

on each time level, cf. [15]. For the backward Euler method we have $\alpha = 1, \beta = 0$, and for the second order backward differentiation method $\alpha = 3/2, \beta = 0$, cf. [7]. F depends on g, k and on the approximate solution at one or more previous time levels.

In a onestep method based on an A-stable rational approximation $r(z) = p(z)/q(z)$ of e^{-z}, such as certain Padé approximations, one similarly has to solve the equation

$$q(-k\Delta_J)u_n = p(-k\Delta_J)u_{n-1} + F, \tag{3.3}$$

where F depends on g and k. After factorization of the denominator $q(z)$ this reduces to a sequence of equations of the form (3.2), since, by A-stability, $r(z)$ has all its poles in the left half-plane. For the Crank-Nicolson method we have $\alpha = 2$, $\beta = 0$. At the end of this section we present an example where $\beta \neq 0$.

Setting

$$A(v, w) = k(\nabla v, \nabla w) + (\alpha + i\beta)(v, w), \quad \text{with } \alpha > 0,\ \beta \in \mathbf{R},$$

our aim is now to check that the assumptions of the previous section are satisfied with constants that are independent of j and k. With

$$[v, w] = S(v, w) = k(\nabla v, \nabla w) + \alpha(v, w), \quad |v| = \left(k\|\nabla v\|^2 + \alpha\|v\|^2\right)^{1/2},$$

it is first of all clear that $S(\cdot,\cdot)$ is symmetric, positive definite and that (2.10) holds.

Let μ_j be the largest eigenvalue of $-\Delta_j$. Then the largest eigenvalue of S_j is $\lambda_j = \alpha + k\mu_j$, where μ_j is bounded above and below by positive multiples of h_j^{-2}. Since $h_{j-1} = Nh_j$ we thus have

$$\frac{\lambda_j}{\lambda_{j-1}} \leqslant \frac{1 + c_2 k h_j^{-2}}{1 + c_1 k h_{j-1}^{-2}} = \frac{1 + c_2 k h_j^{-2}}{1 + N^2 c_1 k h_j^{-2}} \leqslant C,$$

which is (H2) .

We next consider (H1) in the case $k \leqslant h_j^2$. Here $\lambda_j \leqslant 1 + c_2 k h_j^{-2} \leqslant C$ and hence

$$\|(I - P_j)v\| \leqslant \alpha^{-1/2} |(I - P_j)v| \leqslant C\lambda_j^{-1/2} |(I - P_j)v|.$$

For $k \geqslant h_j^2$ we use an adaptation of the standard duality argument. Let $\psi \in L_2$ and let $w \in H_0^1(\Omega)$ be the solution of

$$A(\phi, w) = (\phi, \psi), \quad \forall \phi \in H_0^1(\Omega). \tag{3.4}$$

Then, for any $\chi \in M_j$, and with $C = 1 + |\beta|/\alpha$,

$$((I - P_j)v, \psi) = A((I - P_j)v, w) = A((I - P_j)v, w - \chi) \leqslant C|(I - P_j)v|\, |w - \chi|. \tag{3.5}$$

Here, with suitable χ,

$$|w - \chi|^2 = k\|\nabla(w - \chi)\|^2 + \alpha\|w - \chi\|^2 \leqslant C(kh_j^2 + h_j^4)\|w\|_{H^2(\Omega)}^2 \leqslant Ckh_j^2 \|w\|_{H^2(\Omega)}^2.$$

We now show that $k\|w\|_{H^2(\Omega)} \leqslant C\|\psi\|$ uniformly in k. In fact, since

$$-\Delta w = k^{-1}(\psi - (\alpha + i\beta)w),$$

we have by the standard regularity estimate for elliptic problems

$$k\|w\|_{H^2(\Omega)} \leqslant C(\|w\| + \|\psi\|).$$

But choosing $\phi = w$ in (3.4) we have

$$\alpha\|w\|^2 \leqslant |w|^2 = \mathrm{Re}A(w, w) = \mathrm{Re}(w, \psi) \leqslant \|w\|\, \|\psi\|,$$

so that $\|w\| \leqslant C\|\psi\|$, which completes the proof. Thus, for $\chi \in M_j$ suitable,

$$|w - \chi|^2 \leqslant Ck^{-1}h_j^2\|\psi\|^2 \leqslant C\lambda_j^{-1}\|\psi\|^2,$$

since $\lambda_j \leqslant Ckh_j^{-2}$. Together with (3.5) this shows

$$\|(I - P_j)v\| \leqslant C\lambda_j^{-1/2}|(I - P_j)v|,$$

and thus completes the proof of (H1) .

For $R_j = \mu I$, with μ appropriately chosen, we have thus checked the assumptions of Theorem 2.6, and if $\beta = 0$ the assumptions of Theorems 2.1, 2.2 and 2.4. In the latter (symmetric) case we may also use smoothing iterations of Jacobi and Gauss-Seidel type, see Chapter 5 of [1], where proofs of $A.4$, which is equivalent to our (H3) , may be directly adapted to our situation to show that the constant C_3 is independent of the time step k.

In these situations we thus conclude that the multigrid iteration (2.3) has a rate of convergence which is independent of J and k. Since our results are expressed in the k-dependent norm $|\cdot|$ they may be combined with the results of [6] and [7] on incomplete iterations to obtain estimates of the total error caused both by the discretization and the iterative solution of the algebraic equations.

We finish by presenting a detailed example: the piecewise linear discontinuous Galerkin method, which is related to the (2.1)—Padé approximation of the exponential, cf. [16]. In this method the approximate solution of (3.1) is sought in the space of discontinuous, piecewise linear functions of t with coefficients in M_J corresponding to a subdivision of the t-axis into small intervals not necessarily of the same lengths. If $I=(t_1,t_2)$ is such an interval of length $k=t_2-t_1$, then we introduce

$$\Pi_1(I;M_J)=\Big\{v:v(t)=v_1\frac{t_2-t}{k}+v_2\frac{t-t_1}{k},\ \ v_1,v_2\in M_J\Big\},$$

and represent functions $v\in\Pi_1(I;M_J)$ by their left and right nodal values $v_1,v_2\in M_J$. The approximation is determined on I as the solution of the variational problem: find $u\in\Pi_1(I;M_J)$ such that

$$(u_1,v_1)+\int_I\Big((u_t,v)+(\nabla u,\nabla v)\Big)\mathrm{d}t=(u_0,v_1)+\int_I(g,v)\mathrm{d}t,\quad \forall v\in\Pi_1(I;M_J),$$

where u_0 is the right nodal value from the previous time interval. In order to be able to progress in time we need to compute the right nodal value u_2 on the present time interval. Simple calculations show that

$$\left(-\frac{k}{6}\Delta_J\begin{bmatrix}2&1\\1&2\end{bmatrix}+\frac{1}{2}\begin{bmatrix}1&1\\-1&1\end{bmatrix}\right)\begin{bmatrix}u_1\\u_2\end{bmatrix}\begin{bmatrix}\begin{bmatrix}F_1\\F_2\end{bmatrix}\end{bmatrix}$$

where $F_1=Q_J(u_0+\int_I g(t)(t_2-t)\mathrm{d}t/k)$ and $F_2=Q_J\int_I g(t)(t-t_1)\mathrm{d}t/k$. By elimination of u_1 we obtain

$$((-k\Delta_J)^2+4(-k\Delta_J)+6)u_2=-(2(-k\Delta_J)+6)F_1+(4(-k\Delta_J)+6))F_2,$$

which is of the form (3.3). Dividing the solution into partial fractions we find

$$u_2=\mathrm{Re}\ \Big(\Big(-k\Delta_J+2+i\sqrt{2}\Big)^{-1}\Big((-2+i5\sqrt{2})F_1+(4-i\sqrt{2})F_2\Big)\Big).$$

so that $u_2=\mathrm{Re}\ w$, where

$$(-k\Delta_J+2+i\sqrt{2}\ )w=(-2+i5\sqrt{2}\ )F_1+(4-i\sqrt{2}\ )F_2, \tag{3.6}$$

which is of the form (3.2).

Acknowledgments The authors are indebted to Professor J. H. Bramble for useful comments on an earlier version of the manuscript.

References

[1] J.H. Bramble, *Multigrid Methods*, Pitman, 1993.

[2] R.E. Bank and T. Dupont, An optimal order process for solving finite element equations, *Mathematics of Computation*, 36 (1981), 35-51.

[3] D. Braess and W. Hackbusch, A new convergence proof for the multigrid method including the V-cycle, *SIAM Journal on Numerical Analysys*, 20 (1983), 967-975.

[4] J.H. Bramble and J. E. Pasciak, New estimates for multilevel algorithms including the V-cycle, *Mathematics of Computation*, 60 (1993), 447-471.

[5] J.H. Bramble, J.E. Pasciak and J. Xu, The analysis of multigrid algorithms with nonnested spaces or noninherited quadratic forms, *Mathematics of Computation*, 56 (1991), 1-34.

[6] J. Douglas, Jr., T. Dupont and R. Ewing, Incomplete iterations for time-stepping a Galerkin method for a quasilinear parabolic problem, *SIAM Journal on Numerical Analysis*, 16 (1979), 503-522.

[7] J.H. Bramble, J.E. Pasciak, P.H. Sammon and V. Thomée, Incomplete iterations in multistep backward difference methods for parabolic problems with smooth and nonsmooth data, *Mathematics of Computation*, 52 (1989), 339-367.

[8] A. Brandt, *Multi-level adaptive finite-element methods. I. Variational problems, Special Topics of Applied Mathematics*, J. Frehse, D. Pallaschke and U.Trottenberg, eds., North-Holland, 1980, pp. 91-128.

[9] J. Burmeister and R. Paul, *Schrittweitensteurerung für ein zeitparalleles Mehrgitterverfahren*, Bericht Nr. 9314, Institut für Informatik und Praktische Mathematik, Christian-Albrechts-Universität zu Kiel, 1993.

[10] W. Hackbusch, Parabolic multi-grid methods, Computing Methods, in *Applied Sciences and Engineering, VI* (R. Glowinski and J.-L. Lions, eds.), North-Holland, 1984, pp. 189-197.

[11] J. Janssen and S. Vandewalle, Multigrid waveform relaxation on spatial finite element meshes: the discrete-time case, Preprint CRPC-94-8, California institute of technology, 1994.

[12] Ch. Lubich and A. Ostermann, Multi-grid dynamic iteration for parabolic equations, *BIT*, 27 (1987), 216-234.

[13] S.Z. Zhou and C.B. Wen, Multigrid method for P_1-nonconforming finite element approximation of a parabolic problem, *Chinese Numerical Mathematics*, (to appear).

[14] J.H. Bramble, J.E. Pasciak, J. Wang and J. Xu, Convergence estimates for multigrid algorithms without regularity assumptions, *Mathematics of Computation*, 57 (1991), 23-45.

[15] G. Savaré, $A(\theta)$-stable approximations of abstract Cauchy problems, *Numerishe Mathematik*, 65 (1993), 319-335.

[16] K. Eriksson, C. Johnson and V. Thomée, Time discretization of parabolic problems by the discontinuous Galerkin method, *RAIRO Modélisation Mathématique et Analyse Numérique*, 19 (1985), 611-643.

Nonlinear Stability and D-convergence of Runge-Kutta Methods for Delay Differential Equations*

C.J. Zhang (张诚坚)† S.Z. Zhou (周叔子)†

Abstract This paper deals with the stability and convergence of Runge-Kutta methods with the Lagrangian interpolation (RKLMs) for nonlinear delay differntial equations (DDEs). Some new concepts, such as strong algebraic stability, GDN-stability and D-convergence, are introduced. We show that strong algebraic stability of a RKM for ODEs implies GDN-stability of the corresponding RKLM for DDEs, and that a strongly algebraically stable and diagonally stable RKM with order p, together with a Lagrangian interpolation of order q, leads a D-convergent RKLM of order $\min\{p, q+1\}$.

Keywords Strong algebraic stability, GDN-stability, D-convergence, DDE, Runge-Kutta method.

1 Introduction

Consider the following nonlinear DDEs:

$$\begin{cases} y'(t) = f(t, y(t), y(t-\tau)), & t \in [t_0, T], \\ y(t) = \varphi(t), & t \in [t_0 - \tau, t_0] \end{cases} \tag{1.1}$$

and

$$\begin{cases} z'(t) = f(t, z(t), z(t-\tau)), & t \in [t_0, T], \\ z(t) = \psi(t), & t \in [t_0 - \tau, t_0], \end{cases} \tag{1.2}$$

where $f : [t_0, T] \times C^N \times C^N \to C^N$ and $\varphi, \psi : [t_0 - \tau, t_0] \to C^N$ are continuous functions such that (1.1) and (1.2) has a unique solution, respectively. Moreover, we assume that there exist some inner product $\langle \cdot, \cdot \rangle$ and the induced norm $\|\cdot\|$ such that

$$\mathrm{Re}\,\langle f(t, y_1, u) - f(t, y_2, u), y_1 - y_2 \rangle \leqslant \sigma \|y_1 - y_2\|^2, \tag{1.3}$$

* 发表于: Journal of Computational and Applied Mathematics, Vol. 85, 1997, pp. 225-237.

† Department of Applied Mathematics, Hunan University, Changsha 410082, China.

$$\|f(t,y,u_1)-f(t,y,u_2)\| \leqslant \gamma\|u_1-u_2\|, \quad \forall t\in[t_0,T], \quad y,y_1,y_2,u,u_1,u_2\in C^N, \tag{1.4}$$

where σ, γ are constants with

$$0 \leqslant \gamma \leqslant -\sigma. \tag{1.5}$$

Torelli [1] pointed out that the analytic solutions of (1.1) and (1.2), under the conditions (1.3)–(1.5), satisfy

$$\|y(t)-z(t)\| \leqslant \max_{t_0-\tau\leqslant x\leqslant t_0} \|\varphi(x)-\psi(x)\|, \quad \forall t\in[t_0,T].$$

In recent years, many authors, such as in't Hout, Spijker and Jackiewicz (cf. [1–7]), have adapted RKMs (A,b,c) (for ODEs):

$$\begin{array}{c|c} c & A \\ \hline & b^T \end{array} \tag{1.6}$$

where $A=(a_{ij})\in R^{s\times s}$, $b=(b_1,b_2,\cdots,b_s)^T$ and $c=(c_1,c_2,\cdots,c_s)^T\in R^s$, to DDEs (1.1). Especially, in't Hout [3, 4] present RKLMs for DDEs (1.1), which defined by

$$\begin{cases} y_{n+1}=y_n+h\sum\limits_{j=1}^{s} b_j f(t_n+c_jh, y_j^{(n)}, \tilde{y}_j^{(n)}), \\ y_i^{(n)}=y_n+h\sum\limits_{j=1}^{s} a_{ij} f(t_n+c_jh, y_j^{(n)}, \tilde{y}_j^{(n)}), \quad i=1,2,\cdots,s, \end{cases} \tag{1.7}$$

where coefficients a_{ij}, b_j and c_j satisfy

$$\sum_{j=1}^{s} b_j=1, \quad \sum_{j=1}^{s} a_{ij}=c_i, \quad 0\leqslant c_i\leqslant 1, \quad i=1,2,\cdots,s; \tag{1.8}$$

$t_n=t_0+nh\in[t_0,T]$; $y_n, y_j^{(n)}, \tilde{y}_j^{(n)}$ are approximations to the analytic solution $y(t_n), y(t_n+c_jh), y(t_n+c_jh-\tau)$ of (1.1), respectively, and the argument $\tilde{y}_j^{(n)}$ is determined by

$$\tilde{y}_j^{(n)}=\begin{cases} \varphi(t_n+c_jh-\tau), & t_n+c_jh-\tau\leqslant t_0, \\ \sum\limits_{l=-r}^{v} L_l(\delta) y_j^{n-m+l}, & t_n+c_jh-\tau>t_0, \end{cases} \tag{1.9}$$

with $\tau=(m-\delta)h$, $\delta\in[0,1)$, integer $m\geqslant v+1, r, v\geqslant 0$ and

$$L_l(\delta)=\prod_{q=-r,q\neq l}^{v}\left(\frac{\delta-q}{l-q}\right), \quad l=-r,-r+1,\cdots,v.$$

What we assume $m\geqslant v+1$ is to guarantee that no (unknown) values $y_j^{(i)}$ with $i\geqslant n$ are used in the interpolation procedure. In addition, we always put $y_j^{(n)}=\varphi(t_n+c_jh)$ whenever $n<0$, and $y_n=\varphi(t_n)$ whenever $n\leqslant 0$. in't Hout [3, 4] gave some nice results on asymptotic stability of methods (1.7)–(1.9) for linear DDEs.

For discussing nonlinear stability of numerical methods, Torelli [1] presented the concept of GRN-stability. However, it is difficult to verify that a method for DDEs is GRN-stable, and only the Euler method has been proved to be GRN-stable so far (cf.[1]). In view of this, we gave a new stability concept, i.e. GDN-stability, which is different from GRN-stability by σ, γ in (1.3)–(1.5) are constants independent of t and positive constant C in (2.1) is not necessarily equal to one. A convenient criterion for GDN-stability is presented, and the convergence behaviour of methods (1.7)–(1.9) is revealed by introducing the concept of D-convergence in this paper.

2 GDN-Stability

Some new stability concepts are introduced as follows.

Definition 2.1. *A numerical method (1.7)–(1.9) for DDEs is called GDN-stable if, under the conditions (1.3)–(1.5), numerical approximations y_n and z_n to the solution of (1.1) and (1.2), respectively, satisfy*

$$\|y_n - z_n\| \leqslant C \max_{t_0-\tau\leqslant t\leqslant t_0} \|\varphi(t) - \psi(t)\|, \quad n \geqslant 0, \tag{2.1}$$

where constant $C > 0$ depends only on the method, the parameter c and the interval length $T - t_0$.

Definition 2.2. *A RKM (A,b,c) for ODEs is called strongly algebracially stable if matrices*

$$M_i = (diag\, A_i)A + A^T(diag\, A_i) - A_iA_i^T, \quad i = 0, 1, 2, \cdots, s$$

are nonnegative definite, where

$$A_0 = b, \quad A_i = (a_{i1}, a_{i2}, \cdots, a_{is})^T, i = 1, 2, \cdots, s$$

with $a_{ij} \geqslant 0$ and $b_i \geqslant 0$ $(i, j = 1, 2, \cdots, s)$.

In particular, algebraic stability can be defined by that matrix M_0 which is nonnegative definite together with $b_i \geqslant 0(1, 2, \cdots, s)$.

Let $\left\{y_n, y_j^{(n)}, \tilde{y}_j^{(n)}\right\}_{j=1}^{s}$ and $\left\{z_n, z_j^{(n)}, \tilde{z}_j^{(n)}\right\}_{j=1}^{s}$ be two sequences of approximations to problems (1.1) and (1.2), respectively, by method (1.7)–(1.9) with the same stepsize h, and write

$$T_i^{(n)} = t_n + c_jh, \quad U_i^{(n)} = y_i^{(n)} - z_i^{(n)}, \quad \tilde{U}_i^{(n)} = \tilde{y}_i^{(n)} - \tilde{z}_i^{(n)}, \quad U_0^{(n)} = y_n - z_n,$$

$$Q_i^{(n)} = h[f(T_i^n, y_i^{(n)}, \tilde{y}_i^{(n)}) - f(T_i^n, z_i^{(n)}, \tilde{z}_i^{(n)})], \quad i = 1, 2, \cdots, s.$$

Then (1.7) reads

$$\begin{cases} U_i^{(n)} = U_0^{(n)} + \sum\limits_{j=1}^{s} a_{ij}Q_j^{(n)}, \quad i = 1, 2, \cdots, s, \\ U_0^{(n+1)} = U_0^{(n)} + \sum\limits_{j=1}^{s} b_jQ_j^{(n)}, \end{cases} \tag{2.2}$$

Theorem 2.3. *Assume RKM (1.6) is strongly algebraically stable. Then the corresponding RKLM (1.7)–(1.9) is GDN-stable, and satisfies*

$$\|y_n - z_n\| \leqslant \exp\left[-\frac{1}{2}(T-t_0)\sigma L_0\right] \max_{t_0-\tau\leqslant t\leqslant t_0} \|\varphi(t)-\psi(t)\|, \quad n \geqslant 0, \tag{2.3}$$

where $L_0 = \sup_{\delta\in[0,1)} \left(\sum_{l=-r}^{v} |L_l(\delta)|\right)^2$.

Proof. From (2.2) we get

$$\begin{aligned}
\|U_0^{(n+1)}\|^2 &= \left\langle U_0^{(n)} + \sum_{j=1}^{s} b_j Q_j^{(n)}, U_0^{(n)} + \sum_{j=1}^{s} b_j Q_j^{(n)} \right\rangle \\
&= \|U_0^{(n)}\|^2 + 2\sum_{i=1}^{s} b_i \text{ Re } \left\langle Q_i^{(n)}, U_0^{(n)} \right\rangle + \sum_{i,j=1}^{s} b_i b_j \left\langle Q_i^{(n)}, Q_j^{(n)} \right\rangle \\
&= \|U_0^{(n)}\|^2 + 2\sum_{i=1}^{s} b_i \text{ Re } \left\langle Q_i^{(n)}, U_i^{(n)} - \sum_{j=1}^{s} a_{ij} Q_j^{(n)} \right\rangle + \sum_{i,j=1}^{s} b_i b_j \left\langle Q_i^{(n)}, Q_j^{(n)} \right\rangle \\
&= \|U_0^{(n)}\|^2 + 2\sum_{i=1}^{s} b_i \text{ Re } \left\langle Q_i^{(n)}, U_i^{(n)} \right\rangle - \sum_{i,j=1}^{s} (b_i a_{ij} + b_j a_{ji} - b_i b_j) \left\langle Q_i^{(n)}, Q_j^{(n)} \right\rangle
\end{aligned}$$

It follows from the above equality, algebraic stability of method RKM and Lemma 3.4 in [8] that

$$\|U_0^{(n+1)}\|^2 \leqslant \|U_0^{(n)}\|^2 + 2\sum_{i=1}^{s} b_i \text{ Re } \left\langle Q_i^{(n)}, U_i^{(n)} \right\rangle \tag{2.4}$$

Furthermore, by conditions (1.3)–(1.5) and Schwartz inequality we have

$$\begin{aligned}
Re \left\langle Q_i^{(n)}, U_i^{(n)} \right\rangle &= h[\text{ Re } \left\langle f(T_i^n, y_i^{(n)}, \tilde{y}_i^{(n)}) - f(T_i^n, z_i^{(n)}, \tilde{y}_i^{(n)}), U_i^{(n)} \right\rangle \\
&\quad + \text{ Re } \left\langle f(T_i^n, z_i^{(n)}, \tilde{y}_i^{(n)}) - f(T_i^n, z_i^{(n)}, \tilde{z}_i^{(n)}), U_i^{(n)} \right\rangle] \\
&\leqslant h\sigma\|U_i^{(n)}\|^2 + h\|f(T_i^n, z_i^{(n)}, \tilde{y}_i^{(n)}) - f(T_i^n, z_i^{(n)}, \tilde{z}_i^{(n)})\| \cdot \|U_i^n\| \\
&\leqslant h\sigma\|U_i^{(n)}\|^2 + h\gamma\|\tilde{U}_i^{(n)}\|\|U_i^{(n)}\| \\
&\leqslant h\sigma\|U_i^{(n)}\|^2 + \frac{1}{2}h\gamma[\|\tilde{U}_i^{(n)}\|^2 + \|U_i^{(n)}\|^2] \\
&\leqslant h\sigma\|U_i^{(n)}\|^2 - \frac{1}{2}h\sigma[\|\tilde{U}_i^{(n)}\|^2 + \|U_i^{(n)}\|^2] \\
&\leqslant -\frac{h\sigma}{2}\|\tilde{U}_i^{(n)}\|^2.
\end{aligned} \tag{2.5}$$

Substituting (2.5) in (2.4) yields

$$\|U_0^{n+1}\|^2 \leqslant \|U_0^n\|^2 - h\sigma\sum_{i=1}^{s} b_i\|\tilde{U}_i^{(n)}\|^2. \tag{2.6}$$

In addition, with (1.9) we have

$$\|\tilde{U}_i^{(n)}\|^2 \leqslant \left[\sum_{l=-r}^{v} |L_1(\delta)| \|U_i^{(n-m+l)}\|\right]^2$$
$$\leqslant L_0 \max_{-r\leqslant l\leqslant v} \|U_i^{n-m+l}\|^2. \tag{2.7}$$

Combining (2.6) with (2.7) and using (1.8) we arrive at

$$\|U_0^{(n+1)}\|^2 \leqslant (1 - h\sigma L_0) \max\{\|U_0^{(n)}\|^2, \max_{(i,l)\in E} \|U_i^{(n-m+l)}\|^2\}, \tag{2.8}$$

where $E = \{(i,l)|1 \leqslant i \leqslant s, -r \leqslant l \leqslant v\}$. Similar to (2.8), the inequalities

$$\|U_i^{(n)}\|^2 \leqslant (1 - h\sigma L_0) \max\{\|U_0^{(n)}\|^2, \max_{(i,l)\in E} \|U_i^{(n-m+l)}\|^2\}, \quad i = 1, 2, \cdots, s, \tag{2.9}$$

can be obtained by considering the RKLM corresponding to the schemes (A, A_i^T, c) $(i = 1, 2, \cdots, s)$ and the strong algebraic stability of the method.

In the following, with the help of inequalities (2.8), (2.9) and an induction we shall prove the inequalities

$$\|U_i^{(n)}\|^2 \leqslant (1 - h\sigma L_0)^{n+1} \max_{t_0-\tau\leqslant t\leqslant t_0} \|\varphi(t) - \psi(t)\|^2, \quad n \geqslant 0, \quad i = 1, 2, \cdots, s. \tag{2.10}$$

In fact, it is clear from (2.8), (2.9) and $m \geqslant v + 1$ that

$$\|U_i^{(0)}\|^2 \leqslant (1 - h\sigma L_0) \max_{t_0-\tau\leqslant t\leqslant t_0} \|\varphi(t) - \psi(t)\|^2, \quad i = 0, 1, 2, \cdots, s.$$

Supposed for $n \leqslant k$ $(k \geqslant 0)$ that

$$\|U_i^{(n)}\|^2 \leqslant (1 - h\sigma L_0)^{n+1} \max_{t_0-\tau\leqslant t\leqslant t_0} \|\varphi(t) - \psi(t)\|^2, \quad i = 0, 1, 2, \cdots, s.$$

Then, from (2.8), (2.9), $m \geqslant v + 1$ and $1 - h\sigma L_0 > 1$, we conclude that

$$\|U_i^{(k+1)}\|^2 \leqslant (1 - h\sigma L_0)^{k+2} \max_{t_0-\tau\leqslant t\leqslant t_0} \|\varphi(t) - \psi(t)\|^2, \quad i = 0, 1, 2, \cdots, s.$$

This completes the proof of inequalities (2.10). In view of (2.10), we get for $n \geqslant 0$ that

$$\|U_0^{(n)}\|^2 \leqslant (1 - h\sigma L_0)^{n+1} \max_{t_0-\tau\leqslant t\leqslant t_0} \|\varphi(t) - \psi(t)\|^2$$
$$\leqslant \exp[-(n+1)h\sigma L_0] \max_{t_0-\tau\leqslant t\leqslant t_0} \|\varphi(t) - \psi(t)\|^2$$
$$\leqslant \exp[-(T - t_0)\sigma L_0] \max_{t_0-\tau\leqslant t\leqslant t_0} \|\varphi(t) - \psi(t)\|^2.$$

As a result, we know that method (1.7)–(1.9) is GDN-stable. □

An analogous result for ODEs, based on classical Lipschitz condition, can be seen from Gear [9].

3 D-Convergence

In this section, we start discussing the convergence of RKLMs (1.7)–(1.9) for DDEs (1.1) with conditions (1.3)–(1.5). It is always assumed that the analytic solution $y(t)$ of (1.1) is smooth enough and its derivatives used later are bounded by

$$\left\|\frac{d^i y(t)}{dt^i}\right\| \leqslant \tilde{M}_i, \quad t \in [t_0-\tau, T]. \tag{3.1}$$

For convenience, we review some concepts on RKMs (A,b,c) (cf.[10]). A given RKM (A,b,c) is called *diagonally stable* if there exists a positive diagonal matrix D such that is nonnegative definite. Moreover, positive integer $\max\{q|B(q), C(q), \text{hold}\}$ is called *stage order* of the method (A,b,c), where

$$B(q): b^T c^{j-1} = \frac{1}{j}, \quad j=1,2,\cdots,q,$$

$$C(q): Ac^{j-1} = \frac{1}{j}c^j, \quad j=1,2,\cdots,q.$$

are two simplifying order conditions of the method. The following, notations are adopted:

$$y^{(n)} = \begin{pmatrix} y_1^{(n)} \\ y_2^{(n)} \\ \vdots \\ y_s^{(n)} \end{pmatrix}, \quad \tilde{y}^{(n)} = \begin{pmatrix} \tilde{y}_1^{(n)} \\ \tilde{y}_2^{(n)} \\ \vdots \\ \tilde{y}_s^{(n)} \end{pmatrix}, \quad Y^{(n)} = \begin{pmatrix} y(t_n+c_1h) \\ y(t_n+c_2h) \\ \vdots \\ y(t_n+c_sh) \end{pmatrix},$$

$$\tilde{Y}^{(n)} = \begin{pmatrix} y(t_n+c_1h-\tau) \\ y(t_n+c_2h-\tau) \\ \vdots \\ y(t_n+c_sh-\tau) \end{pmatrix}, \quad F(t,y,\tilde{y}) = \begin{pmatrix} f(t_n+c_1h, y_1, \tilde{y}_1) \\ f(t_n+c_2h, y_2, \tilde{y}_2) \\ \vdots \\ f(t_n+c_sh, y_s, \tilde{y}_s) \end{pmatrix}$$

$e=(1,1,\cdots,1)^T \in R^s$, I_N denotes $N \times N$ identity matrix, $\tilde{G} = G \otimes I_N$ is the Kroneck product of matrix G and I_N. Moreover, we also introduce the set

$$D_\sigma = \{z \in c^{NS\times NS} | z = \text{blockdiag } (z_i) \text{ with } \mu(z_i) \leqslant \sigma, z_i \in C^N, i=1,2,\cdots,s\}.$$

where $\mu(\cdot)$ denotes the logarithmic norm [10] defined by

$$\mu(G) = \max_{x\neq 0} \frac{\text{Re } \langle Gx, x\rangle}{\|x\|}, \quad \forall G \in C^{N\times N}.$$

With the above notations, methods (1.7)–(1.9) can be written as

$$\begin{cases} y_{n+1} = y_n + h\tilde{b}^T F(t_n, y^{(n)}, \tilde{y}^{(n)}), \\ y^{(n)} = \tilde{e} y_n + h\tilde{A} F(t_n, y^{(n)}, \tilde{y}^{(n)}), \\ \tilde{y}^{(n)} \text{ determined by (1.9)} \end{cases} \tag{3.2}$$

and the local errors in (3.2), $Q_n \in C^N$, $r_n = (r_1^{(n)}, r_2^{(n)}, \cdots, r_s^{(n)})$, $\rho_n = (\rho_1^{(n)}, \rho_2^{(n)}, \cdots, \rho_s^{(n)}) \in C^{NS}$ can be defined as

$$\begin{cases} y(t_{n+1}) = y(t_n) + h\tilde{b}F(t_n, Y^{(n)}, \tilde{Y}^{(n)}) + Q_n, \\ Y^{(n)} = \tilde{e}y(t_n) + h\tilde{A}F(t_n, Y^{(n)}, \tilde{Y}^{(n)}) + r_n, \\ \tilde{Y}^{(n)} = (\tilde{Y}_1^{(n)^T}, \tilde{Y}_2^{(n)^T}, \cdots, \tilde{Y}_s^{(n)^T})^T \text{ with} \\ \tilde{Y}_i^{(n)} = \begin{cases} \varphi(t_n + c_j h - \tau), & t_n + c_i h - \tau \leqslant t_0, \\ \sum\limits_{l=-r}^{v} L_l(\delta) Y_i^{(n-m+l)} + \rho_i^{(n)}, & t_n + c_i h - \tau \geqslant t_0. \end{cases} \end{cases} \tag{3.3}$$

According to Taylor formula and Lemma 5.1 in [11], Q_n, r_n and ρ_n can be determined, respectively, as follows:

$$\begin{cases} Q_n = \sum\limits_{l=1}^{p} \dfrac{h^l}{(l-1)!}\left(\dfrac{1}{l} - \sum\limits_{j=1}^{s} b_j c_j^{l-1}\right) y^{(l)}(t_n) + R_0^{(n)}, \\ r_i^{(n)} = \sum\limits_{l=1}^{p} \dfrac{h^l}{(l-1)!}\left(\dfrac{1}{l}c_i^l - \sum\limits_{j=1}^{s} a_{ij} c_j^{l-1}\right) y^{(l)}(t_n) + R_i^{(n)}, \\ \rho_i^{(n)} = \dfrac{h^{v+r+1}}{(v+r+1)!}\left[\prod\limits_{l=-r}^{v}(\delta - l)\right] y^{(v+r+1)}(\xi_i^{(n)}), \quad \xi_i^{(n)} \in (t_{n-m-r} + c_i h, t_{n-m+v} + c_i h), \end{cases} \tag{3.4}$$

where

$$\|R_i^{(n)}\| \leqslant \hat{M}_i h^{p+1}, \quad i = 0, 1, 2, \cdots, s, \quad h \in (0, h_0],$$

h_0 depends only on the method, and $\hat{M}_i (i = 0, 1, 2, \cdots, s)$ depend only the method and some $\tilde{M}_i$ in (3.1).

Definition 3.1. *An interpolation scheme is called being of order q if its local error is $O(h^{q+1})$.*

In accordance with the last equality of (3.4) we know that the interpolation scheme (1.9) is of order $r + v$.

Definition 3.2. *A given RKLM (1.7)–(1.9) with $y_n = y(t_n)(n \leqslant 0)$ and $y_i^{(n)} = y(t_n + c_j h)(n < 0)$ is called D-convergent of order p if this method, when applied to any given DDE (1.1) subject to (1.3)–(1.5), produces an approximation sequence $\{y_n\}$, and the global error satisfies a bound of the form*

$$\|y(t_n) - y_n\| \leqslant c(t_n) h^p, \quad 0 < h \leqslant h_0,$$

where the maximum stepsize h_0 depends only on characteristic parameter σ and the method; the function $c(t)$ depends only on some $\tilde{M}_i$ in (3.1), delay τ, characteristic parameter σ and the method.

It is remarkable that D-convergence for $\tau = 0$ is just B-convergence [10] in ODEs.

To describe the main results, we define the inner product and the induced norm in $(C^N)^S$

as follows:

$$\langle u,v\rangle=\sum_{i=1}^{s}\langle u_i,v_i\rangle,\quad \|u\|=\sqrt{\langle u,v\rangle}=\sqrt{\sum_{i=1}^{s}\|u_i\|^2},$$

where $u=(u_1,u_2,\cdots,u_s)$, $v=(v_1,v_2,\cdots,v_s)\in(C^N)^S$, $u_i,v_i\in C^N$. In addition, we also present two lemmas, which are obtained by a slight alteration to the proofs of the Lemmas 2.1 and 2.2 in [12].

Lemma 3.3. *Assume RKM (1.6) is diagonally stable. Then there exist constants $D_1,D_2>0$, which depend only on the method, such that for any given $h>0$ and $\forall z\in D_\sigma$, with $\sigma\leqslant 0$ we have*

(1) matrix $\tilde{I}_s-h\tilde{A}z$ is nonsingular,

(2) $\|(\tilde{I}_s-h\tilde{A}z)^{-1}\|\leqslant D_1$, $\|h\tilde{A}_iz(\tilde{I}_s-h\tilde{A}z)^{-1}\|\leqslant D_2$, $\quad i=0,1,2,\cdots,s$.

Lemma 3.4. *Assume RKM (1.6) is strongly algebraically stable. Then for $\forall h>0$ and $\forall z\in D_\sigma$ with $\sigma\leqslant 0$ we have*

$$\|k_i(z)\|\leqslant 1,\quad i=0,1,2,\cdots,s.$$

where $k_i(z)=I_N+h\tilde{A}_i^Tz(\tilde{I}_s-h\tilde{A}z)^{-1}\tilde{e}$.

Based on the above lemmas we can derive our main result.

Theorem 3.5. *Assume RKM (1.6) with stage order p is strongly algebraically stable and diagonally stable, and the interpolation scheme (1.9) is of order q(where $q=r+v$). Then the corresponding RKLM (1.7)–(1.9) is D-convergent of order* $\min\{p,q+1\}$.

Proof. Subtraction of (3.2) from (3.3) yields the following recursion scheme:

$$\begin{cases}\varepsilon_0^{(n+1)}=\varepsilon_0^{(n)}+h\tilde{b}^T[z_n\varepsilon_n+F(t_n,y^{(n)},\tilde{Y}^{(n)})-F(t_n,y^{(n)},\tilde{y}^{(n)})]+Q_n,\\ \varepsilon_n=\tilde{e}\varepsilon_0^{(n)}+h\tilde{A}[z_n\varepsilon_n+F(t_n,y^{(n)},\tilde{Y}^{(n)})-F(t_n,y^{(n)},\tilde{y}^{(n)})]+r_n,\end{cases}\tag{3.5}$$

where $\varepsilon_0^{(n)}=y(t_n)-y_n$, $\varepsilon_n=(\varepsilon_1^{(n)^T},\varepsilon_2^{(n)^T},\cdots,\varepsilon_s^{(n)^T})^T=Y^{(n)}-y^{(n)}$, $z_n=$ blockdiag $(z_i^{(n)})$ with

$$z_i^{(n)}=\int_0^1 f_2(t_n+c_ih,y_i^{(n)}+\theta(Y_i^{(n)}-y_i^{(n)}),\tilde{Y}_i^{(n)})\mathrm{d}\theta,\quad i=1,2,\cdots,s,$$

and $f_2(t,u,v)$ is the Jacobian matrix $(\partial f(t,u,v)/\partial u)(t\in R,u,v\in C^N)$. From (3.5) we can get

$$\begin{aligned}\varepsilon_0^{(n+1)}=&k_0(z_n)\varepsilon_0^{(n)}+h\tilde{b}^Tz_n(\tilde{I}_s-h\tilde{A}z)^{-1}r_n+h\tilde{b}^T[hz_n(\tilde{I}_s-h\tilde{A}z)^{-1}\tilde{A}+\tilde{I}_s]\\&\times[F(t_n,y^{(n)},\tilde{Y}^{(n)})-F(t_n,y^{(n)},\tilde{y}^{(n)})]+Q_n.\end{aligned}\tag{3.6}$$

By conditions (1.3),(3.4), $B(p)$ and $C(p)$ we know for $h\in(0,h_0]$ that

$$z_n\in D_\sigma,\quad r_n=(R_1^{(n)},R_2^{(n)},\cdots,R_s^{(n)}),\quad Q_n=R_0^{(n)}.\tag{3.7}$$

A combination of (1.3)–(1.5), (1.8), (3.4), (3.6), (3.7), as well as Lemmas 3.3 and 3.4, yields for

$h \in (0, h_0]$ that

$$\begin{aligned}
\|\varepsilon_0^{(n+1)}\| \leqslant & \|k_0(z_n)\|\|\varepsilon_0^{(n)}\| + \|h\tilde{b}^T z_n(\tilde{I}_s - h\tilde{A}z)^{-1}\|\|r_n\| \\
& + h\{\|h\tilde{b}^T z_n(\tilde{I}_s - h\tilde{A}z)^{-1}\|\|\tilde{A}[F(t_n, y^{(n)}, \tilde{Y}^{(n)}) - F(t_n, y^{(n)}, \tilde{y}^{(n)})]\| \\
& \|\tilde{b}^T[F(t_n, y^{(n)}, \tilde{Y}^{(n)}) - F(t_n, y^{(n)}, \tilde{y}^{(n)})]\|\} + \|Q_n\| \\
\leqslant & \|\varepsilon_0^{(n)}\| + D_2\sqrt{\sum_{i=1}^{s}\|R_i^{(n)}\|^2} \\
& + h\left[D_2\sqrt{\sum_{i=1}^{s}\left\|\sum_{j=1}^{s}a_{ij}[f(t_n + c_jh, y_j^{(n)}, \tilde{Y}_j^{(n)}) - f(t_n + c_jh, y_j^{(n)}, \tilde{y}_j^{(n)})]\right\|^2}\right. \\
& \left. + \|\sum_{j=1}^{s}b_j[f(t_n + c_jh, y_j^{(n)}, \tilde{Y}_j^{(n)}) - f(t_n + c_jh, y_j^{(n)}, \tilde{y}_j^{(n)})]\|\right] + \hat{M}_0h^{p+1} \\
\leqslant & \|\varepsilon_0^{(n)}\| + \left[D_2\sqrt{\sum_{i=1}^{s}\hat{M}_i^2} + \hat{M}_0\right]h^{p+1} + h(-\sigma)\left[D_2\sqrt{\sum_{i=1}^{s}c_i^2} + 1\right] \\
& \times \max_{1\leqslant j\leqslant s}\|\tilde{Y}_j^{(n)} - \tilde{y}_j^{(n)}\|.
\end{aligned} \tag{3.8}$$

It follows from (3.2), (3.3) and (3.4) for $h \in (0, h_0]$, that

$$\begin{aligned}
\|\tilde{Y}_j^{(n)} - \tilde{y}_j^{(n)}\| \leqslant & \sum_{l=-r}^{v}|L_l(\delta)|\|\varepsilon_j^{n-m+l}\| + \|\rho_j^{(n)}\| \\
\leqslant & \sup_{\delta\in[0,1)}\sum_{l=-r}^{v}|L_l(\delta)| \max_{-r\leqslant l\leqslant v}\|\varepsilon_j^{(n-m+l)}\| + \frac{h^{q+1}}{(q+1)!}\sup_{\delta\in[0,1)}\left[\prod_{l=-r}^{v}|\delta - l|\right]\tilde{M}_{q+1}.
\end{aligned} \tag{3.9}$$

where $q = v + r$. Substituting (3.8) in (3.7) we get

$$\|\varepsilon_0^{(n+1)}\| \leqslant \varepsilon_0^{(n)} + h\gamma_1 \max_{(i,l)\in E}\|\varepsilon_i^{(n-m+l)}\| + \gamma_2^{(0)}h^{p+1} + \gamma_3h^{q+1}, \quad h \in (0, h_0), \tag{3.10}$$

with

$$\begin{aligned}
\gamma_1 = & -\sigma\left[D_2\sqrt{\sum_{i=1}^{s}c_i^2} + 1\right]\sup_{\delta\in[0,1)}\sum_{l=-r}^{v}|L_l(\delta)|, \\
\gamma_2^{(0)} = & D_2\sqrt{\sum_{i=1}^{s}\hat{M}_i^2} + \hat{M}_0, \\
\gamma_3 = & -\sigma\frac{\tilde{M}_{q+1}}{(q+1)!}\left[D_2\sqrt{\sum_{i=1}^{s}c_i^2} + 1\right]\sup_{\delta\in[0,1)}\prod_{l=-r}^{v}|\delta - l|.
\end{aligned}$$

By (3.10) we have for $h \in (0, h_0]$ that

$$\|\varepsilon_0^{(n+1)}\| \leqslant (1 + h\gamma_1)\max\{\|\varepsilon_0^{(n)}\|, \max_{(j,l)\in E}\|\varepsilon_j^{(n-m+l)}\|\} + \gamma_2^{(0)}h^{p+1} + \gamma_3h^{q+2}. \tag{3.11}$$

Using Lemma 3.3 and 3.4, similar to (3.11) the inequalities

$$\|\varepsilon_i^{(n)}\| \leqslant (1+h\gamma_1)\max\{\|\varepsilon_0^{(n)}\|, \max_{(j,l)\in E}\|\varepsilon_j^{(n-m+l)}\|\} + \gamma_2^{(i)}h^{p+1} + \gamma_3 h^{q+2},$$
$$i=1,2,\cdots,s, \quad h\in(0,h_0]. \tag{3.12}$$

can be obtained by (3.2) and (3.3), where

$$\gamma_2^{(i)} = D_2\sqrt{\sum_{i=1}^{s}\hat{M}_i^2 + \hat{M}_i^2}, \quad i=1,2,\cdots,s.$$

putting $\gamma_2=\max\{\gamma_2^{(i)}|0\leqslant i\leqslant s\}$ and combining (3.11) with (3.12) we arrive at

$$\|\varepsilon_i^{(n)}\| \leqslant (1+h\gamma_1)\max\{\|\varepsilon_0^{(n)}\|, \max_{(j,l)\in E}\|\varepsilon_j^{(n-m+l)}\|\} + \gamma_2 h^{p+1} + \gamma_3 h^{q+2},$$
$$i=1,2,\cdots,s, \quad h\in(0,h_0]. \tag{3.13}$$

Next, with an induction to (3.13) we shall prove inequalities

$$\|\varepsilon_i^{(n)}\| \leqslant \sum_{j=0}^{n}(1+h\gamma_1)^j(\gamma_2 h^{p+1}+\gamma_3 h^{q+2}), \quad i=0,1,2,\cdots,s, \quad h\in(0,h_0]. \tag{3.14}$$

In fact, it is apparent from (3.13) and $m\geqslant v+1$ that

$$\|\varepsilon_i^{(0)}\| \leqslant \gamma_2 h^{p+1}+\gamma_3 h^{q+2}, \quad i=0,1,2,\cdots,s, \quad h\in(0,h_0].$$

Suppose for $n\leqslant k$ $(k\geqslant 0)$ that

$$\|\varepsilon_i^{(n)}\| \leqslant \sum_{j=0}^{n}(1+h\gamma_1)^j(\gamma_2 h^{p+1}+\gamma_3 h^{q+2}), \quad i=0,1,2,\cdots,s, \quad h\in(0,h_0].$$

Then, from (3.13) and $m\geqslant v+1$ we get

$$\|\varepsilon_i^{(k+1)}\| \leqslant \sum_{j=0}^{k+1}(1+h\gamma_1)^j(\gamma_2 h^{p+1}+\gamma_3 h^{q+2}), \quad i=0,1,2,\cdots,s, \quad h\in(0,h_0].$$

This completes the proof of inequalities (3.14). By (3.14) we arrive at for $h\in(0,h_0]$ that

$$\begin{aligned}\|\varepsilon_0^{(n)}\| &\leqslant \sum_{j=0}^{n}(1+h\gamma_1)^j(\gamma_2 h^{p+1}+\gamma_3 h^{q+2})\\ &= \frac{(1+h\gamma_1)^{n+1}-1}{h\gamma_1}(\gamma_2 h^{p+1}+\gamma_3 h^{q+2})\\ &\leqslant \frac{\exp[(n+1)h\gamma_1]-1}{\gamma_1}(\gamma_2 h^{p+1}+\gamma_3 h^{q+2})\\ &\leqslant c(t_n)h^{\min\{p,q+1\}}\end{aligned}$$

where

$$c(t)=\begin{cases}\dfrac{\exp[(t-t_0)\gamma_1]\exp(h_0\gamma_1)-1}{\gamma_1}(\gamma_2+\gamma_3 h_0^{q+1-p}), & p\leqslant q,\\[2ex] \dfrac{\exp[(t-t_0)\gamma_1]\exp(h_0\gamma_1)-1}{\gamma_2}(\gamma_2 h_0^{p-q-1}+\gamma_3), & p>q.\end{cases}$$

Hence, method (1.7)–(1.9) is D-convergence of order $\min\{p,q+l\}$. □

4 Some Examples

Example 1. The two-stage two-order diagonally implicit method (cf. [13])

$$\begin{array}{c|cc} \lambda & \lambda & 0 \\ 1-\lambda & 1-2\lambda & \lambda \\ \hline & \dfrac{1}{2} & \dfrac{1}{2} \end{array} \tag{4.1}$$

is strongly algebraically stable iff $\dfrac{1}{4} \leqslant \lambda \leqslant \dfrac{1}{2}$, since

$$M_0 = \left(\lambda - \frac{1}{4}\right)\begin{bmatrix} 1 & -1 \\ -1 & 1 \end{bmatrix}, \quad M_1 = \begin{bmatrix} \lambda^2 & 0 \\ 0 & 0 \end{bmatrix}, \quad M_2 = \begin{bmatrix} (1-\lambda)(4\lambda-1) & 0 \\ 0 & \lambda^2 \end{bmatrix}.$$

As to the diagonal stability of the method, we can verify it to be true by taking $D = I_2$, such that

$$DA + A^T D = \left(\lambda - \frac{1}{4}\right)\begin{bmatrix} 2\lambda & 1-2\lambda \\ 1-2\lambda & 2\lambda \end{bmatrix}$$

is positive definite for $\lambda > \dfrac{1}{4}$. Moreover, we find that $B(2)$, $C(1)$ hold. Therefore, it follows from Theorems 2.3 and 3.5 that the RKLM (1.7)–(1.9) corresponding to method (4.1) is GDN-stable and D-convergent of order one if $\dfrac{1}{4} < \lambda \leqslant \dfrac{1}{2}$.

Example 2. (cf. Burrage and Butcher [13]). The s-stage method given by

$$\begin{array}{c|cccccc} \lambda & \lambda & & & & & \\ b_1+\lambda & b_1 & \lambda & & & & \\ \sum\limits_{i=1}^{2} b_i+\lambda & b_1 & b_2 & \lambda & \ddots & & \\ \vdots & \vdots & \vdots & \vdots & & \vdots & \\ \sum\limits_{i=1}^{s-1} b_i+\lambda & b_1 & b_2 & b_3 & \cdots & b_{s-1} & \lambda \\ \hline & b_1 & b_2 & b_3 & \cdots & b_{s-1} & b_s \end{array} \tag{4.2}$$

is strongly algebraically stable iff $0 \leqslant b_i \leqslant 2\lambda (i = 1, 2, \cdots, s)$, since

$$\begin{aligned} M_0 &= \operatorname{diag}\,(b_1(2\lambda - b_1), b_2(2\lambda - b_2), \cdots, b_s(2\lambda - b_s)), \\ M_i &= \operatorname{diag}\,(b_1(2\lambda - b_1), \cdots, b_{i-1}(2\lambda - b_{i-1}), \lambda^2, 0, \cdots, 0). \end{aligned}$$

Further, if we take $D = I_s$, then

$$DA + A^TD = \left(\lambda - \frac{1}{4}\right)\begin{bmatrix} 2\lambda & b_1 & b_1 & \cdots & b_1 & b_1 \\ b_1 & 2\lambda & b_2 & \cdots & b_2 & b_2 \\ b_1 & b_2 & 2\lambda & \cdots & b_3 & b_3 \\ \vdots & \vdots & \vdots & & \vdots & \vdots \\ b_1 & b_2 & b_3 & \cdots & 2\lambda & b_{s-1} \\ b_1 & b_2 & b_3 & \cdots & b_{s-1} & 2\lambda \end{bmatrix}$$

is positive definite whenever

$$2\lambda > \max\left\{(s-1)b_1, \max_{2\leqslant i\leqslant s}\left[\sum_{j=1}^{i-1} b_j + (s-i)b_i\right]\right\} \geqslant 0.$$

Consequently, in terms of Theorems 2.3 and 3.5 we know that the RKLM (1.7)–(1.9) corresponding the method (4.2) is GDN-stable, and D-convergent of order $\min\{p, q+1\}$ if the interpolation is order q and

$$2\lambda > \max\left\{(s-1)b_1, \max_{2\leqslant i\leqslant s}\left\{\left[\sum_{j=1}^{i-1} b_j + (s-i)b_i\right], b_i\right\}\right\} \geqslant 0,$$

$B(p)$, $C(p)$ hold. In particular, setting $s = 2, \lambda = b_1 = b_2 = \frac{1}{2}$, the method (4.2) becomes

$$\begin{array}{c|cc} \frac{1}{2} & \frac{1}{2} & 0 \\ 1 & \frac{1}{2} & \frac{1}{2} \\ \hline & \frac{1}{2} & \frac{1}{2} \end{array} \tag{4.3}$$

It is easy to testify for (4.3) that $B(1), C(1)$ hold. Hence, in accordance with above discussion, the method (4.3) together with the linear interpolation scheme produces a GDN-stable and order one D-convergent RKLM

$$\begin{aligned} y_1^{(n)} &= y_n + \frac{h}{2}f\left(t_n + \frac{h}{2}, y_1^{(n)}, \tilde{y}_1^{(n)}\right), \\ y_2^{(n)} &= y_n + \frac{h}{2}f\left(t_n + \frac{h}{2}, y_1^{(n)}, \tilde{y}_1^{(n)}\right) + \frac{h}{2}f\left(t_{n+1} + \frac{h}{2}, y_2^{(n)}, \tilde{y}_2^{(n)}\right), \\ \tilde{y}_j^{(n)} &= (1-\delta)y_j^{(n-m)} + \delta y_j^{(n-m+l)}, \quad j = 1, 2 \\ y_{n+1} &= y_n + \frac{h}{2}f\left(t_n + \frac{h}{2}, y_1^{(n)}, \tilde{y}_1^{(n)}\right) + \frac{h}{2}f\left(t_{n+1} + \frac{h}{2}, y_2^{(n)}, \tilde{y}_2^{(n)}\right), \end{aligned}$$

where $n > 0$, $(m-\delta)h = \tau$, positive integer $m \geqslant 2$, $\delta \in [0, 1)$.

References

[1] L. Torelli, Stability of numerical methods for delay differential equations, *Journal of Computational and Applied Mathematics*, 25 (1989), 15-26.

[2] A. Bellen, M. Zennaro, Strong contractivity properties of numerical methods for ordinary and delay differential equations, *Applied Numerical Mathematics*, 9 (1992), 321-346.

[3] K.J. in't Hout, A new interpolation procedure for adapting Runge-Kutta methods to delay differential equations, *BIT*, 32 (1992), 634-649.

[4] K.J. in't Hout, Stability analysis of Runge-Kutta methods for systems of delay differential equations, *IMA Journal of Numerical Analysus*, 17 (1997), 17-27.

[5] K.J. in't Hout, M.N. Spijker, Stability analysis of numerical methods for delay differential equations, *Numerische Mathematik*, 59 (1991), 807-814.

[6] Z. Jackiewicz, One step methods of any order for neutral functional equations, *SIAM Journal on Numerical Analysis*, 21 (1984), 486-511.

[7] M. Zemraro, P-stability properties of Runge-Kutta methods for delay differential equations, *Numerische Mathematik*, 49 (1986), 305-318.

[8] G. Dahlquist, G-stability is equivalent to A-stability, *BIT*, 18 (1978), 384-401.

[9] C.W. Gear, *Numerical Initial Value Problems in Ordinary Differential Equations*, Prentice-Hall, Englewood Cliffs, NJ, 1971.

[10] K. Dekker, J.G. Verwer, *Stability of Runge-Kutta Method for Stiff Nonlinear Differential Equations*, North-Holland, Amsterdam, 1984.

[11] P. Hemici, *Discrete Variable Methods in Ordinary Differential Equations*, Wiley, New York, 1962.

[12] A. Xiao, On the order of B-convergence of Runge-Kutta methods, *Natural Science Journal of Xiangtan University*, 14 (1992), 16-19.

[13] K. Burrage, J. C. Butcher, Stability criteria for implicit Runge-Kutta methods, *SIAM Journal on Numerical Analysis*, 16 (1979), 46-57.

On an Axially Symmetric Elastic-Plastic Torsion Problem*

X.P. Yang (杨孝平)† S.Z. Zhou (周叔子)‡ G.Y. Li (李光耀)§

Abstract This paper discussed an axially asymmetric elastic-plastic torsion problem. In virtue penalty method, reflection boundaries, Bernstein estimate and reverse Hólder inequality, on account of studying the corresponding complementary boundary problem which had mixed boundary conditions, the regularity of the solutions was established.

Keywords Complementary boundary problem, penalty method, Bernstein estimate, regularity.

1 Introduction

The elastic-plastic torsion of a shaft is a classical mechanics problem. In the middle of the 1960s, the invention of variational inequalities made the research of this direction quite active and deep. The research about bars with constant cross-sections is relatively complete and intensive ([1, 2]. Cryer [3] first studied the elastic-plastic torsion problem associated with the rotational symmetric shafts of variable cross-sections and derived the existence, uniqueness and regularity. In his paper, there are two fundamental assumptions: (1) the generatrixes of the shafts are monotone curves of C^2 class; (2) so-called "Condition C" are satisfied (see [3, Theorem 5.9]. Afterwards, Zhou Shuzi considered that the generatrixes are generally sectionally continuously differentiable functions, and proved that Haar-Káman Principle was true with respect to this problem under Hencky Condition and this problem was essentially two dimensional problems. This paper employed the method different from that in [3], relaxed the "Condition C", considered the elliptic boundary problem with gradient constrains, and set up $W^{2,\infty}$-regularity of the problem stated in [3]. It should be pointed out that the regularity of solutions of these problems is not only one of important properties of the solutions but also the basis of numerical analysis. Furthermore, two new difficulties appear in the variable cross-

* 发表于: Applied Mathematics & Mechanics, Vol. 18, 1997, pp. 707-720.

† Nanjing University of Science and Technology, Nanjing 210094, China.
‡ Hunan University, Changsha 410082, China.
§ Guizhou University, Guiyang 550025, China.

section case compared with the constant cross-section case. The first one is that the coefficients of the variational inequalities have higher-order degenerate terms, and the second is that the obstacles are not smooth in the corresponding obstacle problem. Those above mean that the discussion about this kind of problem will be meaningful both in theory and practice.

2 Mechanical Problem and Mathematical Statement

The problem to be considered is to seek the final stress distribution when equal and opposite torques T are applied to the ends of a shaft of length L which is ideally elastic-plastic and rotational. This mechanical problem could be stated as follows:

To find out the stress function $u \in C^1(\bar{\Omega}) \cap C^2(\Omega)$, we have

$$|\nabla u| \leqslant k x_2^2 \text{ in } \Omega \quad (k \text{ is a given constant}) \tag{2.1}$$

$$\text{in } \Omega \cap \{(x_1, x_2); |\nabla u| < k x_2^2\}$$

$$Au = -\frac{\partial}{\partial x_1}\left(\frac{1}{x_2^3}\frac{\partial u}{\partial x_1}\right) - \frac{\partial}{\partial x_2}\left(\frac{1}{x_2^3}\frac{\partial u}{\partial x_2}\right) = 0 \tag{2.2}$$

$$u = 0 \quad \text{in } \Gamma_0, \quad u = \frac{T}{2\pi} \quad \text{in } \Gamma_1 \tag{2.3}$$

$$\frac{\partial u}{\partial n} = 0 \quad \text{in } \Gamma_2 = \Gamma_{21} \cup \Gamma_{22}; \;\; n \text{ is the our normal in } \Gamma_2 \tag{2.4}$$

where $\Omega = \{(x_1, x_2):\ 0 < x_1 < L, 0 < x_2 < R(x_1)\}$, $x_2 = R(x_1)$ is the generatrix equation of the shaft, the boundary of Ω is the following

$$\Gamma_0 = \{(x_1, x_2) : 0 \leqslant x_1 \leqslant L, x_2 = 0\}$$
$$\Gamma_1 = \{(x_1, x_2) : 0 \leqslant x_1 \leqslant L, x_2 = R(x_1)\}$$
$$\Gamma_{21} = \{(x_1, x_2) : 0 < x_2 < R(0), x_1 = 0\}$$
$$\Gamma_{22} = \{(x_1, x_2) : 0 < x_2 < R(L), x_1 = L\}.$$

Paper [4] proved that the generalized variational problem of the stress function is to find out $u_0 \in K$, such that

$$\int_\Omega \rho \nabla u_0 \nabla (v - u_0) \mathrm{d}x \geqslant 0, \quad (\forall v \in K) \tag{2.5}$$

where

$$\rho = x_2^{-3}$$

$$K = \left\{v \in H_0^1(\Omega),\ v = \frac{T}{2\pi} \ \text{ in } \Gamma_1,\ |\nabla v| \leqslant k x_2^2 \text{ in } \Omega\right\}$$

$$H_0^1(\Omega) = \left\{v \in H^1(\Omega),\ \int_\Omega \left(\rho v^2 + \rho \sum_{|\alpha| \leqslant 1} |D^\alpha v|^2 \mathrm{d}x\right) < +\infty\right\}.$$

For Γ_1, we assume

$$\text{(a)} \quad R(x_1) \in C^1(0,L), \quad \frac{\mathrm{d}R}{\mathrm{d}x_1} \geqslant 0,$$
$$\text{(b)} \quad \left.\frac{\mathrm{d}R}{\mathrm{d}x_1}\right|_{x_1=0} = \left.\frac{\mathrm{d}R}{\mathrm{d}x_1}\right|_{x_1=L} = 0.$$

Then the following results could be obtained.

Theorem 2.1. *If $k < k_0$, problem (2.5) has no solutions; if $k \geqslant k_0$, problem (2.5) has a unique solution; where $k_0 = \dfrac{3T}{2\pi}R^{-3}(0)$.*

Theorem 2.2. *Let $h_0 = \left(R^3(0) - 3T/(2\pi)\right)^{1/3}$. If $k > k_0$, then in $S_{h_0/2} = \{x \in \Omega :\ 0 < x_2 < h_0/2\}$, we have*

$$\text{(i)} \quad \int_\Omega x_2^{-3}\nabla u_0 \nabla \phi \mathrm{d}x = 0, \quad \forall \phi \in C_0^\infty(S_{h_0/2}),$$
$$\text{(ii)} \quad u_0|S_{h_0/2} \in H^2 \cap C^\infty(S_{h_0/2}),$$
$$\text{(iii)} \quad u_0 = x_2^4 v_0, \text{ where } v_0 \text{ is analytic in } \bar{S}_{h_0/2}.$$

From [3, Theorem 4.3] and [5, Proposition 7.2 and Theorem 8.3], we may always assume $k > k_0$

Theorem 2.3. $u_0 = \max\{u_0, 0\} \geqslant 0$.

3 Auxiliary Problem and Penalized Problem

We construct the functions $\beta_\epsilon(t)$ $(\epsilon > 0)$, $\tau_m(t)$, $(m > 0)$ as follows:

$$\beta_\epsilon(t) \begin{cases} \in C^2(R) \text{ nondecrease}, & \text{convex}, \\ = 0, & (t \leqslant 0), \\ = \dfrac{t-\epsilon}{t}, & t \geqslant 2\epsilon. \end{cases} \tag{3.1}$$

$$\tau_m(t) = \begin{cases} t, & (|t| \leqslant m/2), \\ m, & (t > m), \\ -m, & (t < -m). \end{cases} \qquad (\tau_m(t) \in C^1(R)). \tag{3.2}$$

Let $S_{h_l} = \{(x_1, x_2) \in \Omega, 0 < x_2 < h_1 < h_0/2\}$, $\Omega_0 = \Omega \backslash \bar{S}_{h_l}$. To consider the following penalized problem

$$\left.\begin{aligned} &-\sum_i \frac{\partial}{\partial x_i}\left(\frac{1}{x_2^3}\frac{\partial u_\epsilon}{\partial x_i}\right) - \beta_\epsilon \circ (|\nabla u_\epsilon|^2 - k^2 x_2^4) + \epsilon u_\epsilon = 0, \quad \text{in } \Omega, \\ &u_\epsilon|_{\Gamma_0'} = u_0,\ u_\epsilon|_{\Gamma_1} = \frac{T}{2\pi},\ \left.\frac{\partial u_\epsilon}{\partial n}\right|_{\Gamma_2'} = 0, \end{aligned}\right\} \tag{3.3}$$

where $\Gamma_0' = \bar{S}_{h_1} \cap \overline{\Omega}_0$, $\Gamma_2' = \Gamma_2 \cap \overline{\Omega}$.

We introduce the following auxiliary problem to study (3.3).

Find $u_\epsilon^m \in \hat{K}$ such that

$$\int_{\Omega_0} \frac{1}{x_2^3} \nabla u_\epsilon^m \nabla (v - u_\epsilon^m) \mathrm{d}x - \int_{\Omega_0} [\beta_{\alpha,m} \circ (\omega_m^2(u_\epsilon^m) - k^2 x_2^4) - \epsilon u_\epsilon^m](v - u_\epsilon^m) \mathrm{d}x \geqslant 0, \quad \forall v \in \hat{K}, \tag{3.4}$$

where

$$\hat{K} = \left\{ v \in H^1(\Omega_0) : v|_{\Gamma_0'} = u_0, v_\epsilon|_{\Gamma_1} = \frac{T}{2\pi} \right\},$$
$$\beta_{\epsilon,m} \circ (\omega_m^2(\nabla u_\epsilon^m) - k^2 x_2^4) = \tau_m \circ (\beta_\epsilon \circ (\omega_m^2(\nabla u_\epsilon^m) - k^2 x_2^4)),$$
$$\omega_m(p) = (\tau_m(p_1), \cdots, \tau_m(p_n)), \quad (\forall p \in R^n).$$

First, similar to the proof of [6, Theorem 9.2], we can obtain

Lemma 3.1. *If Ω is a bounded domain of R^n, $\partial\Omega = \Gamma_1^* \cup \Gamma_2^*$, $\Gamma_1^* \cap \Gamma_2^* = \emptyset$, operator*

$$Qu = a^{ij}(x, \nabla u) D_{ij} u + b(x, u, \nabla u)$$

$u, v \in C^0(\overline{\Omega}) \cap C^2(\Omega \cup \Gamma_2^*)$, where

(i) operator Q is elliptic with respect to u;

(ii) the coefficient $b(x,z,p)$ is non-increasing w. r. t. z for every $(x,p) \in \Omega \times R^n$;

(iii) the coefficients a^{ij}, b are continuously differentiable w. r. t. p in $\Omega \times R \times R^n$;

(iv) $Qv \geqslant Qu$ in Ω, $(u-v)|_{\Gamma_1^*} \leqslant 0$, $\dfrac{\partial(v-u)}{\partial n}|_{\Gamma_2^*} = 0$,

then $v \leqslant u$ in Ω.

To be explicit, we give the following definition.

Definition 3.2. *Let u_ϵ be a weak solution of (3.3). If $u_\epsilon \in \widehat{K}$ and satisfies*

$$\int_{\Omega_0} x_2^{-3} \frac{\partial u_\epsilon}{\partial x_i} \cdot \frac{\partial \varphi}{\partial x_i} \mathrm{d}x - \int_{\Omega_0} (\beta_\varepsilon \circ (|\nabla u_\epsilon|^2 - k^2 x_2^4) - \varepsilon u_\epsilon) \varphi \mathrm{d}x = 0, \quad \forall \varphi \in H_0^1(\Omega_0 \setminus \Gamma_2'), \tag{3.5}$$

where

$$H_0^1(\Omega_0 \setminus \Gamma_2') = \left\{ v \in H^1(\Omega_0), v = 0 \ \text{ in } \ \partial\Omega_0 \setminus \Gamma_2' \right\}.$$

The reflection domain of Ω_0, w. r. t. Γ_{21}' is denoted by Ω_0'. Let $\widetilde{\Omega} = \Omega_0 \cup \Gamma_{21}' \cup \Omega_0'$, its boundary be $\widetilde{\Gamma} = \widetilde{\Gamma}_0' \cup \widetilde{\Gamma}_1' \cup \widetilde{\Gamma}_{21}' \cup \widetilde{\Gamma}_{22}'$ (see the gtaph bellow).

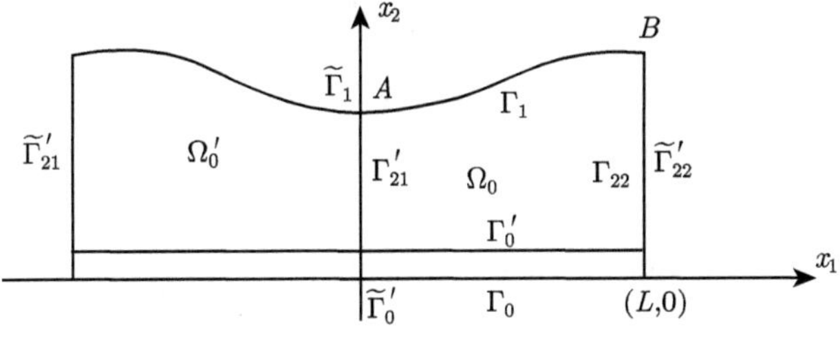

Figure 1

Define the function $\widetilde{u}$ in $\widetilde{\Omega}$ as follows:

$$\widetilde{u}(x) = u(x_1, x_2), \quad x \in \Omega_0; \quad \widetilde{u}(x) = u(-x_1, x_2), \quad x \in \Omega_0'. \tag{3.6}$$

Lemma 3.3. *The extention u_ϵ of the weak solution of problem (3.3) in $\widetilde{\Omega}$ defined by (3.6) belongs to $H^1(\widetilde{\Omega})$, and satisfies*

$$\int_{\widetilde{\Omega}} x_2^{-3}\nabla\widetilde{u}_\varepsilon\nabla\varphi dx - \int_{\widetilde{\Omega}}(\beta_\varepsilon\circ(|\nabla\widetilde{u}_\varepsilon|^2 - k^2x_2^4) - \varepsilon\widetilde{u}_s)\varphi dx$$
$$=0,\quad (\forall\varphi\in H_0^1(\widetilde{\Omega}\setminus\widetilde{\Gamma}_2'),\ \widetilde{\Gamma}_2' = \widetilde{\Gamma}_{22}'\cup\Gamma_{22}') \tag{3.7}$$

$$\widetilde{u}_\varepsilon|_{\Omega_0} = u_\epsilon. \tag{3.8}$$

Now let us discuss the auxiliary problem and penalized problem.

Lemma 3.4. *In variational inequalitiy (3.4) a solution $u_\epsilon^m\in C^0(\widetilde{\Omega}_0)\cap C^2(\Omega_0\cup\Gamma_2')$ exists, and there is the constant M independent of m such that*

$$|u_\epsilon^m(x)|\leqslant M,\quad x\in\widetilde{\Omega}_0 \tag{3.9}$$

$$\|u_\epsilon^m\|_{H^1(\Omega_0)}\leqslant M. \tag{3.10}$$

Proof. For convenience, the ε in β_ε, $\beta_{\varepsilon,m}$, u_ε, u_ε^m would be omitted hereafter when there is not equivocalness.

From the standard theory, for any $v_0\in K$. the following variational inequality has a solution.

Find $u_0^m\in\widehat{K}$, for any $v\in\widehat{K}$

$$\int_{\Omega_0} x_2^{-3}\nabla u_0^m\nabla(v-u_0^m)dx + \int_{\Omega_0}(-\beta_m\circ(\omega_m^2(\nabla v_0) - k^2x_2^4) + \varepsilon u_0^m)(v-u_0^m)dx\geqslant 0. \tag{3.11}$$

Then a mapping T can be defined as

$$Tv_0 = u_0^m.$$

To notice the boundness of β_m and apply the prior estimation in Γ_2', translate the problem into interior case by using Lemma 3.3, then it can be proved that the operator T is continuously compact. Turn $\widehat{K}$ into the bounded set of $\widehat{K}$, so a fixed point can be obtained by fixed point theorem. It can be conclude,d that the variational inequality (3.4) has a solution.

Let $v = u^m\pm\eta$ for any $\eta\in H_0^1(\Omega_0\setminus\Gamma_2')$, and get

$$\int_{\Omega_0} x_2^{-3}\nabla u^m\cdot\nabla\eta dx + \int_{\Omega_0}(-\beta_m\circ(\omega^2(\nabla u^m) - k^2x_2^4) + \varepsilon u^m)\cdot\eta dx = 0. \tag{3.12}$$

This demonstrates that any solution of (3.4) satisfies (3.14). Noticing the construction and boundness of β_m. applying difference quotient. the method of straightening the boundary and Lemma 3.3, we can easily prow that $u^m\in H^{2,2}(\Omega_0)$, furthermore, $u^m\in C(\widetilde{\Omega}_0)$ by using Sobolev's embedding theorem. and $u^m\in C^{1,2}(\Omega_0)$ (some $a>0$), then ∇u^m is locally bounded. In (3.12) replacing η by $D_\epsilon\eta$ $(s=1,2)$. integrating by parts and we get

$$\int_{\Omega_0} x_2^{-3}\cdot\nabla(D_\epsilon u^m)\cdot\nabla\eta dx + \int_{\Omega_0}\left(\frac{-3}{x_2^4}\nabla u^m\cdot\nabla\eta dx\cdot sign(s-1)\right)$$
$$+\int_{\Omega_0}[-\beta_m'\circ(\cdot)\cdot(\tau_m^2 - k^2x_2^4)_{x_2}' + \varepsilon D_\epsilon u^m]\eta dx = 0,\quad\forall\eta\in H_0^1(\Omega_0\setminus\Gamma_2'). \tag{3.13}$$

Use [7, p.219] again to get $D_\epsilon u^m \in C^{1,\alpha}(\Omega_0)$, then $u_m \in C^{2,\alpha}(\Omega_0)$. Employ Lemma 3.3 and the argument above to get

$$\left.\begin{aligned}&-\frac{\partial}{\partial x_i}\left(\frac{1}{x_2^3}\cdot\frac{\partial u^m}{\partial x_i}\right)-\beta_m\circ(\omega^2(\nabla u^m)-k^2x_2^4)+\varepsilon u^m=0, \quad \text{in } \Omega_0,\\ &u^m|_{\Gamma_0'}=u_0, \quad u^m|_{\Gamma_1}=\frac{T}{2\pi}, \quad \left.\frac{\partial u^m}{\partial n}\right|_{\Gamma_2'}=0,\end{aligned}\right\} \tag{3.14}$$

which has a solution that belongs to $C^0(\widetilde{\Omega}_0)\cap C^2(\Omega_0\cup\Gamma_2')$.

We will give the uniform boundness of $\|u^m\|_{C^0(\widetilde{\Omega}_0)}$ and $\|u^m\|_{H^1(\Omega_0)}$.

Because $w_1=0$ and $w_2=T/2\pi$ respectely satisfy

$$\left.\begin{aligned}&-\frac{\partial}{\partial x_i}\left(x_2^{-3}\frac{\partial w_1}{\partial x_i}\right)-\beta_m\circ(\omega^2(\nabla w_1)-k^2x_2^4)+\epsilon w_1\leqslant 0,\\ &w_1|_{\Gamma_0'}=w_1|_{\Gamma_1'}=0, \qquad \left.\frac{\partial w_1}{\partial n}\right|_{\Gamma_1'}=0\end{aligned}\right\} \tag{3.15}$$

and

$$\left.\begin{aligned}&-\frac{\partial}{\partial x_i}\left(x_2^{-3}\frac{\partial w_1}{\partial x_i}-\right)-\beta_m\circ(\omega^2(\nabla w_1)-k^2x_2^4)+\epsilon w_1\geqslant 0,\\ &w_1|_{\Gamma_2'}=w_2|_{\Gamma_0'}=T/2\pi, \qquad \left.\frac{\partial w_2}{\partial n}\right|_{\Gamma_2'}=0.\end{aligned}\right\} \tag{3.16}$$

Consider $u_0=x_2^4v$ in $S_{h_0/2}$ (Theorem 2.2), let h_1 appropriately small to guarantee $u_0|_{\Gamma_0'}<T/2\pi$, then apply Lemma 3.1 to obtain

$$0\leqslant u^m\leqslant T/2\pi. \tag{3.17}$$

Now in analogy with the deduction of [8], we will estimate $\|u^m\|_{H^1(\Omega_0)}$, Let $\xi_0(x)=|u^m(x)-u_0(x)|^{q-1}$ $(q\geqslant 1)$, it is reasonable to choose $\delta>0$ such that $\delta|\xi_0(x)|<1$, because of the boundness of $u_0(x)$, and $u^m(x)$. In (3.4) let $v=u^m-\delta\xi_0(u^m-u_0)$ and get

$$\begin{aligned}&q\int_{\Omega_0}x_2^{-3}|\nabla(u^m-u_0)|^2\cdot|u^m-u_0|^{q-1}\mathrm{d}x\\ \leqslant{}& M_1+M_2\int_{\Omega_0}|\nabla(u^m-u_0)|^2\cdot|u^m-u_0|^q\mathrm{d}x\\ &+\delta_0\int_{\Omega_0}|u^m-u_0|^{q-1}\cdot|\nabla(u^m-u_0)|^2\mathrm{d}x\\ &+M_3(\delta_0)\int_{\Omega_0}|\nabla u_0|^2\cdot|u^m-u_0|^{q-1}\mathrm{d}x.\end{aligned} \tag{3.18}$$

If allow $q=l,\delta_0=\alpha/2$ (α is some constant), then

$$\begin{aligned}(\alpha-\delta_0)\int_{\Omega_0}|\nabla(u^m-u_0)|^2\mathrm{d}x\leqslant{}& M_1+M_4\int_{|u^m-u_0|\geqslant l}|\nabla(u^m-u_0)|^2\cdot|u^m-u_0|\mathrm{d}x\\ &+M_5\cdot l\int_{|u^m-u_0|\leqslant l}|\nabla(u^m-u_0)|^2\mathrm{d}x+M_0\int_{\Omega_0}|\nabla u_0|^2\mathrm{d}x.\end{aligned}$$

To choose l such that $M_5 \cdot l < 1$, then

$$\int_{\Omega_0} |\nabla(u^m - u_0)|^2 \mathrm{d}x \leqslant M_7 + M_8 \int_{|u^m - u_0| \geqslant l} |\nabla(u^m - u_0)|^2 \cdot |u^m - u_0| \mathrm{d}x. \tag{3.19}$$

If $q = \overline{q} > 1$, from (3.18) we can deduce

$$\int_{|u^m - u_0| \geqslant l} |\nabla(u^m - u_0)|^2 |u^m - u_0| \mathrm{d}x \leqslant l^{2-\overline{q}} \cdot M_9. \tag{3.20}$$

So the uniform boundness of $\|u^m\|_{H^1(\Omega)}$ which is independent of m can be derived from the boundness of u^m and (3.19), (3.20). The proof of the lemma is complete. □

In order to discuss the existence of solutions of (3.3), it is necessary to prove the following.

Lemma 3.5. *The solution u^m of problem (3.4) belongs to $H^{1,p}(\Omega_0)$ (some $p > 2$), and there exists a constant c independent of m such that*

$$\|u^m\|_{H^{1,p}(\Omega_0)} \leqslant C. \tag{3.21}$$

Proof. Let us consider the equivalent weak form of (3.4)

$$\int_{\Omega_0} x_2^{-3} \nabla u^m \cdot \nabla \eta \mathrm{d}x \int_{\Omega_0} (-\beta_m \circ (\cdot) + \varepsilon u^m) \cdot \eta \mathrm{d}x = 0, \quad \forall \eta \in H_0^1(\Omega_0 \setminus \Gamma_2').$$

Let ξ be the smooth cutoff function in $B_{2R} \subset \Omega_0$, $\xi \equiv 1$ in B_R. Allow $\eta = (u^m - u_R^m)\xi^2 \exp[t|u^m - u_R^m|^2]$, where $u_R^m = ⨍_{B_R} u^m \mathrm{d}x = \dfrac{\int_{B_R} u^m \mathrm{d}x}{\int_{B_R} \mathrm{d}x}$. Notice the construction of $\beta_{\varepsilon,m}$, for fixed ε, choose t which is big enough and apply Lemma 3.4 to estimate

$$\int_{B_R} |\nabla u^m|^2 \mathrm{d}x \leqslant \frac{c}{R^2} \left(\int_{B_{2R}} |\nabla u^m|^q \mathrm{d}x \right)^{2/q} + cR^2 \int_{B_{2R}} |\nabla u^m|^2 \mathrm{d}x, \tag{3.22}$$

where $q = \dfrac{2n}{n+2}$. Here the Sobolev-Poincare inequality is used.

In the following we will give the boundary estimate. First consider the points near by Γ_0'. B_{R_0} is a circle in R^2 and $B_{R_0} \cap \Omega_0 \neq \phi_0$ for $x_0 \in B_R$, $R < \dfrac{1}{2}\mathrm{dist}(x_0, \partial B_{R_0})$, there are three possibilities:

(1) $B_{3R/2}(x_0) \cap \Omega_0 = \phi$;

(2) $B_{3R/2}(x_0) \cap (B_{R_0} \setminus \Omega_0) = \phi$;

(3) $B_{3R/2}(x_0) \cap \Omega_0 \neq \phi$, $B_{3R/2}(x_0) \cap (B_{R_0} \setminus \Omega_0) \neq \phi$.

To discuss the case (3). In (3.14) let $\eta = (u^m - u_0)\xi \exp[t|u^m - u_0|^2]$, where $\xi \in C_0^1(B_{3R/2}(x_0))$, $\xi \equiv 1$ in $B_R(x_0)$. then by calculation to get

$$⨍_{B_R} |\nabla u^m|^2 \mathrm{d}x \leqslant c \left\{ \left(⨍_{B_{2R} \cap \Omega_0} |\nabla(u^m - u_0)|^q \mathrm{d}x \right)^{2/q} + ⨍_{B_{2R} \cap \Omega_0} (1 + |\nabla u_0|)^2 \mathrm{d}x \right\}.$$

In order to prove the lemma, here we temporarily extend u^m and u_0 to be zero outside of $B_{2R_0}\cap\Omega_0$. Define the functions $g(x)$ and $f(x)$ as follows

$$g(x)=\begin{cases}|\nabla u^m|^q, & x\in B_{R_0}\cap\Omega_0,\\ 0, & x\in B_{R_0}\setminus\Omega_0,\end{cases}\qquad f(x)=\begin{cases}(1+|\nabla u_0|)^q, & x\in B_{R_0}\cap\Omega_0,\\ 0, & x\in B_{R_0}\setminus\Omega_0,\end{cases}$$

Then for any $x\in B_R$, $R<\dfrac{1}{2}\mathrm{dist}(x,\partial B_{R_0})$,

$$⨍_{B_R(x)}g^{2/q}\mathrm{d}x\leqslant c\left\{\left(⨍_{B_{2R}(x)}g\mathrm{d}x\right)^{2/q}+⨍_{B_{2R}(x)}f^{2/q}\mathrm{d}x\right\}. \tag{3.23}$$

Applying the reverse Holder inequality of [7] and the finite covering theorem to (3.22) and (3.23), taking notice of using Lemma 3.3, one can conclude that estimation (3.23) is also true for the points near Γ_1 and Γ_2'; finally one can obtain that $\exists p>2$ such that

$$\left(\int_{\Omega_0}|\nabla u^m|^p\mathrm{d}x\right)^{1/p}\leqslant c\left\{\left(\int_{\Omega_0}|\nabla u^m|^2\mathrm{d}x\right)^{1/2}+\left(\int_{\Omega_0}(1+|\nabla u_0|^p)\mathrm{d}x\right)^{1/p}\right\}.$$

where c is independent of m. Now the conclusion of the lemma is obvious. □

In the following we will prove

Theorem 3.6. *In the mixed problem (3.3) the solution exists, which is an element of the space* $C^0(\widetilde{\Omega}_0)\cap C^3(\Omega_0\cup\Gamma_2')$.

Proof. By Lemma 3.5, using Sobolev's compact embedding theorem to get that $\{u^m\}$ is compact in $C^0(\widetilde{\Omega}_0)$, combining with Lemma 3.4, one can extract a subsequence from $\{u^m\}$ (still denoted by u^m) s. t.

$$\lim_{m\to\infty}u^m=u\ \text{ in }\ C^0(\overline{\Omega}_0), \tag{3.24}$$

$$\lim_{m\to\infty}u^m=u\ \text{ in the weak sense of }\ H^1(\Omega_0). \tag{3.25}$$

Additionally, because β_m is bounded in $L^1(\Omega_0)$ (see (3.12)), one can select a subsequence such that [8]

$$\lim_{m\to\infty}\beta_m\circ(\omega^2(\nabla u^m)^2-k^2x_2^4)=\mu\ \text{ in the measure space }\ \mu(\Omega_0),$$

then

$$\begin{aligned}&\lim_{n,m\to\infty}\sup\int_{\Omega_0}x_2^{-3}|\nabla(u^m-u^n)|^2\mathrm{d}x\\ &\leqslant\lim_{n,m\to\infty}\left\{\int_{\Omega_0}[-\beta_m\circ(\omega^2(\nabla u^m)-k^2x_2^4)+\beta_n\circ(\omega^2(\nabla u^n)-k^2x_2^4)]\right.\\ &\quad\left.\times(u^m-u^n)\mathrm{d}x-\varepsilon\int_{\Omega_0}|u^m-u^n|^2\mathrm{d}x\right\}\\ &=0.\end{aligned}$$

So we get that u^m strongly converges u in $H_0^1(\Omega_0)$, and

$$\lim_{m\to\infty}\beta_m\circ(\omega^2(\nabla u^m)-k^2x_2^4)=\beta\circ(|\nabla u|^2-k^2x_2^4)\ \text{ a.e. in } \Omega.$$

Here the equi-integrablity of $\beta_m\circ(\omega^2(\nabla u^m)-k^2x_2^4)$ is used. Then

$$\lim_{m\to\infty}\beta_m\circ(\omega^2(\nabla u^m)-k^2x_2^4)=\beta_\varepsilon\circ(\omega^2(|\nabla u|^2-k^2x_2^4)\ \text{ in } L^1(\Omega_0).$$

Now we obtain that u is the solution of the following variational inequality

$$\left.\begin{array}{l}\text{Find } u\in\widehat{K}\cap C^0(\Omega),\quad \forall v\in\widehat{K}\ \text{ s.t.}\\ \displaystyle\int_{\Omega_0}x_2^{-3}\nabla u\cdot\nabla(v-u)\mathrm{d}x+\int_{\Omega_0}[-\beta\circ(|\nabla u|^2--k^2x_2^4)+\varepsilon u](v-u)\mathrm{d}x\geqslant 0.\end{array}\right\}\tag{3.26}$$

In analogy with the proof of Lemma 3.4. noticing that $\beta_\varepsilon\in C^2$, one can get that any bounded solution of (3.30) belongs to $C^3(\Omega_0\cup\Gamma_2')\cap C^0(\widetilde{\Omega}_0)$. Combining (3.26) in which the continuous solution exists, one may obtain that the classical mixed problem (3.3) is solvable in $C^3(\Omega_0\cup\Gamma_2')\cap C^0(\widetilde{\Omega}_0)$. Now the theorem is proved. □

Let domain $T_{h_2}=\{(x_1,x_2)\in\Omega,0<h_2<x_2<h_1\}$, consider the boundary value problem:

$$\left.\begin{array}{l}Au_\varepsilon^h+\varepsilon u_\varepsilon^h=0\\ u_\varepsilon^h|_{\partial T_{h_2}}=u_0\end{array}\quad \text{in } T_{h_2}\right\}\tag{3.27}$$

where u_0 is the solution of (2.5).

Theorem 3.7. *In (3.27) there is a solution* $u_\varepsilon^h\in C^2(\overline{T}_h)\cap C^2(T_h)$, *and*

(i) $u_\varepsilon^h=x_2^4\cdot v_\varepsilon^h$, *where* v_ε^h *satisfies*

$$\left.\begin{array}{l}Lv_\varepsilon^h\triangleq x^2\dfrac{\partial^2v_\varepsilon^h}{\partial x_1^2}+5\dfrac{\partial v_\varepsilon^h}{\partial x_2}-\varepsilon x_2^4v_\varepsilon^h=0\\ u_\varepsilon^h|_{\partial T_{h_2}}=u_0\end{array}\right\}\tag{3.28}$$

(ii) there exists the constant c *independent of* ε,h_2 *such that*

$$|u_\varepsilon^h(x)|\leqslant c,\quad \forall x\in\overline{T}_{h_2},\tag{3.29}$$

$$|\nabla u_\varepsilon^h(x)|\leqslant kx_2^2,\quad \forall x\in T_{h_2},\tag{3.30}$$

where h_1 *is suitably small.*

Proof. The existence of the solution of problem (3.28) is obvious. Because of the boundness of v_0, with help of the maximal value principle one may derive the following

$$|v_\varepsilon^h(x)|\leqslant\max_{x\in\partial T_{h_2}}|v_0|\leqslant c,\tag{3.31}$$

where c is independent of ε,h_2.

By constructing a barrier in ∂T_{h_2} with respect to the operator L and v_ε^h, one may get $|\nabla v_\varepsilon^h(x)| \leqslant c$ ($x \in \partial T_{h_2}$, where c is independent of ε, h_2). Now

$$x_2 \frac{\partial^3 v_\varepsilon^h}{\partial x_2^2 \partial x_1} + 5 \frac{\partial^2 v_\varepsilon^h}{\partial x_2 \partial x_1} - \varepsilon x_2^4 \frac{\partial u_\varepsilon^h}{\partial x_1} = 0.$$

Apply the maximal value principle to $\frac{\partial v_\varepsilon^h}{\partial x_1}$ and get

$$\left|\frac{\partial v_\varepsilon^h}{\partial x_1}\right| \leqslant \max_{x \in \partial T_{h_2}} \left|\frac{\partial v_\varepsilon^h}{\partial x_1}\right| \leqslant c, \tag{3.32}$$

where c is independent of ε, h_2.

Using Bernstein technique to $\dfrac{\partial v_\varepsilon^h}{\partial x_2}$ we may give its uniform bound which is independent of ε, h. In fact, let $v = \left|\dfrac{\partial v_\varepsilon^h}{\partial x_2}\right|^2$, then

$$v_{x_i} = 2 v_{\varepsilon, x_2}^h \cdot v_{\varepsilon, x_2 x_i}^h,$$

$$v_{x_i x_i} = 2(v_{\varepsilon, x_2 x_i})^2 + 2 v_{x_2}^h \cdot v_{\varepsilon, x_2 x_1 x_i}^h, \quad (i = 1, 2),$$

where $v_{x_i} = \partial v / \partial x_i$. If v gets its maximal value in $x_0 \in T_{h_2}$, then in x_0,

$$\left.\begin{aligned} &v_{x_k} = 0, \quad k = 1, 2, \\ &Lv \leqslant 0. \end{aligned}\right\} \tag{3.33}$$

From the above, using (3.28), putting v_{x_i}, $v_{x_i x_i}$ into (3.33) and calculating, one can derive that $|v_{\varepsilon, x_2}^h| \leqslant c$ (independent of ε, h) in the point x_0. So it can be concluded that $|\nabla v_\varepsilon^h| \leqslant c$ in T_{h_2} (c independent of ε, h).

On the other hand, if u_ε^h is the solution of (3.27), $v_\varepsilon^h = x_2^{-4} u_\varepsilon^h$ satisfies (3.33). When h is chosen to be small enough, (ii) can be got due to uniform boundness of $v_\varepsilon^h(x)$ and ∇v_ε^h. Now Theorem 3.7 is proved. □

It is easy to prove that the following function defined by the solution u_ε, of (3.3) and the solution u_ε^h of (3.27)

$$u_\varepsilon(x) = \begin{cases} u_\varepsilon, & x \in \Omega_0, \\ u_\varepsilon^h, & x \in T_{h_2} \cup \Gamma_0', \end{cases}$$

belongs to $C^0(\overline{\Omega_0 \cup T_{h_2}}) \cap H^{1,2}(\Omega_0 \cup T_{h_2})$.

Then for suitably small h_2, u_ε is a $H^{1,2}(\Omega_1)$, which is weak solution of the following

$$\left.\begin{aligned} &A u_\varepsilon - \beta_\varepsilon \circ (|\nabla u_\varepsilon|^2 - k^2 x_2^4) + \varepsilon u_\varepsilon = 0 \quad \text{in } \Omega, \\ &u_\varepsilon|_{\partial T_h \backslash \Gamma_0'} = u_0, \quad u_\varepsilon|_{\Gamma_1} = T/2\pi, \quad \frac{\partial u_\varepsilon}{\partial n}|_{\Gamma_2'} = 0. \end{aligned}\right\} \tag{3.34}$$

where $\Omega_1 = \Omega_0 \cup T_{h_2} \cup \Gamma_0'$, $\Gamma_2'' = \partial\Omega_1 \cap \Gamma_2$, $\Gamma_2'' = \{(x_1, h_2) \in \overline{\Omega}\}$. In analogy with the behind part of the proof of the Theorem 3.6, one may obtain that $u_\varepsilon \in C^3(\Omega_0 \cup T_{h_2} \cup \Gamma_2'' \cup \Gamma_0') \cap C^0(\overline{\Omega}_1)$.

4 The Uniform Estimations for the Gradients

In this section, we will give the uniform boundness for the gradients of the solutions of the penalized problem (3.34). The method used is so-called Bernstein [2, 6, 9].

Apply lemma 3.1 and (3.3), construct barriers to get

Lemma 4.1. *There exist the constants* c_1, c_2 *independent of* ε *such that*

$$0 \leqslant u_\varepsilon(x) \leqslant c_1 \qquad (x \in \overline{\Omega}_1, 0 < \varepsilon < 1), \tag{4.1}$$

$$|\nabla u_\varepsilon(x)| \leqslant c_2 \qquad (x \in \Gamma_0' \cap \Gamma_1, 0 < \varepsilon < \varepsilon_0 < 1). \tag{4.2}$$

Lemma 4.2. *There exists the constants* c_3 *such that*

$$|\nabla u_\varepsilon(x)| \leqslant c_3 \qquad (\forall x \in \Omega_1, 0 < \varepsilon < \varepsilon_0 < 1, \ \textit{for some } \varepsilon_0). \tag{4.3}$$

Proof. It is convenient to omit ε of u_ε in the proof. Let the auxiliary function $v = |\nabla u|^2 + \lambda u$ ($\lambda > 0$ be settled). Then

$$v_{x_i} = 2u_{x_k} u_{x_k x_i} + \lambda u_{x_i} \quad (k, i = 1, 2), \tag{4.4}$$

$$v_{x_i x_j} = 2u_{x_k x_j} u_{x_k x_i} + 2u_{x_k} u_{x_k x_i x_j} + \lambda u_{x_i x_j} \quad (j = 1, 2), \tag{4.5}$$

Make derivation of equation (3.34) with respect to x_1, x_2 respectively

$$Au_{x_1} - \beta' \circ (|\nabla u|^2 - k^2 x_1^4) \cdot (|\nabla u|^2 - k^2 x_1^4)_{x_1} + \varepsilon u_{x_1} = 0, \tag{4.6}$$

$$Au_{x_2} - \beta' \circ (|\nabla u|^2 - k^2 x_2^4) \cdot (|\nabla u|^2 - k^2 x_2^4)_{x_2} + \varepsilon u_{x_2} = -\frac{3}{x_2^4} u_{x_i x_i} + \frac{12}{x_2^5} u_{x_1}. \tag{4.7}$$

Let v reach its maximal value in $x_0 \in \overline{\Omega}_1$, then there are three possibilities:

1° the point $x_0 \in \Omega_1$, then in the point x_0,

$$\left.\begin{aligned} &v_k = 0 \quad (k = 1, 2), \\ &Av = -\frac{\partial}{\partial x_i}\left(x_2^{-3} \frac{\partial v}{\partial x_i}\right) \geqslant 0, \end{aligned}\right\} \tag{4.8}$$

By (3.34), (4.6), (4.7) and Young inequality, one may get

$$\begin{aligned} 0 &\leqslant Av \\ &\leqslant \beta' \circ (\cdot)(-2\lambda |\nabla u|^2 + c|\nabla u|^2 + c) + cu_{x_2}^2 + \lambda \beta \circ (\cdot). \end{aligned}$$

$\beta(t) \leqslant \beta'(t) \cdot t$ ($t \in R$), due to the convexity of $\beta(t)$. One may assume $\beta' \circ (|\nabla u|^2 - k^2 x_2^4) \geqslant 1$ in x_0, then get the bound of v, so $0 \leqslant \beta' \circ (\cdot)(-\lambda |\nabla u|^2 + c|\nabla u|^2 + c)$ that is $\lambda |\nabla u|^2 \leqslant c|\nabla u|^2 + c$ in the point x_0.

Now let λ big enough, one may get the uniform boundness of $|\nabla u|^2$ in x_0, furthermore the uniform boundness of v in x_0.

2° the point $x_0 \in \Gamma_2' = \Gamma_{21}' \cap \Gamma_{22}'$. In this case, with help of Lemma 3.3, one may turn x_0 into the interior point and get the uniform boundness of v in x_0.

3° the point x_0 is another point $\partial\overline{\Omega}_1$, and one may easily get (4.3) by applying Lemmas 3.3 and 4.1.

Combining the conclusions above we know that (4.3) is true. The proof of the lemma is complete. □

5 $W^{2,p}$, $W^{2,\infty}$ Estimation and the Stress Function

Lemma 5.1. *There exists a constant c_4 independent of ε s.t.*

$$|Au_\varepsilon(x)| \leqslant c_4(\Omega_1') \qquad (\forall x \in \Omega_1' \subset\subset \Omega_1, 0 < \varepsilon < \varepsilon_0). \tag{5.1}$$

and for $1 \leqslant p < \infty, \Omega_1' \subset\subset \Omega_1$,

$$||u_\varepsilon||_{H^{2,\infty}(\Omega_1')} \leqslant c(p, \Omega_1') \quad (0 < \varepsilon < \varepsilon_0). \tag{5.2}$$

Proof. We will omit ε in $u_\varepsilon, \beta_\varepsilon$ in the proof for convenience. Construct an auxiliary function $V = \xi^2\beta \circ (|\nabla u|^2 - k^2 x_2^4)$, where ξ is a smooth function which vanishes near $\partial\Omega_1$. Then

$$V_{x_i} = \xi^2\beta' \circ (\cdot)(2u_{x_k} \cdot u_{x_k x_i} - (k^2x_2^4)_{x_i}) + 2\xi_{x_i} \cdot \beta \circ (\cdot),$$

$$\begin{aligned}
V_{x_ix_j} =& \xi^2\beta'' \circ (\cdot)(|\nabla u|^2 - k^2x_1^4)_{x_i} \cdot (|\nabla u|^2 - k^2x_2^4)_{x_j} \\
&+\xi^2\beta' \circ (\cdot) \cdot (2u_{x_kx_j}u_{x_kx_i} + 2u_{x_k}u_{x_kx_ix_j} - (k^2x_2^4)_{x_ix_j}) \\
&+2\xi\xi_{x_j}\beta' \circ (\cdot)(2u_{x_k}u_{x_kx_i} - (k^2x_2^4)_{x_i}) \\
&+2\xi\xi_{x_i}\beta' \circ (\cdot) \cdot (2u_{x_k}u_{x_kx_j} - (k^2x_2^4)_{x_j}) \\
&+\beta \circ (\cdot) \cdot (2\xi_{x_i}\xi_{x_j} + 2\xi\xi_{x_ix_j}).
\end{aligned}$$

From equation (3.34). $\xi^2 Au = V - \varepsilon\xi^2 \cdot u$, then $\xi^2 Au_k = V_{x_k} + \xi\widetilde{D}_k^2 u$ $(k = 1, 2)$, where $\widetilde{D}_k^2 u = \sum\limits_{|\alpha|\leqslant 2} \sigma_k^\alpha D^\alpha u$, σ_k^α is bounded.

If V attains its maximal value in $x_0 \in \Omega_1$, then in x_0,

$$V_{x_k} = 0 \ \ (k = 1, 2, AV \geqslant 0).$$

Now in the point x_0, employing Lemma 4.2, we may carry out the following estimation

$$\begin{aligned}
0 \leqslant AV =& -x_2^{-3}V_{x_ix_i} + 3x_2^{-4}V_{x_2} \\
=& -x_2^{-3}\xi^2\beta' \circ (\cdot) \cdot (|\nabla u|^2 - k^2x_2^4)_{x_i}^2 + c \cdot \beta \circ (\cdot) \\
&+\beta' \circ (\cdot)\{-2x_2^{-3}\xi^2u_{x_kx_i}^2 + 2u_{x_k}\xi^2(-x_2^{-3}u_{x_kx_ix_i} + 3x_2^{-4}u_{x_2x_k} + cu_{x_k}) + c_\varsigma^c|\nabla^2 u| + c\}.
\end{aligned}$$

Because β is convex, $\beta(t) \leqslant t\beta'(t)$.

$$0 \leqslant \beta' \circ (\cdot)\{-x_2^{-3}\xi^2|\nabla^2 u|^2 + 2u_{x_k}\xi^2 A(u_k) + (\xi^2|\nabla^2 u| + c)\}$$
$$\leqslant \beta' \circ (\cdot) \cdot \{-x_2^{-3}\xi^2|\nabla^2 u|^2 + c\xi|\nabla^2 u| + c)\}.$$

From these, one may derive $\xi^2|\nabla^2 u|^2 \leqslant c$, c is independent of ε, so $V \leqslant c$. The lemma is proved. □

Furthermore similar to the proof of the theorem of paper [2] we can get the result bellow.

Lemma 5.2. *There exists a constant $c_\delta(\Omega')$ such that*

$$||\nabla^2 u_\varepsilon||_{L^\infty(\Omega')} \leqslant c_\delta \Omega' \quad (\forall \Omega' \subset\subset \Omega_1) \tag{5.3}$$

In the following we will prove the theorem of the stress function.

Theorem 5.3. *The classical problem (2.1)–(2.4) has a unique weak solution $u \in H^2(\Omega_{\Gamma_1}) \cap W^{2,\infty}_{loc}(\Omega \cup \Gamma_2 \cup \Gamma_0)$, that is, $u \in C^1(\overline{\Omega} \setminus \Gamma_1) \cap C^{1,1}(\Omega \cap \Gamma_2)$, where $\Omega_{\Gamma_1} \subset\subset \overline{\Omega} \setminus \Gamma_1$.*

Proof. We first prove the existence and uniqueness of the solution for the following boundary value problem.

$$\left.\begin{aligned} &\max\{-A\overline{u}, |\nabla\overline{u}| - kx_2^2\} = 0 \quad \text{in } \Omega_1, \\ &\overline{u}|_{\Gamma_0} = u_0, \overline{u}|_{\Gamma_1} = T/2\pi, \left.\frac{\partial \overline{u}}{\partial n}\right|_{\Gamma_2} = 0, \end{aligned}\right\} \tag{5.4}$$

In fact, by Lemmas 4.1 and 5.2, there exists $\varepsilon_m \to 0$ and the function $\overline{u} \in C^{0,1}(\overline{\Omega}_1) \cap W^{2,\infty}_{loc}(\Omega_1)$ such that

$$\text{in } C(\Omega_1) \lim_{m\to\infty} u_{\varepsilon_m} = \overline{u}, \tag{5.5}$$

$$\text{in } C(\Omega_1) \lim_{m\to\infty} u_{\varepsilon_m, x_i} = \overline{u}_{x_i}, \tag{5.6}$$

$$\text{in } W^{2,\infty}_{loc}(\Omega_1) \lim_{m\to\infty} u_{\varepsilon_m, x_i x_j} = \overline{u}_{x_i x_j} \text{ (in weak sense)}, \quad (i,j,m = 1,2). \tag{5.7}$$

By Lemma 5.1, $\beta_{\varepsilon_m} \cdot (|\nabla u_m|^2 - k^2 x_2^4)$ is locally bounded. Applying the continuity of $|\nabla \overline{u}|$ and ∇u_{ε_m} and the construction of β_ε, we can derive

$$\text{in } \Omega_1 \quad |\nabla \overline{u}|^2 \leqslant k^2 x_2^4, \tag{5.8}$$

Now because $-Au_\varepsilon \leqslant 0$ in Ω_1, then

$$-A\overline{u} \leqslant 0, \quad a.e. \text{ in } \Omega_1. \tag{5.9}$$

For $x_0 \in \Omega_1$, $|\nabla \overline{u}(x_0)| < kx_2^2$, and due to the continuity of $|\nabla \overline{u}|$ and the function $g = kx_2^2$. It is certain that there exists a neighborhood $N(x_0)$ of x_0 such that $|\nabla \overline{u}(x)| < kx_2^2$ for $x \in N(x_0)$. By (5.6), there exists m_0 which is big enough so that $|\nabla u_{\varepsilon_m}(x)| < kx_2^2$, $\forall x \in N(x_0)$ when $m > m_0$. Then one may derive by limitation that

$$A\overline{u}(x) = 0, \quad \text{when } x \in N(x_0). \tag{5.10}$$

Putting (5.8)–(5.10) together one may conclude that $\overline{u} \in C^{0,1}(\overline{\Omega}_1) \cap W^{2,\infty}_{loc}(\Omega_1)$ is a weak solution of (5.4) and the regularity of $\overline{u}$ in the boundary $\Gamma_2^{'}$, i.e. $\overline{u} \in C^{1,1}(\Omega_1 \cap \Gamma_2^{''})$ by using Lemma 3.3 again.

Now the rest is to verify the uniqueness of the solution of (5.4). Let $\overline{u}$ and $\widehat{u}$ be two distinct solutions of (5.4), $\rho = \max\limits_{\Omega_1}(\widehat{u} - \overline{u}) > 0$ and $u_\tau = \widehat{u} + (\tau - 1)\overline{u}$ $(0 < \tau < 1)$. Choose suitable τ such that $\max\limits_{\overline{\Omega}_0} u_\tau \geqslant \rho/2$. Make function $V(x) = a\exp(bx_1)$, where $a, b > 0$, then

$$AV = -x_2^{-3} b^2 V < 0, \quad \text{in } \Omega_1. \tag{5.11}$$

If $u_\tau + V$ obtains non-negative maximal value in $x^\tau \in \overline{\Omega}_0$, then there are two possibilities: (i) $x^\tau \in \Omega_1^{'} \subset\subset \Omega_1$, by Bony maximal value principle

$$\text{in the point } x^\tau, \quad \nabla(u^\tau + V) = 0, \tag{5.12}$$

$$\lim_{x \to x^\tau} \text{ ess sup } A(u^\tau + V) \geqslant 0. \tag{5.13}$$

Let $\delta = \min k(x_2^\tau)^2, a_1$ small enough such that when $a \in (0, a_1), bV < \tau\delta, \max\limits_{\overline{\Omega}_1} V < \rho/4$ in Ω_1, then for $a \in (0, a_1)$,

$$|\nabla \widehat{u}(x^\tau)| \leqslant k(x_2^\tau)^2 - \varepsilon\delta + |\nabla V| < k(x_2^\tau)^2, \tag{5.14}$$

and $A\widehat{u}(x) = 0$ a.e. in a neighborhood of x^τ, furthermore, by (5.13) and (5.14)

$$0 \leqslant \lim_{x \to x^\tau} \text{ ess sup } ((\tau - 1)A\widehat{u} + AV) < 0.$$

This is a contradiction. It is not possible for the case (i) to be true.

(ii) $x^\tau \in \Gamma_2^{''}$. Now applying Lemma 3.3, one may translate (ii) into (i), so (ii) can not be true. In a word. the solution of (5.4) exists and is unique. Now we will extend the solution of (5.4) so that it becomes the solution of (2.1)–(2.4).

In fact, from Lemma 3.3 and the above, $A\overline{u} = 0$, in T_{h_2}. Use Theorem 2.2, and construct the following function:

$$u(x) = \begin{cases} \overline{u}(x), & (x \in \Omega_1), \\ u_0(x), & (x \in \Gamma_0^{''} \cup (\Omega \setminus \Omega_1)), \end{cases} \tag{5.15}$$

then $u \in W^{1,\infty}(\Omega) \cap W^{2,\infty}(\Omega \cup \Gamma_2 \cup \Gamma_0)$ and it satisfies

$$\left.\begin{aligned} &\max\{-Au, |\nabla u| - kx_2^2\} = 0 \quad \text{in } \Omega, \\ &u|_{\Gamma_0} = u_0, u|_{\Gamma_1} = T/2\pi, \left.\frac{\partial u}{\partial n}\right|_{\Gamma_1} = 0. \end{aligned}\right\} \tag{5.16}$$

Now we finish the proof of the theorem. □

Remark 5.1. In general. we can not expect that the solutions belong to C^2 for obstacle problems or the variational problem with constraints. In this sense $C^{1,1}$-regularity is optimal.

References

[1] T.W. Ting. Elastic-plastic torsion problem over multiple connected domains, *Annali Della Scuola Normale Superiore Di Pisa*, 4 (1977), 291-312.

[2] M. Wiegner. The $C^{1,1}$-character of solutions of second order elliptic equations with gradient constraint, *Communications in Partial Differential Equations*, 6 (1981), 361-371.

[3] C.W. Cryer, The solution of the axisymmetric elastic-plastic torsion of a shaft using variational inequalities. *Journal of Mathematics and Applications*, 76 (1980), 535-570.

[4] L.C. Evans. A second order elliptic equation with gradient constraint, *Communications in Partial Differential Equations*, 49 (1979), 557-572.

[5] S.Z. Zhou, On an axisymmetric free boundary problem, *Journal of Mathematical Analysis & Applications*, 121 (1983), 465-486.

[6] D. Gilbary and N.S. *Trudinger, Elliptic Pariial Differential Eqations of Second Order*, Spring-Verlag, New York, 1983.

[7] M. Giaquinta, *Multiple Integrals in the Calculus of Variations and Nonlinear Elliptic Systems*, Princeton University Press, Princeton, 1983.

[8] M. Biroli and U. Mosco. Stability and homogenization for nonlinear variational inequalities with irregular obstacles and quadratic growth, *Nonlinear Analysis Theory Methods & Applicatio*, 7 (1983), 41-60.

[9] H. Ishii and S. Koike, Boundary regularity and uniqueness for an elliptic equation with gradient constraint, *Communications in Partial Differential Equations*, 8 (1983), 317-346.

On Monotone and Geometric Convergence of Schwarz Methods for Two-Sided Obstacle Problems*

J.P. Zeng (曾金平)† S.Z. Zhou (周叔子)†

Abstract Convergence of Schwarz methods is discussed for certain discretizations of two-sided obstacle problems. Monotone and especially geometrical convergence are established if the stiffness matrix is a kind of M-matrix, and an h-independent convergence rate is proved for uniformly overlapping decomposition.

Keywords Variational inequalities, obstacle problems, Schwarz algorithms, monotone convergence, geometrical convergence, h-independent convergence rate.

1 Introduction

Schwarz algorithms have been studied extensively for linear elliptic problems; cf. [1–8] and the references therein. But very few works have been done on Schwarz algorithms for variational inequalities. The first work concerning this topic appears to be the papers of P.L. Lions [5, 9] in which a multiplicative Schwarz algorithm was proposed for an obstacle problem of second order. Subsequent works on obstacle problems include [10] on the multiplicative Schwarz algorithm for fourth order and [11] on a new multiplicative and an additive Schwarz algorithm.

Recently, more attention has been paid to Schwarz algorithms for discrete (e.g., finite element) problems of variational inequalities. An algorithm of this kind for solving these problems was proposed in [6, 12] for the case when the convex subset K is a cone. Multiplicative and additive Schwarz algorithms were given in [13] and [14], respectively, for the obstacle problem with one or two obstacles. Monotone convergent Schwarz algorithms with special choice of initial values were proposed in [15, 16] for the single obstacle problem. The same monotone convergent algorithms for single obstacle problems were discussed in [17] from a different viewpoint, and an h-independent convergence result was claimed there without proof.

* 发表于：SIAM Journal on Numerical Analysis, Vol. 35, 1998, pp. 600-616.

† Department of Applied Mathematics, Hunan University, Changsha 410082, China.

In this paper, monotone convergent methods for two-sided obstacle problems are proposed. Geometrical convergence is first proved for the Schwarz algorithm solving obstacle problems. In order to prove the geometrical convergence, a kind of strong discrete maximum principle is established. For the additive algorithm solving the single obstacle problem, the iterative solution is constructed by using the weighted average of the solutions of the subproblems. This scheme cannot guarantee the monotonicity of the iterative solution sequence in the two-side obstacle case. So the minimum of the solution of the subproblems is used to construct an iterative solution for the additive algorithm.

The paper is organized as follows. The obstacle problem is introduced in section 2. The monotone multiplicative and additive Schwarz algorithms for obstacle problems are discussed in sections 3 and 4, respectively. The geometrical convergence is presented in section 5, and the h-independent convergence rate is proved in section 6. Finally, some numerical examples are given in section 7.

2 The Obstacle Problem

We assume, for simplicity, that Ω is a convex polygonal domain in R^2 with a boundary $\Gamma = \partial\Omega$. Consider the two-sided obstacle problem: find $u \in K$ such that

$$a(u, v-u) \geqslant (f, v-u), \quad \forall v \in K, \tag{2.1}$$

where

$$K = \{v \in H_0^1 : \phi \leqslant v \leqslant \psi \text{ in } \Omega\},$$

$$a(u,v) = \int_\Omega [\nabla u \nabla v + (\overrightarrow{\beta} \cdot \nabla u)v + \gamma uv]\mathrm{d}x, \quad (f,v) = \int_\Omega fv\mathrm{d}x,$$

$\overrightarrow{\beta} = (\beta_1, \beta_2)^T \in (L^\infty(\Omega))^2$, $\gamma \in L^\infty(\Omega)$, $\gamma \geqslant 0$ in Ω, $f \in L^2(\Omega)$, $\phi, \psi \in H^2(\Omega)$, $\phi < \psi$ in Ω, $\phi \leqslant 0 \leqslant \psi$ on Γ.

It can be shown (cf. [18]) that (2.1) is equivalent to the following problem: find $u \in H^2(\Omega)$ such that

$$\begin{cases} \phi \leqslant u \leqslant \psi, & \text{in } \Omega, \quad u = 0 \text{ on } \Gamma, \\ Łu - f = 0, & \text{if } \phi < u < \psi, \\ Łu - f \geqslant 0, & \text{if } u = \phi, \\ Łu - f \leqslant 0, & \text{if } u = \psi, \end{cases} \tag{2.2}$$

where operator Ł is defined by $Łv = -\Delta v + \overrightarrow{\beta} \cdot \nabla v + \gamma v$.

Now we consider the discretization for (2.1). Assume that $V^h \subset H^1(\Omega)$ is a linear finite element space. The finite element approximation of (2.1) is: find $u_h \in K^h$ such that

$$a(u_h, v-u_h) \geqslant (f, v-u_h) \quad \forall v \in K^h, \tag{2.3}$$

where

$$K^h = \{v \in V^h : \phi_h \leqslant v \leqslant \psi_h \quad \text{in } \Omega, \ v = 0 \text{ on } \Gamma_h\},$$

ϕ_h and ψ_h are the finite element approximations of the functions ϕ and ψ, respectively, and Γ_h is the set of the nodes belonging to Γ [19].

The problem (2.3) is equivalent to the following algebraic problem: find $W \in C$, such that

$$(AU, W-U) \geqslant (F, W-U) \quad \forall W \in C, \tag{2.4}$$

where $A = (a_{ij}) \in R^{N\times N}$ is the stiffness matrix,

$$C = \{W \in R^N : \Phi \leqslant W \leqslant \Psi\}.$$

Componentwise,

$$a_{ij} = a(\theta_h^{(i)}, \theta_h^{(j)}),$$
$$U_i = u_h(x_i), \quad \Phi_i = \phi_h(x_i), \quad \Psi_i = \psi_h(x_i) \quad \forall x_i \in \Omega_h,$$
$$F_i = \int_\Omega f\theta_h^{(i)} \mathrm{d}x,$$

where Ω_h is the set of the nodes belonging to Ω. $\theta_h^{(i)}$ $(i = 1, \cdots, N)$ are the nodal basis functions.

The algebraic problem (2.4) has the following equivalent problem similar to (2.2): find $U \in R^N$ such that

$$\begin{cases} \Phi \leqslant U \leqslant \Psi, & \\ (AU-F)_j = 0, & \text{if } \Phi_j < U_j < \Psi_j, \\ (AU-F)_j \geqslant 0, & \text{if } U_j = \Phi_j, \\ (AU-F)_j \leqslant 0, & \text{if } U_j = \Psi_j. \end{cases} \tag{2.5}$$

In the sequel, we assume that A is an M-matrix (cf. [20]) and, therefore, (2.5) has a unique solution. To obtain this property, we need some other assumptions on the finite element triangulation. For example, we can assume that the triangulation is a Delauney triangulation and h is small enough.

For M-matrix A, it is easy to verify the following comparison theorem.

Lemma 2.1. *If A is an M-matrix and $I \subset \{1, 2, \cdots, N\}$, then for any $V, W \in R^N$ satisfying*

$$(AW)_i \geqslant 0, \quad (AV)_i \leqslant 0 \quad \text{for } \ i \in I$$

and

$$V_i \leqslant W_i \quad \text{for } \ i \notin I,$$

we have that $V \leqslant W$.

We need a lemma in [21] for our convergence theory. We give a proof of this lemma here for convenience.

Lemma 2.2. *If A is an M-matrix, then the minimal element of T solves the problem (2.5), where*

$$T = \{W \in R^N : \Phi \leqslant W \leqslant \Psi, (AW - F_k) \geqslant 0 \text{ or } W_k = \Psi_k \text{ for } k = 1, \cdots, N\}.$$

Here, the minimal element U of T means $U \in T$ and $U \leqslant W$ for all $W \in T$.

Proof. Let U be the solution of (2.5). For any $W \in T$ let $J = \{i : U_i = \Phi_i \text{ or } W_i = \Psi_i\}$, $I = \{1, 2, \cdots, N\} \setminus J$. Then, $U_i \leqslant W_i$ for $i \in J$ and

$$(AU - F)_i \leqslant 0, \quad (AW - F)_i \geqslant 0 \quad \text{for } i \in I.$$

From Lemma 2.1 we have that $U \leqslant W$. This proves that U is the minimal element of T. □

In order to construct a Schwarz algorithm for solving (2.3), we decompose Ω into overlapping open subdomains Ω_i, i.e.,

$$\Omega = \bigcup_{i=1}^{m} \Omega_i.$$

For h-level finite element triangulation, we assume that any finite element does not cross the boundaries of Ω_i $(i = 1, 2, \cdots, m)$.

In the sequel, we let Ω_h and Ω_{ih} $(i = 1, \cdots, m)$ be the sets of the nodes belonging to domain Ω and subdomain Ω_i, respectively, N_i be the subset of index set $\{1, 2, \cdots, N\}$ as follows: $N_i = \{j \in \{1, 2, \cdots, N\} : x_j \in \Omega_{ih}\}$, and for $I, J \subset \{1, 2, \cdots, N\}$, we let $A_{I,J}$ be the submatrix of A with elements a_{ij} $(i \in I, j \in J)$ and U_I be the subvector of U with elements U_i $(i \in I)$.

3 A Multiplicative Schwarz Algorithm

In this section, we discuss the following multiplicative Schwarz algorithm for solving (2.3).

Algorithm I.

Let $u_h^0 \in K^h$ be given such that

$$u_h^0(x_k) = \psi_h(x_k) \quad \text{or} \quad \sum_{j=1}^{N} a(\theta_h^{(k)}, \theta_h^{(j)}) u_h^0(x_j) - F_k \geqslant 0$$

hold for $k = 1, \cdots, N$. For $n = 0, 1, \cdots$, do Step 1 until convergence.

Step 1. For $i = 1, \cdots, m$ solve the following subproblems successively: find $u^{n+i/m} \in (u_h^{n+(i-1)/m} + V_i) \cap K^h$ such that

$$a(u_h^{n+i/m}, v - u_h^{n+i/m}) \geqslant (f, v - u_h^{n+i/m}) \quad \forall v \in (u_h^{n+(i-1)/m} + V_i) \cap K^h,$$

where $V_i = H_0^1(\Omega_i) \cap V^h$.

It is easy to prove that Algorithm I is equivalent to the following algorithm in matrix form, which solves (2.4).

Algorithm II.

Let $U^0 \in T$ be given. For $n = 0, 1, \cdots$, do Step 1 until convergence.

Step 1. For $i = 1, \cdots, m$ solve the following subproblems successively: find $U^{n+i/m} \in R^N$ such that

$$\begin{cases} \Phi_{N_i} \leqslant U_{N_i}^{n+i/m} \leqslant \Psi_{N_i}, \\ (A_{N_i,N_i} U_{N_i}^{n+i/m} - F^{n,i})_k = 0, & \text{if } (\Phi_{N_i})_k < (U_{N_i}^{n+i/m})_k < (\Psi_{N_i})_k, \\ (A_{N_i,N_i} U_{N_i}^{n+i/m} - F^{n,i})_k \geqslant 0, & \text{if } (U_{N_i}^{n+i/m})_k = (\Phi_{N_i})_k, \\ (A_{N_i,N_i} U_{N_i}^{n+i/m} - F^{n,i})_k \leqslant 0, & \text{if } (U_{N_i}^{n+i/m})_k = (\Psi_{N_i})_k \end{cases} \tag{3.1}$$

and

$$U_{N\backslash N_i}^{n+i/m} = U_{N\backslash N_i}^{n+(i-1)/m}, \tag{3.2}$$

where $F^{n,i} = F_{N_i} - A_{N_i,N\backslash N_i} U_{N\backslash N_i}^{n+(i-1)/m}$ (here we assume that $u_h^{n+i/m} = u_h^{n+(i-1)/m}$ in $\Omega\backslash\Omega_i$).

Remark 3.1. The simplest choice for $U^0 \in T$ is $U^0 = \Psi$. Similarly, the simplest choice of u_h^0 in Algorithm I is $u_h^0 = \psi_h$.

Remark 3.2. Since Algorithms I and II are equivalent, we only need to analyze the convergence of Algorithm II.

Since the principal submatrix of an M-matrix is also an M-matrix, we immediately obtain by Lemma 2.2 the following lemma.

Lemma 3.1. *Assume A is an M-matrix and $U_{N_i}^{n+i/m}$ is the solution of (3.1). Then $U_{N_i}^{n+i/m}$ is the minimal element of $T^{n,i}$, where*

$$T^{n,i} = \{\Phi_{N_i} \leqslant W \leqslant \Psi_{N_i} : (A_{N_i,N_i} W - F^{n,i})_k \geqslant 0 \ \ or \ \ W_k = (\Psi_{N_i})_k, k \in N_i\}.$$

Theorem 3.2. *Assume A is an M-matrix and U is the solution of problem (2.4). Then the iterative sequence $\{U^n\}$ generated by Algorithm II is in T and it is monotone; that is, $U^n \in T$ and*

$$U \leqslant U^{n+1} \leqslant U^n \leqslant \cdots \leqslant U^0.$$

Proof. We prove the conclusion by induction. From Step 1 of Algorithm I we know that $U^0 \in T$. Noting that

$$A_{N_1,N_1} U_{N_1}^0 - F^{0,1} = A_{N_1,N} U^0 - F_{N_1} = (AU^0 - F)_{N_1},$$

we know that if $(U_{N_1}^0)_k < (\Psi_{N_1})_K$ then $((AU^0 - F)_{N_1})_k \geqslant 0$, that is, $(A_{N_1,N_1} U_{N_1}^0 - F^{0,1})_k \geqslant 0$. Therefore $U_{N_1}^0 \in T^{0,1}$. From Lemma 3.1 we know that $U_{N_1}^{1/m}$ is the minimal element of $T^{0,1}$. So $\Phi_{N_1} \leqslant U_{N_1}^{1/m} \leqslant \Psi_{N_1}$ and $U_{N_1}^0 \geqslant U_{N_1}^{1/m}$. This combined with (3.2) yields that $U^{1/m} \leqslant U^0$. From (3.1) we have

$$((AU^{1/m} - F)_{N_1})_k = (A_{N_1,N_1} U_{N_1}^{1/m} - F^{0,1})_k \geqslant 0 \quad \text{for } (U_{N_1}^{1/m})_k < (\Psi_{N_1})_k.$$

Now look at the other components of $U^{1/m}$. Assume that $(U_{N\backslash N_1}^{1/m})_k < (\Psi_{N\backslash N_1})_k$. Since $(U_{N\backslash N_1}^0)_k = (U_{N\backslash N_1}^{1/m})_k < (\Psi_{N\backslash N_1})_k$ and $U^{1/m} \leqslant U^0$, we get (noting that the off-diagonal element of M-matrix are nonpositive)

$$\begin{aligned}((AU^{1/m}-F)_{N\backslash N_1})_k &= (A_{N\backslash N_1,N\backslash N_1})_{kk}(U^0_{N\backslash N_1})_k+\sum_{j\neq k}(A_{N\backslash N_1,N\backslash N_1})_{kj}(U^{1/m}_{N\backslash N_1})_j\\ &\quad+\sum_{j\notin N_1}(A_{N\backslash N_1,N})_{kj}(U^{1/m})_j-(F_{N\backslash N_1})_k\\ &\geqslant ((AU^0-F)_{N\backslash N_1})_k\geqslant 0.\end{aligned}$$

Therefore, we have $U^{1/m}_k=\Psi_k$ or $(AU^{1/m}-F)_k\geqslant 0$ for all $k=1,\cdots,N$, i.e., $U^{1/m}\in T$. By induction, for index i we obtain

$$U^1\leqslant U^{(m-1)/m}\leqslant\cdots\leqslant U^{1/m}\leqslant U^0$$

and $U^{i/m}\in T, i=0,1,\cdots,m$. Hence $U^1\leqslant U^0$ and $U^1\in T$. By using induction for index n we get $U^{n+1}\leqslant U^n, n=0,1,\cdots$, and $U^{n+1}\in T$. Since U is the minimal element of T we have that $U^n\geqslant U, n=0,1,\cdots$. □

The following result is immediately obtained from the proof of Theorem 3.2.

Corollary 3.3. *Under the conditions of Theorem 3.2 we have*

$$U^{n+i/m}\in T,\ U^{n+1}\leqslant U^{n+i/m}\leqslant U^{n+(i-1)/m}\leqslant U^n,\quad i=1,\cdots,m,\quad n=0,1,\cdots.$$

Theorem 3.4. *Assume the conditions of Theorem 3.2 hold. Let $\epsilon^{n+i/m}=U^{n+i/m}-U$. Then for all $n=0,1,\cdots$ and $i=1,\cdots,m$, we have*

$$0\leqslant \epsilon^{n+i/m}\leqslant G_i\epsilon^{n+(i-1)/m}. \tag{3.3}$$

Thereby

$$0\leqslant \epsilon^{n+1}\leqslant G\epsilon^n, \tag{3.4}$$

where

$$G_i=P_i\begin{pmatrix}0 & -A^{-1}_{N_i,N_i}A_{N_i,N\backslash N_i}\\ 0 & I_{N\backslash N_i,N\backslash N_i}\end{pmatrix}P_i^T,$$

$G=\Pi_{i=1}^m G_i$, I is the unit matrix, and P_i is the permutation matrix such that

$$U=P_i\begin{pmatrix}U_{N_i}\\ U_{N\backslash N_i}\end{pmatrix}.$$

Moreover, $G_i(i=1,\cdots,m)$ and G are nonnegatives.

Proof. By Theorem 3.2 and Corollary 3.3, we have $\epsilon^{n+i/m}\geqslant 0$, $\epsilon^n\geqslant 0$. For $i=1,\cdots,m$ and index k we have the following three cases.

(a) $(U^{n+i/m}_{N_i})_k>(\Phi_{N_i})_k$ and $(U_{N_i})_k<(\Psi_{N_i})_k$. Then from (3.1) we get that

$$(A_{N_i,N_i}U^{n+i/m}_{N_i}-F_{n,i})_k\leqslant 0. \tag{3.5}$$

On the other hand, from (2.5) we have

$$(A_{N_i,N_i}U_{N_i} + A_{N_i,N\backslash N_i}U_{N\backslash N_i} - F_{N_i})_k \geqslant 0. \tag{3.6}$$

Subtracting (3.6) from (3.5), we get

$$(A_{N_i,N_i}\epsilon_{N_i}^{n+i/m})_k \leqslant (-A_{N_i,N\backslash N_i}\epsilon_{N\backslash N_i}^{n+(i-1)/m})_k. \tag{3.7}$$

(b) $(U_{N_i}^{n+i/m})_k > (\Phi_{N_i})_k$ and $(U_{N_i})_k = (\Psi_{N_i})_k$. From Corollary 3.3 we have $(U_{N_i})_k \leqslant (U_{N_i}^{n+i/m})_k \leqslant (\Psi_{N_i})_k = (U_{N_i})_k$, i.e., $(\epsilon_{N_i}^{n+i/m})_k = 0$. Since A is an M-matrix we know that $-A_{N_i,N\backslash N_i} \geqslant 0$. So (3.7) holds in this case.

(c) $(U_{N_i}^{n+i/m})_k = (\Phi_{N_i})_k$. Then we also have $(\epsilon_{N_i}^{n+i/m})_k = 0$ because $U^{n+i/m} \geqslant U \geqslant \Phi$ by Corollary 3.3. Therefore, (3.7) holds again.

Combining (a)–(c) we obtain

$$A_{N_i,N_i}\epsilon_{N_i}^{n+i/m} \leqslant -A_{N_i,N\backslash N_i}\epsilon_{N\backslash N_i}^{n+(i-1)/m}. \tag{3.8}$$

Since principal matrices of an M-matrix are also M-matrices, so we have

$$\epsilon_{N_i}^{n+i/m} \leqslant -A_{N_i,N_i}^{-1}A_{N_i,N\backslash N_i}\epsilon_{N\backslash N_i}^{n+(i-1)/m}. \tag{3.9}$$

Using (3.2) we get

$$\epsilon_{N\backslash N_i}^{n+i/m} = \epsilon_{N\backslash N_i}^{n+(i-1)/m}, \tag{3.10}$$

which together with (3.9) yields (3.3) and (3.4). □

Now we can obtain the following theorem, which is the first monotone convergence theorem for the two-sided obstacle problem.

Theorem 3.5. *Under the conditions of Theorem 3.2, the iterative sequence $\{U^n\}$ monotonically converges to U.*

Proof. By Theorem 3.2 we know that $\{U^n\}$ is a decreasing sequence and has a low bound U. Then there exists $U^A \in R^N$ such that $\lim\limits_{n\to\infty} U^n = U^A$. By Lemma 2.1 we have that

$$U \leqslant U^A \leqslant U^n, \quad n = 0, 1, \cdots,$$

which together with Corollary 3.3 yields

$$\lim_{n\to\infty} U^{n+i/m} = U^A, \quad i = 1, \cdots, m.$$

Therefore, letting $n \to \infty$ in (3.1) we obtain that $\Phi_{N_i} \leqslant U_{N_i}^A \leqslant \Psi_{N_i}$ and

$$\begin{cases} (A_{N_i,N_i}U_{N_i}^A - F^{A,i})_k = 0, & \text{if } (\Phi_{N_i})_k < (U_{N_i}^A)_k < (\Psi_{N_i})_k, \\ (A_{N_i,N_i}U_{N_i}^A - F^{A,i})_k \geqslant 0, & \text{if } (U_{N_i}^A)_k = (\Phi_{N_i})_k, \\ (A_{N_i,N_i}U_{N_i}^A - F^{A,i})_k \leqslant 0, & \text{if } (U_{N_i}^A)_k = (\Psi_{N_i})_k, \end{cases}$$

where $F^{A,i} = F_{N_i} - A_{N_i,N\backslash N_i}U_{N\backslash N_i}^A, i = 1, \cdots, m$, which means U^A satisfies (2.5) and then $U^A = U$. □

4 An Additive Schwarz Algorithm

In this section, we discuss the following additive Schwarz algorithm for solving (2.3).

Algorithm III.

Let $u_h^0 \in K^h$ be given such that

$$u_h^0(x_k) = \psi_h(x_k) \quad \text{or} \quad \sum_{j=1}^{N} a(\theta_h^{(k)}, \theta_h^{(j)}) u_h^0(x_j) - F_k \geqslant 0$$

hold for $k = 1, \cdots, N$. For $n = 0, 1, \cdots$, do Step 1 and 2 until convergence.

Step 1. For $i = 1, \cdots, m$ solve the following subproblems independently: find $u_h^{n,i} \in (u_h^n + V_i) \cap K^h$ such that

$$a(u_h^{n,i}, v - u_h^{n,i}) \geqslant (f, v - u_h^{n,i}) \quad \forall v \in (u_h^n + V_i) \cap K^h,$$

where $V_i = H_0^1(\Omega_i) \cap V^h$ (it is easy to see that $u_h^{n,i} = u_h^n$ in $\Omega \backslash \Omega_i$).

Step 2. Find $u_h^{n+1} \in V^h$ such that $u_h^{n+1}(x_j) = \inf_{1 \leqslant i \leqslant m} u_h^{n,i}(x_j), j = 1, \cdots, N$.

Algorithm III for solving (2.3) is equivalent to the following algorithm for solving (2.4).

Algorithm IV.

Let $U^0 \in T$ be given. For $n = 0, 1, \cdots$, do Steps 1 and 2 until convergence.

Step 1. For $i = 1, \cdots, m$ solve the following subproblems independently: find $U^{n,i} \in R^N$ such that

$$\begin{cases} \Phi_{N_i} \leqslant U_{N_i}^{n,i} \leqslant \Psi_{N_i}, \\ (A_{N_i,N_i} U_{N_i}^{n,i} - \hat{F}^{n,i})_k = 0, & \text{if } (\Phi_{N_i})_k < (U_{N_i}^{n,i})_k < (\Psi_{N_i})_k, \\ (A_{N_i,N_i} U_{N_i}^{n,i} - \hat{F}^{n,i})_k \geqslant 0, & \text{if } (U_{N_i}^{n,i})_k = (\Phi_{N_i})_k, \\ (A_{N_i,N_i} U_{N_i}^{n,i} - \hat{F}^{n,i})_k \leqslant 0, & \text{if } (U_{N_i}^{n,i})_k = (\Psi_{N_i})_k. \end{cases} \tag{4.1}$$

and

$$U_{N \backslash N_i}^{n,i} = U_{N \backslash N_i}^n, \tag{4.2}$$

where $\hat{F}^{n,i} = F_{N_i} - A_{N_i, N \backslash N_i} U_{N \backslash N_i}^n$.

Step 2.

$$U^{n+1} = \inf_i \{U^{n,i}\}. \tag{4.3}$$

As before, we need only to analyze the convergence of Algorithm IV. The following lemma is similar to Lemma 3.1.

Lemma 4.1. *Assume A is an M-matrix and $U_{N_i}^{n,i}$ is the solution of (4.1). Then $U_{N_i}^{n,i}$ is the minimal element of $\hat{T}^{n,i}$, where*

$$\hat{T}^{n,i} = \{\Phi_{N_i} \leqslant W \leqslant \Psi_{N_i} : (A_{N_i,N_i} W - \hat{F}^{n,i})_k \geqslant 0 \quad \text{or } W_k = (\Psi_{N_i})_k, k \in N_i\}.$$

Theorem 4.2. *Assume A is an M-matrix and U is the solution of problem (2.4). Then the iterative sequence $\{U^n\}$ generated by Algorithm IV is in T and it is monotone; that is, $U^n \in T$ and*

$$U \leqslant U^{n+1} \leqslant U^n \leqslant \cdots \leqslant U^0.$$

Proof. As in the proof of Theorem 3.2, we prove the conclusion by induction. From Step 1 of Algorithm IV we have that $U^0 \in T$. Noting that for every $i \in \{1, \cdots, m\}$,

$$A_{N_i,N_i} U^0_{N_i} - \hat{F}^{0,i} = (AU^0 - F)_{N_i},$$

we know that if $(U^0_{N_i})_k < (\Psi_{N_i})_k$ then $((AU^0 - F)_{N_i})_k \geqslant 0$, i.e., $(A_{N_i,N_i} U^0_{N_i} - \hat{F}^{0,i})_k \geqslant 0$. Therefore, $U^0_{N_i} \in \hat{T}^{0,i}$. From Lemma 4.1 we get that $\Phi_{N_i} \leqslant U^{0,i}_{N_i} \leqslant \Psi_{N_i}$, and $U^{0,i}_{N_i} \leqslant U^0_{N_i}$, which combined with (4.2) yields that $U^{0,i} \leqslant U^0$. It follows from (4.1) that

$$((AU^{0,i} - F)_{N_i})_k = (A_{N_i,N_i} U^{0,i}_{N_i} - \hat{F}^{0,i})_k \geqslant 0 \quad \text{for } (U^{0,i}_{N_i})_k < (\Psi_{N_i})_k.$$

Now consider another part of $U^{0,i}$; that is, $U^{0,i}_{N\backslash N_i} (= U^0_{N\backslash N_i})$. Assume that $(U^{0,i}_{N\backslash N_i})_k < (\Psi_{N\backslash N_i})_k$. Then by (4.2), $U^{0,i} \leqslant U^0$, and the fact that the off-diagonal elements of the M-matrix are nonpositive, we obtain that

$$\begin{aligned}((AU^{0,i} - F)_{N\backslash N_i})_k &= (A_{N\backslash N_i, N\backslash N_i})_{kk} (U^0_{N\backslash N_i})_k + \sum_{j\neq k} (A_{N\backslash N_i, N\backslash N_i})_{kj} (U^{0,i}_{N\backslash N_i})_j \\ &\quad + \sum_{j\in N_i} (A_{N\backslash N_i, N})_{kj} (U^{0,i})_j - (F_{N\backslash N_i})_k \\ &\geqslant ((AU^0 - F)_{N\backslash N_i})_k \geqslant 0.\end{aligned}$$

Therefore,

$$U^{0,i} \in T, \quad U^{0,i} \leqslant U^0. \tag{4.4}$$

Obviously $\Phi \leqslant U^1 \leqslant \Psi$ and $U^1 \leqslant U^{0,i} \leqslant U^0$ by (4.3). Now we prove $U^1 \in T$. It suffices to prove that $U^1_k < \Psi_k$ implies $(AU^1 - F)_k \geqslant 0$.

Assume $U^1_k < \Psi_k$. From (4.3) there exists an index i such that $U^1_k = U^{0,i}_k < \Psi_k$. By (4.4) and $a_{kj} \leqslant 0$ for $j \neq k$ we have

$$\begin{aligned}(AU^1 - F)_k &= a_{kk} U^1_k + \sum_{j\neq k} a_{kj} U^1_j - F_k = a_{kk} U^{0,i}_k + \sum_{j\neq k} a_{kj} U^1_j - F_k \\ &\geqslant a_{kk} U^{0,i}_k + \sum_{j\neq k} a_{kj} U^{0,i}_j - F_k = (AU^{0,i} - F)_k \geqslant 0.\end{aligned}$$

Hence $U^1 \in T$. It is easy to prove by induction that $U^n \in T$ and $U^{n+1} \leqslant U^{n,i} \leqslant U^n$. At last we know by Lemma 2.1 that $U^n \geqslant U$. □

Theorem 4.3. *Assume that the conditions of Theorem 4.1 hold and* $\epsilon^n = U^n - U$. *Then*

$$0 \leqslant \epsilon^{n+1} \leqslant G_i \epsilon^n, \quad i = 1, \cdots, m,$$

where G_i *is defined as in Theorem 3.4.*

Proof. It follows from Theorem 4.2 and (4.3) that $\epsilon^n \geqslant 0$ and $\epsilon^{n,i} = U^{n,i} - U \geqslant \epsilon^{n+1} \geqslant 0$. As in the proof of Theorem 3.4 we may discuss the following three cases:

(a) $(U^{n,i}_{N_i})_k > (\Phi_{N_i})_k$ and $(U_{N_i})_k < (\Psi_{N_i})_k$,
(b) $(U^{n,i}_{N_i})_k > (\Phi_{N_i})_k$ and $(U_{N_i})_k = (\Psi_{N_i})_k$,
(c) $(U^{n,i}_{N_i})_k = (\Phi_{N_i})_k$,

and obtain

$$A_{N_i,N_i}\epsilon^{n,i}_{N_i} \leqslant -A_{N,N\setminus N_i}\epsilon^{n}_{N\setminus N_i}. \tag{4.5}$$

Hence,

$$\epsilon^{n,i}_{N_i} \leqslant -A^{-1}_{N_i,N_i}A_{N_i,N\setminus N_i}\epsilon^{n}_{N\setminus N_i}, \quad i=1,\cdots,m,$$

which combined with (4.2) implies that

$$\epsilon^{n,i} \leqslant G_i\epsilon^n, \quad i=1,\cdots,m.$$

Hence we obtain

$$0 \leqslant \epsilon^{n+1} \leqslant G_i\epsilon^n, \quad i=1,\cdots,m.$$

□

As in Theorem 3.5 we have the following result.

Theorem 4.4. *Assume the conditions of Theorem 4.2 hold. Then $U^n \to U$ $(n\to\infty)$.*

5 The Geometrical Convergence

In this section we establish the so-called strong maximum principle for a kind of M-matrices and then prove the geometrical convergence for the above algorithms.

For matrix A we will consider the following conditions.

(i) $a_{ii} > 0, a_{ij} \leqslant 0 (j \neq i)$;
(ii) A is irreducible and weakly diagonally dominant;
(iii) A_{N_i,N_i} are irreducible, $i=1,\cdots,m$;
(iv) for every $i=1,\cdots,m$, there exists $l_i \in N_i$ such that $\sum_{j=1}^{N} a_{l,j} > 0$.

Remark 5.1. Conditions (i)–(iv) are satisfied for Delauney triangulation provided that mesh-size h is small enough and that the boundary of every subdomain has a common part with the boundary of the whole domain.

Lemma 5.1. *If matrix A satisfies condition (i)–(ii), then A is an M-matrix.*

Proof. Since $a_{ii} > 0, a_{ij} \leqslant 0 (j \neq i)$, it is sufficient to prove that

$$\rho(D^{-1}C) < 1,$$

where $D = diag(A)$, $C = D - A$ [20]. We know that $\rho(D^{-1}C) \leqslant 1$ because A is irreducible and weakly diagonally dominant. Obviously $(I - D^{-1}C)$ is nonsingular and $D^{-1}C$ is nonnegative. So, if $\rho(D^{-1}C) = 1$, then there exists a vector $V \neq 0$ such that $D^{-1}CV = V$. This means that $I - D^{-1}C$ is singular. This contradiction proves the lemma. □

Now we prove the discrete strong maximum principle which is the base for proving the geometrical convergence.

Lemma 5.2 (discrete strong maximum principle). *Assume matrix A satisfies conditions (i)–(iv), $V \in R^N$. If $V \geqslant 0$, $(AV)_{N_i} \leqslant 0$, then there exists constants $k_i \in (0,1)$ such that*

$$\sup_{j\in N_i} V_j \leqslant k_i \sup_{j\notin N_i} V_j. \tag{5.1}$$

Proof. Let $M_1 = \sup_{j\in N_i} V_j$, $M = \sup_{j\notin N_i} V_j$. We may suppose $M_1 \neq 0$. At first we prove

$$M_1 < M \tag{5.2}$$

If (5.1) is not true then there exists $j_0 \in N_i$ such that $V_{j_0} = M_1 \leqslant M$. Hence, we have (noting $a_{ii} > 0, a_{ij} \leqslant 0$ for $j \neq i$)

$$0 \geqslant \sum_{j=1}^{N} a_{j_0 j} V_j \geqslant \sum_{j=1}^{N} a_{j_0 j} M_1 \geqslant 0,$$

which implies that

$$\sum_{j\neq j_0} a_{j_0 j}(V_j - M_1) = 0.$$

Therefore,

$$V_j = M_1 \quad \text{if } \ a_{j_0 j} \neq 0. \tag{5.3}$$

Since A_{N_i,N_i} is irreducible there exist $j_1, \cdots, j_s \in N_i$ such that

$$a_{j_0 j_1} a_{j_1 j_2} \cdots a_{j_s j_i} \neq 0.$$

We know by (5.3) that $V_{j_1} = M_1$. Similarly, we get

$$V_{j_2} = \cdots = V_{j_s} = V_{l_i} = M_1.$$

Hence, we have

$$0 \geqslant \sum_{j=1}^{N} a_{l_i j} V_j \geqslant \sum_{j=1}^{N} a_{l_i j} M_1 > 0.$$

This contradiction proves (5.2).

Now we prove (5.1). If it is not true then there exist $V^n \in R^N, n = 1, 2, \cdots$, such that

$$V^n \geqslant 0, \quad (AV^n)_{N_i} \leqslant 0, \quad \sup_{j\in N_i} V_j^n > (1-1/n) \sup_{j\notin N_i} V_j^n, \quad n = 1, 2, \cdots.$$

Let $W^n = V^n / \sup_{j\notin N_i} V_j^n$. Then noting (5.2) we have

$$0 \leqslant W_j^n \leqslant 1, \quad (AW_j^n)_{N_i} \leqslant 0, \quad n = 1, 2, \cdots, \tag{5.4}$$

$$\sup_{j\in N_i} W_j^n > (1-1/n) \sup_{j\notin N_i} W_j^n, \quad n = 1, 2, \cdots. \tag{5.5}$$

$\{W^n\}$ is bounded and has a subsequence convergent to $\bar{W} \in R^N$. It follows from (5.4) and (5.5) that

$$0 \leqslant \bar{W} \leqslant 1, \quad (A\bar{W})_{N_i} \leqslant 0, \tag{5.6}$$

$$\sup_{j\in N_i} \bar{W}_j \geqslant \sup_{j\notin N_i} \bar{W}_j, \tag{5.7}$$

But (5.2) and (5.6) imply that

$$\sup_{j\in N_i} \bar{W}_j < \sup_{j\notin N_i} \bar{W}_j,$$

which contradicts (5.7). □

The following two theorems present the geometrical convergence of the algorithm given in this paper.

Theorem 5.3. *Suppose matrix A satisfies conditions (i)–(iv), $\{U^n\}$ is produced by Algorithm II, $\epsilon^n = U^n - U$. Then there exists $k \in (0,1)$ such that*

$$\sup_{1\leqslant j\leqslant N} \epsilon_j^{n+1} \leqslant k \sup_{1\leqslant j\leqslant N} \epsilon_j^n, \quad n = 0, 1, 2, \cdots.$$

Proof. By Lemma 5.1 we know that A is an M-matrix. Hence it follows from Theorem 3.2 and (3.8) that for $i = 1, \cdots, m,\quad n = 1, 2, \cdots$.

$$0 \leqslant \epsilon^{n+1} \leqslant \epsilon^{n+i/m} \leqslant \epsilon^{n+(i-1)/m} \leqslant \epsilon^n, \quad (A\epsilon^{n+i/m})_{N_i} \leqslant 0, \tag{5.8}$$

which combined with Lemma 5.2 implies that there exists a constant $k_i \in (0,1)$ such that

$$\begin{aligned}\sup_{j\in N_i} \epsilon_j^{n+i/m} &\leqslant k_i \sup_{j\notin N_i} \epsilon_j^{n+i/m} \leqslant k_i \sup_{j\notin N_i} \epsilon_j^{n+(i-1)/m} \\ &\leqslant k_i \sup_{1\leqslant j\leqslant N} \epsilon_j^{n+(i-1)/m} \leqslant k_i \sup_{1\leqslant j\leqslant N} \epsilon_j^n, \quad i = 1, \cdots, m, \quad n = 0, 1, 2, \cdots.\end{aligned}$$

Hence, we have

$$\begin{aligned}\sup_{1\leqslant j\leqslant N} \epsilon_j^{n+1} &= \sup_{1\leqslant i\leqslant m} \left\{ \sup_{j\in N_i} \epsilon_j^{n+1} \right\} \leqslant \sup_{1\leqslant i\leqslant m} \left\{ \sup_{j\in N_i} \epsilon_j^{n+i/m} \right\} \\ &\leqslant \sup_{1\leqslant i\leqslant m} k_i \sup_{1\leqslant j\leqslant N} \epsilon_j^n = k \sup_{1\leqslant j\leqslant N} \epsilon_j^n, \quad n = 0, 1, \cdots.\end{aligned}$$

with $k = \sup_{1\leqslant i\leqslant m} k_i \in (0,1)$. □

Theorem 5.4. *Suppose matrix A satisfies conditions (i)–(iv), $\{U^n\}$ is produced by Algorithm IV, $\epsilon^n = U^n - U$. Then there exists $k \in (0,1)$ such that*

$$\sup_{1\leqslant j\leqslant N} \epsilon_j^{n+1} \leqslant k \sup_{1\leqslant j\leqslant N} \epsilon_j^n, \quad n = 0, 1, 2, \cdots.$$

Proof. It follows from Theorem 4.2 and (4.5) that

$$0 \leqslant \epsilon^{n+1} \leqslant \epsilon^{n,i} \leqslant \epsilon^n, \quad (A\epsilon^{n,i})_{N_i} \leqslant 0, \quad i = 1, \cdots, m, \quad n = 0, 1, \cdots,$$

which together with Lemma 5.2 yields that there exists a constant $k_i \in (0,1)$ such that

$$\sup_{j\in N_i} \epsilon_j^{n,i} \leqslant k_i \sup_{1\leqslant j\leqslant N} \epsilon_j^n,$$

$$\sup_{1\leqslant j\leqslant N} \epsilon_j^{n+1} = \sup_{1\leqslant i\leqslant m} \left\{ \sup_{j\in N_i} \epsilon_j^{n,i} \right\} \leqslant k \sup_{1\leqslant j\leqslant N} \epsilon_j^n, \quad n=0,1,2,\cdots.$$

where $k = \sup_{1\leqslant i\leqslant m} k_i \in (0,1)$. □

Remark 5.2. If A is strictly diagonally dominant then we can derive a value for k as follows:

$$k = \sup_{1\leqslant i\leqslant N} \{-\sum_{j\neq i} a_{ij}/a_{ii}\}.$$

6 The h-Independent Convergence

In the previous section we established the geometrical convergence of our algorithm, but we have not proved that the convergence rate is independent of h. Now we prove this result for the case of uniform overlapping decomposition. For simplicity, we only discuss the case of $m=2$. Then the uniform overlapping means that

$$\text{dist}\,(\Gamma_1,\Gamma_2) = \rho > 0, \tag{6.1}$$

where $\Gamma_i = \partial\Omega_i \cap \Omega$ $(i=1,2)$. In the sequel, we also assume that conditions posed in the previous section hold. Moreover, we assume that Ω_i, $i=1,2$ are convex.

Using (6.1), we immediately have the following lemma.

Lemma 6.1. *There exists a function $g_1 \in C^\infty(\bar{\Omega})$ such that*

$$g_1(x) = \begin{cases} 0, & \text{if } x \in \Omega_1 \setminus (\Gamma_1 \cup (\partial\Omega_1 \cap \partial\Omega_2)), \\ 1, & \text{if } x \in \Gamma_1, \end{cases}$$

and

$$0 \leqslant g_1(x) \leqslant 1, \quad \text{if } x \in \partial\Omega_1 \cap \partial\Omega_2.$$

For the function g_1, we have the following lemma.

Lemma 6.2. *If $w \in H^1(\Omega)$ satisfies*

$$Lw = 0, \quad \text{if } x \in \Omega_1 \tag{6.2}$$

and

$$w = g_1, \quad \text{if } x \in \partial\Omega_1 \tag{6.3}$$

then

$$\bar{k}_1 = \sup\{w : x \in \Gamma_2\} \in (0,1).$$

Proof. In our case we have $w \in H^2(\Omega) \cap C(\bar{\Omega})$. Then the lemma is a result of the strong maximum principle. □

By the use of the above lemma and the convergence theorem of finite element with maximal norm, the finite element approximation w_h for w satisfies that

$$k_{1h} = \sup\{w_h : x \in \Gamma_2\} \leqslant k_1 \in (0,1) \quad \text{for sufficiently small } h. \tag{6.4}$$

Let

$$L_h w_h = 0 \tag{6.5}$$

be the finite element approximation of (6.1). Define $w_h(x_j) = 0$ for $x_j \in \Omega_h \setminus \bar{\Omega}_1$ and denote $W_j = w_h(x_j)$, $x_j \in \Omega_h$. Remove the term in (6.5) corresponding to $x_j \in \partial\Omega_1 \cap \Gamma_h$ into the right-hand side. Then we have for $x_k \in \Omega_{1h}$ that

$$(AW)_k = -\sum_{x_j \in \partial\Omega_1 \cap \Gamma_h} a(\theta_h^{(k)}, \theta_h^{(j)}) w_h(x_j). \tag{6.6}$$

Since $g_1(x_j) \geqslant 0$ for $x_j \in \partial\Omega_1 \cap \Gamma_h, g_1(x_j) = 1$ for $x_j \in \Gamma_1 \cap \Omega_h$, and due to (6.3), we have that

$$(AW)_k \geqslant 0, \quad \text{if } x_k \in \Omega_{1h} \tag{6.7}$$

and

$$W_k = 1 \quad \text{if } x_k \in \Gamma_1 \bigcap \Omega_h. \tag{6.8}$$

Let

$$\tilde{\epsilon}^{n+1/2} = \frac{\epsilon^{n+1/2}}{\sup\limits_{x_j \in \Gamma_1 \cap \Omega_h} \epsilon_j^{n+1/2}} = \frac{\epsilon^{n+1/2}}{\sup\limits_{x_j \in \Gamma_1 \cap \Omega_h} \epsilon_j^n}.$$

Since every finite element does not cross the boundary of Ω_1, we have that $x_j \in \Gamma_1 \cap \Omega_h$ for each $k \in N_1, j \in N \setminus N_1$ with $a_{kj} \neq 0$. Therefore, from (5.8), (6.7), and (6.8), we get that

$$0 \leqslant \tilde{\epsilon}_j^{n+1/2} \leqslant 1, \quad \text{if } x_j \in \Omega_{1h},$$

$$(A(\tilde{\epsilon}^{n+1/2} - W))_{N_1} = (A\epsilon^{n+1/2})_{N_1} / \sup_{x_j \in \Gamma_1 \cap \Omega_h} \epsilon_j^n - (AW)_{N_1} \leqslant 0.$$

Thereby,

$$A_{N_1,N_1}(\tilde{\epsilon}^{n+1/2} - W)_{N_1} \leqslant -A_{N_1,N\setminus N_1}(\tilde{\epsilon}^{n+1/2} - W)_{N\setminus N_1} \leqslant 0.$$

Hence,

$$\tilde{\epsilon}_{N_1}^{n+1/2} \leqslant W_{N_1}.$$

By (6.4), we get

$$\sup_{x_j \in \Gamma_2 \cap \Omega_h} \epsilon_j^{n+1/2} \leqslant \sup_{x_j \in \Gamma_2 \cap \Omega_h} W_j \cdot \sup_{x_j \in \Gamma_1 \cap \Omega_h} \epsilon_j^n \leqslant k_1 \sup_{x_j \in \Gamma_1 \cap \Omega_h} \epsilon_j^n. \tag{6.9}$$

Theorem 6.3. *Under the assumptions above, for sufficiently small h, the convergence rate of Algorithm II is independent of h; i.e., there exists a positive constant $k_1 < 1$ independent of h such that*

$$\sup_{1 \leqslant j \leqslant N} \epsilon_j^{n+1} \leqslant k_1 \sup_{1 \leqslant j \leqslant N} \epsilon_j^n.$$

Proof. Again from (5.8), we have that

$$\sup_{j\in N\backslash N_2} \epsilon_j^{n+1/2} \leqslant \sup_{x_j\in\Gamma_2\cap\Omega_h} \epsilon_j^{n+1/2}$$

and

$$\sup_{j\in N_2} \epsilon_j^{n+2/2} \leqslant \sup_{x_j\in\Gamma_2\cap\Omega_h} \epsilon_j^{n+2/2} = \sup_{x_j\in\Gamma_2\cap\Omega_h} \epsilon_j^{n+1/2}.$$

Hence from (6.9) we get

$$\begin{aligned}
\sup_{1\leqslant j\leqslant N} \epsilon_j^{n+1} &= \sup\left\{\sup_{j\in N\backslash N_2} \epsilon_j^{n+2/2}, \sup_{j\in N_2} \epsilon_j^{n+2/2}\right\} \\
&\leqslant \sup\left\{\sup_{j\in N\backslash N_2} \epsilon_j^{n+1/2}, \sup_{x_j\in\Gamma_2\cap\Omega_h} \epsilon_j^{n+1/2}\right\} \\
&\leqslant \sup_{x_j\in\Gamma_2\cap\Omega_h} \epsilon_j^{n+1/2} \leqslant k_1 \sup_{x_j\in\Gamma_1\cap\Omega_h} \epsilon_j^{n} \leqslant k_1 \sup_{1\leqslant j\leqslant N} \epsilon_j^{n}.
\end{aligned}$$

□

Similarly, for Algorithm IV, there exists a positive constant $k_2 \in (0,1)$ independent of h, such that

$$\sup_{x_j\in\Gamma_2\cap\Omega_h} \epsilon_j^{n,1} \leqslant k_2 \sup_{x_j\in\Gamma_1\cap\Omega_h} \epsilon_j^{n}$$

and

$$\sup_{x_j\in\Gamma_1\cap\Omega_h} \epsilon_j^{n,2} \leqslant k_2 \sup_{x_j\in\Gamma_2\cap\Omega_h} \epsilon_j^{n}.$$

Then, we have that

$$\begin{aligned}
\sup_{1\leqslant j\leqslant N} \epsilon_j^{n+2} &= \sup\left\{\sup_{1\leqslant j\leqslant N} \epsilon_j^{n+1,1}, \sup_{1\leqslant j\leqslant N} \epsilon_j^{n+1,2}\right\} \\
&\leqslant \sup\left\{\sup_{j\in N\backslash N_1} \epsilon_j^{n+1,1}, \sup_{j\in N\backslash N_2} \epsilon_j^{n+1,2}\right\} \\
&= \sup\left\{\sup_{j\in N\backslash N_1} \epsilon_j^{n+1}, \sup_{j\in N\backslash N_2} \epsilon_j^{n+1}\right\} \\
&\leqslant \sup\left\{\sup_{j\in N\backslash N_1} \epsilon_j^{n,2}, \sup_{j\in N\backslash N_2} \epsilon_j^{n,1}\right\} \\
&\leqslant \sup\left\{\sup_{x_j\in\Gamma_1\cap\Omega_h} \epsilon_j^{n,2}, \sup_{x_j\in\Gamma_2\cap\Omega_h} \epsilon_j^{n,1}\right\} \\
&\leqslant \sup\left\{k_2 \sup_{x_j\in\Gamma_2\cap\Omega_h} \epsilon_j^{n}, k_2 \sup_{x_j\in\Gamma_1\cap\Omega_h} \epsilon_j^{n}\right\} \\
&\leqslant k_2 \sup_{1\leqslant j\leqslant N} \epsilon_j^{n}.
\end{aligned}$$

Therefore, we have proved the following theorem.

Theorem 6.4. *Under the assumptions above, for sufficiently small h, the convergence rate of Algorithm IV is independent of h.*

7 Numerical Examples

We have carried out numerical experiments for the following examples. Find a $u \in K$, such that

$$a(u, v-u) \geqslant f(v-u), \quad \forall v \in K.$$

Where $K = \{v \in H_0^1(\Omega) : 0 \leqslant v \leqslant 1\}$, $f = 4x - 2x^2 + 4y - 2y^2$, $\Omega = (0,2) \times (0,2)$.

We use linear Lagrange finite element space and solve the corresponding finite element approximation problem by our Schwarz algorithms for different meshsizes, different overlappings, different numbers of subdomains, and different initial values.

Let h be the meshsize, δ the overlapping, m the number of subdomains. For $m = 2$ we take $\Omega_1 = (0, 1+\delta/2) \times (0,2)$, $\Omega_2 = (1-\delta/2, 2) \times (0,2)$. For $m = 4$ we take $\Omega_1 = (0, 1+\delta/2) \times (0, 1+\delta/2)$, $\Omega_2 = (0, 1+\delta/2) \times (1-\delta/2, 2)$, $\Omega_3 = (1-\delta/2, 2) \times (0, 1+\delta/2)$, $\Omega_4 = (1-\delta/2, 2) \times (1-\delta/2, 2)$. For $m = 8, 16$, we take similar coarse triangulation. The stopping criterion is $\|u^{n+1} - u^n\|_\infty \leqslant \epsilon = 10^{-6}$. The subproblems are solved by project successive overrelaxation (PSOR), and the stopping criterion is that the maximal norm of difference between the successive iterative solutions is less than 10^{-8}. The numbers of iterations in different cases are shown in Tables 1–6.

Table 1 $m = 2$, Algorithm IV, $U^0 = (1, \cdots, 1)^T$

$\delta \setminus h$	$1/2^3$	$1/2^4$	$1/2^5$
1.5	6	6	7
1	9	8	8
0.5	16	14	14

Table 2 $m = 2$, Algorithm II, $U^0 = (1, \cdots, 1)^T$

$\delta \setminus h$	$1/2^3$	$1/2^4$	$1/2^5$
1.5	4	5	7
1	5	5	8
0.5	9	9	10

Table 3 Algorithm IV, $h = 1/2^5$, $U^0 = (1, \cdots, 1)^T$

$m \setminus \delta$	1.5	1	0.5	0.25
2	7	8	14	30
4	11	13	21	37

Table 4 $m = 2$, Algorithm IV, $h = 1/2^5$

$U^0 \setminus \delta$	1.5	1	0.5	0.25
$U^0 = (1, \cdots, 1)^T$	7	8	14	30
$U^0 = (3, \cdots, 3)^T$	7	8	14	30
$U^0 = (0, \cdots, 0)^T$	10*	7*	5*	3*

Table 5 $m = 2$, **Algorithm II,** $h = 1/2^5$

$U^0 \setminus \delta$	1.5	1	0.5	0.25
$U^0 = (1, \cdots, 1)^T$	7	8	10	15
$U^0 = (3, \cdots, 3)^T$	7	8	10	15
$U^0 = (0, \cdots, 0)^T$	8	9	13	19

Table 6 Algorithm II, $h = 0.02$, $U^0 = (1, \cdots, 1)^T$

m	2	4	8	16
Iteration number	24	29	52	75

These results indicate the h-independent convergence of the algorithm and

(1) the convergence is faster when the overlapping is larger (this is a well-known fact for the equation case);

(2) the convergence is slower for the triangulation with inner cross-point (here it is $m = 4, 8, 16$; see Table 6);

(3) the upper obstacle is a better initial value.

Acknowledgments The authors are grateful to the referee and Professor J. C. Xu for their many valuable comments and suggestions.

References

[1] J.H. Bramble, J.E. Pasciak, J. Wang and J. Xu, Convergence estimates for product iterative methods with applications to domain decomposition and multigrid, *Mathematics of Computation*, 57 (1991), 1-22.

[2] X.C. Cai and O.B. Widlund, Multiplicative Schwarz algorithms for nonsymmetric and indefinite elliptic problems, *SIAM Journal on Numerical Analysis*, 30 (1993), 936-952.

[3] M. Dryja, An additive Schwarz algorithm for two- and three-dimensional finite element elliptic problems, in: *Domain Decomposition Methods*, T. Chan, R. Glowinski, J. Périaux, and O. B. Widlund, eds., SIAM, Philadelphia, 1989, pp. 168-172.

[4] M. Dryja and O.B. Widlund, Towards a unified theory of domain decomposition algorithms for elliptic problems, in: *Domain Decomposition Methods for PDEs*, T. F. Chan, R. Glowinski, J. Périaux, and O. B. Widlund, eds., SIAM, Philadelphia, 1990, pp. 340-349.

[5] P.L. Lions, On the Schwarz alternating method I, in: *Proceedings of Domain Decomposition Methods for PDEs*, R. Glowinski, G. H. Golub, G. A. Meurant, and J. Périaux, eds., SIAM, Philadelpha, 1988, pp. 1-40.

[6] T. Lü, T.M. Shih and C.B. Liem, *Domain Decomposition Methods-New Numerical Technique for Solving PDE*, Science Press, Beijing, 1992.

[7] B. Smith, P. Bjϕrstad and W. Gropp, *Domain Decomposition*, Cambridge University Press, Cambridge, UK, 1996.

[8] J.C. Xu, Iterative methods by space decompositiion and subspace correction, *SIAM Review*, 34 (1992), 581-613.

[9] P.L. Lions, On the Schwarz alternating method II, in: *Proceedings of Domain Decomposition Methods for PDEs*, T. F. Chan, R. Glowinski, J. Périaux, and O. B. Widlund, eds., SIAM, Philadelphia, 1989, pp. 47-70.

[10] F. Scarpini, The alternative Schwarz method applied to some biharmonic variational inequalities, *Calcolo*, 27 (1990), 57-72.

[11] X.J. Xu and S.M. Shen, Domain decomposition method for obstacle problems, Numerical Mathematics, *A Journal of Chinese Universities*, 16 (1994), 186-194.

[12] T. Lü, T.M. Shih and C.B. Liem, Parallel algorithms for variational inequalities based on domain decomposition, *Journal of Computational Mathematics*, 9 (1991), 340-349.

[13] L. Bada, On the Schwarz alternating method with more than two subdomains for nonlinear monotone problems, *SIAM Journal on Numerical Analysis*, 28(1991), 179-204.

[14] S.Z. Zhou, An additive Schwarz algorithm for variational inequality, in: *Proceedings of Domain Decomposition Methods in Science and Engineering*, R. Glowinski, J. Périaux, Z. Shi, and O. B. Widlund, eds., John Wiley and Sons, New York, 1997, pp. 133-138.

[15] Y. Kuznetsov, P. Neittaanmäki and P. Tarvainen, Overlapping domain decomposition methods for the obstacle problem, in: *Domain Decomposition Methods in Science and Engineering*, Y. Kuznetsov, J. Périaux, A. Quarteroni, and O. Widlund, eds., AMS, Providence, RI, 1994, pp. 271-277.

[16] Y. Kunetsov, P. Neittaanmäki and P. Tarvainen, Schwarz methods for obstacle problems with convection-diffusion operators, in: *Proc. Domain Decomposition Methods in Scientifical and Engineering Computing*, D. E. Keyes and J. C. Xu, eds., AMS, Providence, RI, 1995, pp. 251-256.

[17] J.P. Zeng, *Monotone convergence and convergent rate analysis of DDM for the solution of LCP*, in Proceedings of DDM8 (Abstract), Chinese Academy of Science, Beijing, 1996, p. 72.

[18] H.C. Huang and L.H. Wang, On the equivalence of the variational iniquality and the difference form for obstacle problem, *Mathematica Numerica Sinica*, 4 (1982), 436-439.

[19] P. G. Ciarlet, *The Finite Element Method for Elliptic Problems*, North-Holland, Amsterdam, 1978.

[20] R. Varga, Matrix Iterative Analysis, Prentice-Hall, Englewood Cliffs, NJ, 1962.

[21] J.P. Zeng and S.Z. Zhou, Two-side obstacle problem and its equivalent linear complementarity problem, *Chinese Science Bulletin*, 39 (1994), 1057-1062.

A Domain Decomposition Method for a Kind of Optimization Problems*

J.P. Zeng (曾金平)† S.Z. Zhou (周叔子)†

Abstract In this paper, we are concerned with the monotone convergence of a multiplicative method for solving a kind of optimization problems. We show that the iterate sequence produced by the method converges to the solution of the problem monotonically.

Keywords Optimization, domain decomposition, monotone convergence.

1 Introduction

Let Ω be a bounded domain in R^2, $V \subset H^1(\Omega)$ be a Hilbert space. We consider the following minimization problem: find a $u^* \in K$, such that

$$F(u^*) = \min_{v \in K} F(v), \tag{1.1}$$

where $F : V \to \mathcal{R}$ is a convex lower semi-continuous twice Gâteau differentiable functional, and

$$K = \{v \in V \mid v \geqslant \phi\} \tag{1.2}$$

for some given function ϕ.

Problem (1.1) is also called obstacle problem and usually equivalent to a variational inequality (see next section). Variational inequalities (VIs) arise from many physical and economical problems such as the fluid flow in porous media, the behavior of elasto-plastic materials and lubrication phenomena in mechanics, and the diffusion problem involving Michaelis-Menten or second-order irreversible reactions (see e.g., [1, 2]).

Recently, more attention has been paid to domain decomposition methods for the solution of VI*s*. For problems with a linear operator, we refer to [1, 3–10] and the references therein. For problems with a nonlinear operator, we refer to [11–13] and the references therein. Some of these topics are related to the monotone convergence property (see e.g., [9, 11–13]). The

* 发表于：Journal of Computational and Applied Mathematics, Vol. 146, 2002, pp. 127-139.

† Department of Applied Mathematics, Hunan University, Changsha 410082, China.

aim of this paper is to study furthermore the monotone convergence property of multiplicative domain decomposition method for solving problem (1.1) with more general operator.

Denote the level set

$$L(F,u^0)=\{v\in V|F(v)\leqslant F(u^0)\}.$$

We assume that F' is Lipschitz continuous and strongly monotone in $L(F,u^0)$. That is, there exist constants $L\geqslant 0$ and $K>0$, such that

$$\|F'(w)-F'(v)\|_{v'}\leqslant L\|w-v\|_V,\quad \forall w,v\in L(F,u^0), \tag{1.3}$$

$$\langle F'(w)-F'(v),w-v\rangle\geqslant K\|w-v\|_V^2,\quad \forall w,v\in L(F,u^0), \tag{1.4}$$

where $\langle\cdot,\cdot\rangle$ is the inner product. To prove the monotonicity of the algorithm, we need to assume that F' is strictly T-monotone over a subset $\tilde{V}$ of V. That is

$$\left\langle F'(w)-F'(v),(w-v)^+\right\rangle\geqslant 0,\quad \forall w,v\in\tilde{V} \tag{1.5}$$

and the equality holds if and only if $(w-v)^+=0$, where $w^+\in V$ is defined for any $w\in V$ as $w^+=\max\{0,w\}$.

T-monotone operators are a kind of important operators, which include many linear and nonlinear elliptic operators (see e.g., [13, 14]). One of the advantages of T-monotone operators is that: the relevant algorithms often possess monotone convergence property.

2 Multiplicative Domain Decomposition Method

In this section, we will present a multiplicative domain decomposition method. First, we need some assumptions as follows (see e.g., [15, 16]):

(S1). There exist Hilbert subspaces $V_1,V_2,\cdots,V_m$ of V, such that $V=V_1+V_2+\cdots+V_m$.

(S2). There exist $\phi_i\in V_i(i=1,2,\cdots,m)$, such that $\sum\limits_{i=1}^m\phi_i=\phi$, and

$$K=\sum_{i=1}^m K_i,$$

where

$$K_i=\{v_i\in V_i|v_i\geqslant\phi_i\}.$$

(S3). There exists a constant $C>0$ such that for any $w_i\in K_i(i=1,2,\cdots,m)$, and $v\geqslant w=\sum\limits_{i=1}^m w_i$, we can find a decomposition of v: $v=\sum\limits_{i=1}^m v_i, v_i\in K_i(i=1,2,\cdots,m)$ satisfying

$$v_i\geqslant w_i\ (i=1,2,\cdots,m),\quad \sqrt{\sum_{i=1}^m\|w_i-v_i\|_V^2}\leqslant C\|w-v\|_V. \tag{2.1}$$

The algorithm we give below is called MDM in the rest of the paper.

Algorithm MDM.

Step 1. Set $u_i^0 \in K_i, i = 1, 2, \cdots, m, n := 0$.

Step 2. For $i = 1, \cdots, m$, solve the following optimization problems successively: find $u_i^{n+1} \in K_i$, such that

$$F\left(\sum_{j\leqslant i} u_j^{n+1} + \sum_{j>i} u_j^n\right) \leqslant F\left(\sum_{j<i} u_j^{n+1} + v_i + \sum_{j>i} u_j^n\right), \quad \forall v_i \in K_i. \tag{2.2}$$

Step 3. Let $n := n + 1$, $u^n = \sum\limits_{i=1}^{m} u_i^n$. Go to Step 2.

It is well known that (1.1) can be reformulated as variational inequality problem (cf. [2, 17]).

Lemma 2.1. *u^* is the solution of (1.1) iff u^* is the solution of the following variational inequality: find $u^* \in K$, such that*

$$\langle F'(u^*), v - u^*\rangle \geqslant 0, \quad \forall v \in K. \tag{2.3}$$

Accordingly, for problem (2.2), the following corollary is true.

Corollary 2.2. *$u_i^{n+1} \in K_i$ is the solution of (2.2) iff*

$$\left\langle F'\left(\sum_{j\leqslant i} u_j^{n+1} + \sum_{j>i} u_j^n\right), v_i - u_i^{n+1}\right\rangle \geqslant 0, \quad \forall v_i \in K_i. \tag{2.4}$$

Denote

$$w_i^n = \sum_{j\leqslant i} u_j^{n+1} + \sum_{j>i} u_j^n, \quad i = 0, 1, \cdots, m, \tag{2.5}$$

and

$$K_i^n = \left\{ v_i + \sum_{j<i} u_j^{n+1} + \sum_{j>i} u_j^n | v_i \in K_i \right\}, \quad i = 1, 2, \cdots, m.$$

Then $w_i^n \in K_i^n$ and satisfies

$$\langle F'(w_i^n), w_i - w_i^n\rangle \geqslant 0, \quad \forall w_i \in K_i^n. \tag{2.6}$$

The following two inequalities will be useful in the sequel. They can be obtained by using (1.3), (1.4), the twice Gâteau differentiability of F, and Taylor's expansion.

$$F(w) - F(v) \leqslant \langle F'(v), w - v\rangle + \frac{L}{2}\|w - v\|_V^2, \quad \forall w, v \in L(F, u^0), \tag{2.7}$$

$$F(w) - F(v) \geqslant \langle F'(v), w - v\rangle + \frac{K}{2}\|w - v\|_V^2, \quad \forall w, v \in L(F, u^0). \tag{2.8}$$

3 Monotone Convergence for Continuous Problems

In this section, we let $V = H_0^1(\Omega)$, $\langle w, v\rangle = \displaystyle\int_\Omega wvdx$ and assume that ϕ, u^* and $u_i^n(i = 1,2,\cdots,m; n = 1,2,\cdots.) \in H^2(\Omega)$. We also assume that F' is strictly T-monotone over $\tilde{V} = V \cap H^2(\Omega)$. Define

$$S = \{v \in \tilde{K} |\min\{F'(v), v-\phi\} \leqslant 0, \text{ a.e. in } \Omega\},$$

where $\tilde{K} = K \cap H^2(\Omega)$. Then we have the following lemmas.

Lemma 3.1. *$u^* \in \tilde{K}$ is the solution of (1.1) iff*

$$F'(u^*) \geqslant 0, \quad \textit{a.e. in } \Omega,$$

$$F'(u^*)(u^*-\phi) = 0, \quad \textit{a.e. in } \Omega.$$

Proof. We need only to prove the suffciency of the lemma. Since $\Omega \in R^2$, we have that $\phi, u^* \in C(\bar{\Omega})$ by embedding theorem. Denote

$$I = \{x | u^*(x) = \phi(x)\}, \quad \Omega_1 = \Omega \setminus I. \tag{3.1}$$

Then Ω_1 is an open set. For any $\psi \in C_0^\infty(\Omega_1)$, denote $E = \text{supp }(\psi)$ and $m_0 = \min_E\{u^* - \phi\}$. Since $E \subset\subset \Omega_1$ (that is $E \in \Omega_1$, and $\text{dist}(E, \partial\Omega_1) > 0$), we obtain that $m_0 > 0$. Then, there exists a positive number $\varepsilon_0 > 0$, such that

$$|\varepsilon\psi| \leqslant m_0, \quad \forall |\varepsilon| < \varepsilon_0. \tag{3.2}$$

Therefore, by (3.1) and (3.2) we can conclude that $u^* + \varepsilon\psi \in K$. Then,

$$\langle F'(u^*), u^* + \varepsilon\psi - u^*\rangle = \varepsilon \langle F'(u^*), \psi\rangle \geqslant 0$$

Hence

$$\varepsilon \langle F'(u^*), \psi\rangle \geqslant 0, \quad \forall |\varepsilon| < \varepsilon_0,$$

which means that

$$\langle F'(u^*), \psi\rangle = 0, \quad \psi \in C_0^\infty(\Omega_1).$$

By variational principle, $F'(u^*) = 0$ a.e. in Ω_1, which implies that $F'(u^*)(u-\phi) = 0$ a.e. in Ω. Similarly, for any $\psi \in C_0^\infty(\Omega)$, with $\psi \geqslant 0$, we have that $u^* + \psi \in K$ and then

$$\langle F'(u^*), u^* + \psi - u^*\rangle = \langle F'(u^*), \psi\rangle \geqslant 0, \quad \forall \psi \in C_0^\infty(\Omega), \quad \psi \geqslant 0,$$

which implies $F'(u^*) \geqslant 0$ a.e. in Ω. Then the proof is completed. □

Lemma 3.2. *The solution u^* of (1.1) is the maximal element of S. That is, $u^* \in S$ and $v \leqslant u^*$ for any $v \in S$.*

Proof. By Lemma 3.1, it is obvious that $u^* \in S$. For any $v \in S$, let $\psi = (v - u^*)^+$. Then we have that

$$\begin{cases} \psi(x) = 0, & \text{if } v(x) = \phi(x), \\ F'(v)(x) \leqslant 0, & \text{if } v(x) > \phi(x), \end{cases}$$

Therefore,

$$\begin{aligned} \langle F'(v), \psi \rangle &= \int_{\{x|v(x)>\phi(x)\}} F'(v)\psi dx + \int_{\{x|v(x)=\phi(x)\}} F'(v)\psi dx \\ &= \int_{\{x|v(x)>\phi(x)\}} F'(v)\psi dx \leqslant 0 \end{aligned} \tag{3.3}$$

and $u^* + \psi \in K$. Using (2.3), we have

$$\langle F'(u^*), \psi \rangle = \langle F'(u^*), (u^* + \psi) - u^* \rangle \geqslant 0. \tag{3.4}$$

Combining (3.3) with (3.4), we have

$$\left\langle F'(v) - F'(u^*), (v - u^*)^+ \right\rangle \leqslant 0.$$

Therefore, (1.5) means $(v - u^*)^+ = 0$ and then $v \leqslant u^*$. □

Lemma 3.3. *Let $\{u^n\}$ be produced by MDM. Then when $u^n \in S$, $u^{n+1} \in S$ and $u^n \leqslant u^{n+1} \leqslant u^*$.*

Proof. Define for each $i = 1, 2, \cdots, m$ that

$$S_i^n = \left\{ w_i \in \tilde{K}_i^n \,\middle|\, \min\left\{ F'(w_i), w_i - \left(\sum_{j<i} u_j^{n+1} + \phi_i + \sum_{j>i} u_j^n \right) \right\} \leqslant 0 \quad \text{a.e. in } \Omega \right\},$$

where $\tilde{K}_i^n = K_i^n \cap H^2(\Omega)$. Since $u^n \in S$,

$$\begin{aligned} \min\left\{ F'(u^n), u^n - \left(\phi_1 + \sum_{j>1} u_j^n \right) \right\} &\leqslant \min\left\{ F'(u^n), u^n - \left(\phi_1 + \sum_{j>1} \phi_j \right) \right\} \\ &= \min\{F'(u^n), u^n - \phi\} \leqslant 0. \end{aligned}$$

That is, $u^n \in S_1^n$. By Lemma 3.2, w_1^n is the maximal element of S_1^n and therefore, $u^n \leqslant w_1^n$. Let

$$I = \{x \in \Omega \mid w_1^n(x) = \phi(x)\}, \quad \Omega_1 = \Omega \setminus I,$$

$$I_1 = \{x \in \Omega_1 \mid u_1^{n+1}(x) > \phi_1(x)\}, \quad I_2 = \{x \in \Omega_1 \mid u_1^{n+1}(x) = \phi_1(x)\}.$$

Then Ω_1 is an open set. In a way similar to the proof of Lemma 3.1, for any $\psi \in C_0^\infty(\Omega_1)$ with $\psi \geqslant 0$, we have $E = \text{supp}(\psi) \subset\subset \Omega_1$, $m_0 = \min_E\{w_1^n - \phi\} > 0$ and therefore, there exists a positive number ε_0, such that $0 \leqslant \varepsilon\psi \leqslant m_0$ holds for any $0 < \varepsilon < \varepsilon_0$. If $x \in I$, we have $w_1^n(x) = \phi(x) = u^n(x)$ and then

$$F'(w_1^n)(x)(w_1^n(x) - \phi(x)) = 0, \quad x \in I. \tag{3.5}$$

If $x \in I_1$, we have $w_1^n(x) = u_1^{n+1}(x) + \sum\limits_{j>1} u_j^n(x) > \phi_1(x) + \sum\limits_{j>1} u_j^n(x)$. Hence,

$$F'(w_1^n)(x) = 0, \quad x \in I. \tag{3.6}$$

If $x \in I_2$, we have $w_1^n(x) = u^(x) > \phi(x)$ and then

$$F'(u^n)(x) \leqslant 0, \quad x \in I_2. \tag{3.7}$$

We define $w_{1,\varepsilon}^n = w_1^n - \varepsilon\psi$ $(0 < \varepsilon < \varepsilon_0)$. Then, $w_{1,\varepsilon}^n \geqslant \phi$. Since $u^n \leqslant w_1^n$, we have that

$$\int_I (F'(u^n) - F'(w_{1,\varepsilon}^n))(u^n - w_{1,\varepsilon}^n)^+ \mathrm{d}x = \int_I (F'(u^n) - F'(w_{1,\varepsilon}^n))(u^n - w_1^n)^+ \mathrm{d}x = 0. \tag{3.8}$$

By (3.7) we have that

$$\int_{I_2} (F'(u^n) - F'(w_{1,\varepsilon}^n))(u^n - w_{1,\varepsilon}^n)^+ \mathrm{d}x = \varepsilon \int_{I_2} (F'(u^n) - F'(w_{1,\varepsilon}^n))\psi \mathrm{d}x \leqslant \varepsilon \int_{I_2} (-F'(w_{1,\varepsilon}^n))\psi \mathrm{d}x. \tag{3.9}$$

Let $I_{11} = I_1 \cap \{x \mid u^n(x) = \phi(x)\}$, $I_{12} = I_1 \cap \{x \mid u^n(x) > \phi(x)\}$. If $x \in I_{11}$, we have

$$0 \leqslant (u^n(x) - w_{1,\varepsilon}^n(x))^+ = (\phi(x) - w_1^n(x) + \varepsilon\psi(x))^+ \leqslant (-m_0 + \varepsilon\psi(x))^+ = 0.$$

Then

$$\int_{I_{11}} (F'(u^n) - F'(w_{1,\varepsilon}^n))(u^n - w_{1,\varepsilon}^n)^+ \mathrm{d}x = 0. \tag{3.10}$$

If $x \in I_{12}$, we have that $F'(u^n)(x) \leqslant 0$ and

$$(u^n(x) - w_{1,\varepsilon}^n(x))^+ = (u^n(x) - w_1^n(x) + \varepsilon\psi(x))^+ \leqslant \varepsilon\psi(x).$$

Therefore,

$$\begin{aligned}
&\int_{I_{12}} (F'(u^n) - F'(w_{1,\varepsilon}^n))(u^n - w_{1,\varepsilon}^n)^+ \mathrm{d}x \\
\leqslant &\int_{I_{12}} (-F'(w_{1,\varepsilon}^n))(u^n - w_{1,\varepsilon}^n)^+ \mathrm{d}x \\
\leqslant &\int_{I_{12}} |-F'(w_{1,\varepsilon}^n)|(u^n - w_{1,\varepsilon}^n)^+ \mathrm{d}x \leqslant \epsilon \int_{I_{12}} |-F'(w_{1,\varepsilon}^n)|\psi \mathrm{d}x.
\end{aligned} \tag{3.11}$$

Denote

$$G_\varepsilon(x) = \begin{cases} |F'(w_{1,\varepsilon}^n)(x)|, & x \in I_1, \\ -F'(w_{1,\varepsilon}^n)(x), & x \in I_2, \end{cases}$$

Then by (3.8)–(3.11) and T-monotonicity, we have

$$0 \leqslant \langle F'(u^n) - F'(w_{1,\varepsilon}^n), (u^n - w_{1,\varepsilon}^n)^+ \rangle \leqslant \varepsilon \int_{\Omega_1} G_\varepsilon \psi \mathrm{d}x.$$

Hence, the continuity of F' implies that

$$\int_{\Omega_1} G\psi \mathrm{d}x \geqslant 0, \quad \forall \psi \in C_0^\infty(\Omega_1), \quad \psi \geqslant 0,$$

where

$$G(x) = \begin{cases} |F'(w_1^n)(x)|, & x \in I_1, \\ -F'(w_1^n)(x), & x \in I_2, \end{cases}$$

By variational principle, we have that $G \geqslant 0$ and then

$$F'(w_1^n)(x) \leqslant 0, \quad x \in I_2. \tag{3.12}$$

Together (3.5), (3.6) with (3.12), we get $w_1^n \in S$.

By the principle of induction, we get

$$w_i^n \in S(i = 1, 2, \cdots, m), \quad u^n \leqslant w_1^n \leqslant w_2^n \leqslant \cdots \leqslant w_m^n = u^{n+1} \leqslant u^*.$$

□

By Lemmas 3.2 and 3.3 we get

Lemma 3.4. *Let $u^0 \in S$ and $\{u^n\}$ be produced by MDM. Then*

$$u^n \in S, \qquad u^n \leqslant u^{n+1} \leqslant u^*.$$

Lemma 3.5. *Let $u^0 \in S$, $\{u^n\}$ be produced by MDM. Then*

$$F(u^{n+1}) \leqslant F(u^n) \leqslant \cdots \leqslant F(u^0), \tag{3.13}$$

$$u^n \in L(F, u^0), \quad n = 1, 2, \cdots$$

and

$$\|u^{n+1} - u^*\|^2 \leqslant \frac{2mL^2C^2}{K^3}(F(u^n) - F(u^{n+1})). \tag{3.14}$$

Proof. By (2.2) and (2.5), we have

$$F(u^n) - F(u^{n+1}) = \sum_{i=1}^{m}(F(w_{i-1}^n) - F(w_i^n)) \geqslant 0, \tag{3.15}$$

which implies (3.13) and $u^n \in L(F, u^0)$. By (1.4), we have

$$K\|u^{n+1} - u^*\|_V^2 \leqslant \left\langle F'(u^{n+1}) - F'(u^*), u^{n+1} - u^* \right\rangle \leqslant \left\langle F'(u^{n+1}), u^{n+1} - u^* \right\rangle. \tag{3.16}$$

By Lemma 3.4 and assumption (S3), for each u^{n+1}, there exists a decomposition of u^*: $u^* = \sum\limits_{i=1}^{m} u_i^*$ ($u_i^* \in K_i$) satisfying

$$u_i^* \geqslant u_i^{n+1} \ (i = 1, 2, \cdots, m), \quad \sqrt{\sum_{i=1}^{m} \|u_i^{n+1} - u_i^*\|_V^2} \leqslant C\|u^{n+1} - u^*\|_V.$$

Moreover, by (2.6), we have

$$-\langle F'(w_i^n), u_i^{n+1} - u_i^* \rangle = \left\langle F'(w_i^n), \sum_{j<i} u_j^{n+1} + u_i^* + \sum_{j>i} u_j^n - w_i^n \right\rangle \geqslant 0. \tag{3.17}$$

Then by (3.16), (3.17) and (1.3), we get

$$\begin{aligned}
K\|u^{n+1} - u^*\|_V^2 \leqslant & \langle F'(u^{n+1}), u^{n+1} - u^* \rangle = \sum_{i=1}^m \langle F'(u^{n+1}), u_i^{n+1} - u_i^* \rangle \\
\leqslant & \sum_{i=1}^m \langle F'(u^{n+1}) - F'(w_i^n), u_i^{n+1} - u_i^* \rangle \\
= & \sum_{i=1}^m \sum_{j>i} \langle F'(w_j^n) - F'(w_{j-1}^n), u_i^{n+1} - u_i^* \rangle \\
\leqslant & L \sum_{i=1}^m \sum_{j>i} \|w_j^n - w_{j-1}^n\|_V \|u_i^{n+1} - u_i^*\|_V \\
\leqslant & L \sum_{i=1}^m \|u_i^{n+1} - u_i^*\|_V \sum_{j=1}^m \|w_j^n - w_{j-1}^n\|_V \\
\leqslant & L\sqrt{m} \sqrt{\sum_{i=1}^m \|u_i^{n+1} - u_i^*\|_V^2} \sum_{j=1}^m \|w_j^n - w_{j-1}^n\|_V .
\end{aligned} \tag{3.18}$$

Noting that $w_{j-1}^n \in K_j^n$, (2.6) and (2.8), we get

$$\begin{aligned}
F(u^n) - F(u^{n+1}) = & \sum_{j=1}^m (F(w_{j-1}^n) - F(w_j^n)) \\
\geqslant & \sum_{j=1}^m (\langle F'(w_j^n), w_{j-1}^n - w_j^n \rangle + \frac{K}{2} \|w_{j-1}^n - w_j^n\|_V^2) \\
\geqslant & \frac{K}{2} \sum_{j=1}^m \|w_{j-1}^n - w_j^n\|_V^2 .
\end{aligned} \tag{3.19}$$

Combining (3.18) and (3.19), we obtain (3.14) □

Note. In (3.14), the constant m can be replaced by the maximal overlapping number $\tau \leqslant m$. In practical computations; the domain Ω is often decomposed so that τ is independent of the number m of subdomains.

By (3.13), Lemmas 3.4 and 3.5, we get the following result.

Theorem 3.6. *Let $u^0 \in S$ and $\{u^n\}$ be produced by MDM. Then $\{u^n\}$ converges to the solution u^* of (1.1) monotonically, and*

$$u^n \in S, \quad u^n \leqslant u^{n+1} \leqslant u^*. \tag{3.20}$$

4 Monotone Convergence for Finite-Dimensional Problems

In this section, we let $V = R^N$. Define $\langle w, v\rangle = \sum_{i=1}^{N} w_i, v_i$ and assume that F' is strictly T-monotone over K. That is

$$\langle F'(w) - F'(v), (w-v)^+\rangle \geqslant 0, \quad \forall w, v \in K,$$

and the equality holds if and only if $(w-v)^+ = 0$, where we define $w^+ = (\max\{0, w_k\})_{k=1}^N \in R_+^N$ for any $w = (w_k)_{k=1}^N$. Let

$$S^N = \{v \in K \mid \min\{F_k'(v), (v-\phi)_k\} \leqslant 0, \quad \forall k = 1, 2, \cdots, N\}.$$

Similarly, we have the following results.

Lemma 4.1. *Let u^* be the solution of (1.1). Then $u^* \in S^N$ and $v \leqslant u^*$ for any $v \in S^N$, where $\leqslant$ means elementwise.*

Lemma 4.2. *Let $\{u^n\}$ be produced by MDM. Then when $u^n \in S^N$, $u^{n+1} \in S^N$ and $u^n \leqslant u^{n+1} \leqslant u^*$.*

Proof. Define for each $i = 1, 2, \cdots, m$ that

$$S_i^{N,n} = \left\{ w_i \in K_i^n \mid \min\left\{ F_k'(w_i), \left(w_i - \left(\sum_{j<i} u_j^{n+1} + \phi_i + \sum_{j>i} u_j^n \right)\right)_k \right\} \leqslant 0 \quad \forall k = 1, 2, \cdots, m \right\}.$$

Since $u^n \in S$, for each $k = 1, 2, \cdots, N$, we have

$$\begin{aligned}\min\left\{ F_k'(u^n), \left(u^n - \left(\phi_1 + \sum_{j>1} u_j^n\right)\right)_k \right\} &\leqslant \min\left\{ F_k'(u^n), \left(u^n - \left(\phi_1 + \sum_{j>1} \phi_j\right)\right)_k \right\} \\ &= \min\{F_k'(u^n), (u^n - \phi)_k\} \leqslant 0.\end{aligned}$$

That is $u^n \in S_1^{N,n}$. By Lemma 4.1, w_1^n is the maximal element of $S_1^{N,n}$ and therefore $u^n \leqslant w_1^n$. For any $k = 1, 2, \cdots, N$, we have the following three cases:

Case A: $(w_1^n)_k > (\phi)_k$ and $(u_1^{n+1})_k > (\phi_1)_k$,

Case B: $(w_1^n)_k > (\phi)_k$ but $(u_1^{n+1})_k = (\phi_1)_k$,

Case C: $(w_1^n)_k = (\phi)_k$.

For *Case* A, we get $(w_1^n)_k = \left(u_1^{n+1} + \sum_{j>1} u_j^n\right)_k > (\phi_1 + \sum_{j>1} u_j^n)_k$. Since $w_1^n \in S_1^{N,n}$, we have $F_k'(w_1^n) \leqslant 0$. For *Case* B, we have $(u_1^{n+1})_k = (u_1^n)_k > (\phi)_k$ and hence $F_k'(u^n) \leqslant 0$. Let $w_{1,\varepsilon}^n = w_1^n - \varepsilon e_k$, where $e_k \in R^N$ is the ith coordinate vector. Then for sufficiently small number $\varepsilon > 0$, we have that $w_{1,\varepsilon}^n \in K$. By strict T-monotonicity, we get

$$0 \leqslant \sum_{j=1}^{N} (F_j'(u^n) - F_j'(w_{1,\varepsilon}^n))(u^n - w_{1,\varepsilon}^n)_j^+ = \varepsilon(F_k'(u^n) - F_k'(w_{1,\varepsilon}^n)).$$

Then

$$F_k'(w_{1,\varepsilon}^n) \leqslant F_k'(u^n) \leqslant 0.$$

By the continuity of F', we have $F_k'(w_1^n) \leqslant 0$. For *Case* C, it is easy to see that

$$\min\{F'(w_1^n), (w_1^n - \phi)_k\} \leqslant (w_1^n - \phi)_k = 0.$$

Then we obtain the conclusion of the lemma. □

By Lemmas 4.1 and 4.2 we get the following results.

Lemma 4.3. *Let $u^0 \in S^N$ and $\{u^n\}$ be produced by MDM. Then*

$$u_i^n \leqslant u_i^{n+1}, \quad i = 1, 2, \cdots, m,$$

and then

$$u^n \in S^N, \quad u^n \leqslant u^{n+1} \leqslant u^*.$$

Theorem 4.4. *Let $u^0 \in S$ and $\{u^n\}$ be produced by MDM. Then $\{u^n\}$ converges to the solution u^* of (1.1) monotonically, and*

$$u^n \in S, \quad u^n \leqslant u^{n+1} \leqslant u^*. \tag{4.1}$$

Proof. By the monotonicity of u^n and $u_i^n (i = 1, 2, \cdots, m)$, we get that there exist $\tilde{u}_i \in K$ $(i = 1, 2, \cdots, m)$, such that $u_i^n \to \tilde{u}_i (i = 1, 2, \cdots, m)$ and

$$u^n \to \tilde{u} = \sum_{i=1}^{m} \tilde{u}_i, \quad w_i^n \to \tilde{u}.$$

Obviously,

$$\min\left\{F_k'(w_i^n), \left(w_i^n - \left(\sum_{j<i} u_j^{n+1} + \phi_i + \sum_{j>i} u_j^n\right)\right)_k\right\} = 0, \quad i = 1, 2, \cdots, m.$$

Take the limit in above equalities and take the sum for $i = 1, 2, \cdots, m$:

$$\min\{F_k(\tilde{u}), (\tilde{u} - (\tilde{u} + \phi_i - \tilde{u}_i))_k\} = \min\{F_k(\tilde{u}), (\tilde{u}_i - \phi_i)_k\} = 0, \quad i = 1, 2, \cdots, m,$$

$$\min\{F_k(\tilde{u}), (\tilde{u} - \phi)_k\} = 0, \quad i = 1, 2, \cdots, N.$$

That is, $\tilde{u} = u^*$ is the solution of (1.1). □

Note. In the proof of above theorem, we see that we can replace F' by operator G, where G may not be a differential operator. That is, we can develop our algorithm and convergence theorem to more general variational inequalities.

5　Applications

Example 1. Let $V = H_0^1(\Omega)$, $f \in L^2(\Omega)$ and

$$K = \{v \in V \mid v \geqslant \phi \text{ in } \Omega\},$$

$$F(v) = \frac{1}{2}\int_\Omega |\nabla v|^2 \mathrm{d}x + \int_\Omega f v \mathrm{d}x.$$

It is easy to see that

$$\langle F'(w), v\rangle = \int_\Omega \nabla w \nabla v \mathrm{d}x + \int_\Omega f v \mathrm{d}x.$$

It is obvious that (1.3), (1.4) hold. Since for $w, v \in \tilde{V} = H_0^1(\Omega) \cap H^2(\Omega)$, we have that

$$\begin{aligned}\left\langle F'(w) - F'(v), (w-v)^+\right\rangle &= \int_\Omega \nabla(w-v)\nabla(w-v)^+ \mathrm{d}x \\ &= \int_\Omega |\nabla(w-v)^+|^2 \mathrm{d}x \geqslant \alpha \|(w-v)^+\|_V,\end{aligned}$$

where α is some positive constant. Hence F' is strictly T-monotone over $\tilde{V}$.

We let subdomains $\Omega_i (i = 1, 2, \cdots, m)$ satisfy $\Omega = \bigcup_{i=1}^m \Omega_i$ and $\mathrm{dist}(\Omega_i, \Omega_j) \geqslant \gamma$ $(i \neq j)$, where γ is a positive constant, and let

$$V_i = \{v \in H_0^1(\Omega_i) \mid v_{\Omega\backslash\Omega_i} = 0\}.$$

Then

$$V = V_1 + V_2 + \cdots + V_m.$$

Let functions $\theta_i \in C_0^\infty(\Omega_i)$ $(i = 1, \cdots, m)$ be a unit decomposition of $\Omega_1, \cdots, \Omega_m$, that is,

$$\sum_{i=1}^m \theta_i = 1, \quad \theta_1, \cdots, \theta_m \geqslant 0. \tag{5.1}$$

Define $\phi_i = \theta_i \phi$, then $\theta_i \phi \in V$ and $\sum_{i=1}^m \theta_i \phi = \phi$. Therefore, $K = K_1 + \cdots + K_m$. We can also verify assumption (S3) holds. Hence we can get the monotone convergence if we use the MDM to above problem for the initial $u^0 \in S$ especially for $u^0 = \phi$.

Example 2. Let

$$\mathcal{K} = \{v \in H_0^1(\Omega) | v \geqslant \phi \ \text{ a.e. in } \Omega\},$$

$$\langle \mathcal{F}(w), v\rangle = \int_\omega \nabla w \nabla v \mathrm{d}x + \int_\Omega f(x, w) v \mathrm{d}x,$$

where Ω is a polygonal domain, $f(x, v)$ is twice continuously differentiable and $\partial f/\partial v$ on $\bar{\Omega} \times \{v \geqslant \phi\}$. Generally, $\mathcal{F}'$ is not Lipschitz on the whole $\mathcal{K}$ while locally Lipschitz constant exists (cf. [18]). We use the conforming linear finite element to discretize the problem, in which Lump region is introduced to deal with the nonlinear term $f(x, v)$, (cf. [11, 12]). Let $u^* \in R^N = V$ be the vector consisting of the inner nodal values of the finite element solution, $K = \{v \in V | v \geqslant \Phi\}$, where $\Phi \in R^N$ is the vector consisting of the inner nodal values of ϕ. The finite element problem can be written as a variational problem with operator F' as follows

$$F'(w) = Aw + D(w) = Aw + (d_k(w_k))_1^N,$$

where $A = (a_{ij})$ is a symmetric and positive definite M-matrix ($a_{ii} > 0$, $a_{ij} \leqslant 0$ $(i \neq j)$, $A^{-1} \geqslant 0$), $d_k(w_k)$ $(k = 1, 2, \cdots, N)$ are smooth and nondecreasing functions. Then

$$\begin{aligned}\langle Aw, w^+\rangle &= \sum_{i,j=1}^{N} a_{ij} w_j w_i^+ = \sum_{i,j=1}^{N} a_{ij}(w_j^+ + w_j^-) w_i^+ \\ &= (w^+)^T A w^+ + \sum_{i=1}^{N} a_{ii} w_i^- w_i^+ + \sum_{i \neq j} a_{ij} w_j^- w_i^+ \\ &= (w^+)^T A w^+ + \sum_{i \neq j} a_{ij} w_j^- w_i^+ \\ &\geqslant (w^+)^T A w^+,\end{aligned}$$

where $w^- = w - w^+ = \min\{0, w\}$. For any $w, v \in K$, we have

$$\begin{aligned}\langle D(w) - D(v), (w-v)^+\rangle &= \sum_{k=1}^{N} (d_k(w_k) - d_k(v_k))(w_k - v_k)^+ \\ &= \sum_{k=1}^{N} \int_0^1 d_k'(w_k + \theta(v_k - w_k))(w_k - v_k)(w_k - v_k)^+ \mathrm{d}\theta \\ &= \sum_{k=1}^{N} \int_0^1 d_k'(w_k + \theta(v_k - w_k))((w_k - v_k)^+)^2 \mathrm{d}\theta \geqslant 0.\end{aligned}$$

Then we can easily verify (1.3), (1.4) as well as the strictly T-monotonicity of F' over K. Decompose Ω in a similar way to Example 1 and assume that each finite element does not cross the boundary of every subdomains. Then, we have the same monotone convergence result as in Example 1.

Example 3. Let $\mathcal{F}'(w)$ be replaced by $\mathcal{G}(w)$, where

$$\langle \mathcal{G}(w), v\rangle = \int_\Omega \nabla w \nabla v \mathrm{d}x + \int_\Omega \partial w v \mathrm{d}x + \int_\Omega f v \mathrm{d}x,$$

and K, V etc. be the same as that in Example 2. Then a similar conclusion as those in Example 2 can be got.

In this example, we see that $\mathcal{G}(w)$ is no longer a differential of some $\mathcal{F}$ (also true for the corresponding discreted problem). As a result, in algorithm MDM, (2.2) is replaced by solving (2.4) and the algorithm is constructed to solve variational inequality (2.3) directly.

Acknowledgments The authors gratefully acknowledge the detailed comments and suggestions of the referees. Their efforts have improved the content of the paper.

References

[1] L. Badea and J. Wang, An additive Schwarz method for variational inequalities, *Mathematics of Computation*, 69 (1999), 1341-1354.

[2] U. Mosco, *An Introduction to the Approximate Solution of Variational Inequalities*, Cremonese, Roma, 1973.

[3] Y. Kuznetsov and P. Neittaanmäki, P. Tarvainen, Schwarz methods for obstacle problem with convection-diffusion operator, in: D.E. Keyes, J.C. Xu (Eds.), *Domain Decomposition Methods in Scientific and Engineering Computing*, American Mathematical Society, Providence, RI, 1995, pp. 251-256.

[4] P.L. Lions, On the Schwarz alternating method I, in: R. Glowinski, G.H. Golub, G.A. Meurant, J. Périaux (Eds.), *Proceedings of the Domain Decomposition Method*, SIAM, Philadelphia, PA, 1988, pp. 1-40.

[5] T. Lü, T.M. Shib, C.B. Liem, *Domain Decomposition Methods*, Science Press, Beijing, 1992. (in Chinese)

[6] T. Lü, T.M. Shih, C.B. Liem, Parallel algorithms for variational inequalities based on domain decomposition, *Journal of Computational Mathematics*, 9 (1991) 340-349.

[7] P. Tarvainen, Two-level Schwarz method for unilateral variational inequalities, *IMA Journal of Numerical Analysis*, 19 (1999), 273-290.

[8] J.P. Zeng and D.H. Li, M. Fukushima, Weighted max norm estimate of additive Schwarz iteration scheme for solving linear complementarity problems, *Journal of Computational Mathematics*, 131 (2001), 1-14.

[9] J.P. Zeng and S.Z. Zhou, On monotone and geometric convergence of Schwarz methods for two-side obstacle problems, *SIAM Journal on Numerical Analysis*, 35 (1998), 600-616.

[10] S.Z. Zhou, An additive Schwarz algorithm for a variational inequality, in: R. Glowinski, J. Périaux, Z.-C. Shi, O. Widlund (Eds.), *Domain Decomposition Methods in Scientific Engineering*, Wiley, NewYork, 1997, pp. 133-138.

[11] K.H. Hoffmann and J. Zou, Parallel solution of variational inequality problems with nonlinear source terms, *IMA Journal of Numerical Analysis*, 16 (1996), 31-45.

[12] J.P. Zeng, S.Z. Zhou, Schwarz algorithm for the solution of variational inequalities with nonlinear source terms, *Applied Mathematics & Computation*, 97 (1998), 23-35.

[13] S.Z. Zhou and J.P. Zeng, Schwarz algorithm for obstacle problems with a type of nonlinear operators, *Acta Mathematicae Applicatae Sinica*, 20 (1997), 521-531 (in Chinese).

[14] R.H.W. Hoppe, Multigrid algorithms for variational inequalities, *SIAM Journal on Numerical Analysis*, 24 (1987), 1046-1065.

[15] X.C. Tai, Parallel function and space decomposition methods: Part II, space decomposition, *Beijing Mathematics*, 1 (1995), Part 2, 135-152.

[16] X.C. Tai and M. Espedal, Rate of convergence of some space decomposition method for linear and nonlinear problems, *SIAM Journal on Numerical Analysis*, 35 (1998), 1558-1570.

[17] S.Z. Zhou, *Variational Inequalities and its Finite Element Methods*, Hunan University Press, Changsha, 1988. (in Chinese).

[18] S.Z. Zhou, J.P. Zeng and G.H. Shan, On the convergence of a space decomposition method for nonlinear problems, *Advances in Mathematics*, 28 (1999), 541-542.

Analysis of Generalized Schwarz Algorithms for Solving Obstacle Problems with T-monotone Operator*

C.L. Li (李郴良)† J.P. Zeng (曾金平)‡ S.Z. Zhou (周叔子)‡

Abstract This paper proves the convergence of some generalized Schwarz algorithms for solving the obstacle problems with a T-monotone operator. Numerical results show that the generalized Schwarz algorithms converge faster than the classical Schwarz algorithms.

Keywords Variational inequalities, T-monotone operator, obstacle problems, generalized Schwarz algorithms, convergence.

1 Introduction

Schwarz algorithms are well-known iterative methods for solving partial differential equations, see, e.g., [1], [2] and the references therein. Since the calculation processes of Schwarz algorithms can be easily implemented in parallel, the algorithms have been widely applied to solve problems including obstacle problems in the last two decades. We refer to [3–7] for earlier work and to [8–13] for some new progress.

In this paper, we consider some generalized Schwarz algorithms for solving obstacle problems with a T-monotone operator. This kind of obstacle problems have many applications (see [14–17] and the references therein for details). In [18–20], some classical Schwarz algorithms were presented for solving obstacle problems with a T-monotone operator. Monotone convergence and global convergence of the algorithms were obtained. In [21], a so-called generalized Schwarz algorithm (GSA) was presented to solve elliptic boundary value problems. Compared with the classical Schwarz algorithms, in which the subproblems are coupled by the Dirichlet boundary condition, the generalized Schwarz algorithm replaces the inner boundary condition by a Robin condition with a parameter. Numerical experiments show that the performance of

* 发表于：Computer and Mathematics with Applications, Vol. 48, 2004, pp. 373-386.

† School of Mathematics and Computing Science, Guilin University of Electronic Technology, Guilin, P.R. China.

‡ Institute of Applied Mathematics, Hunan University, P.R. China.

GSA with appropriate parameters is much better than the performance of the classical Schwarz algorithm. In [22], this technique was extended to solve the discrete obstacle problems with a linear self-adjoint elliptic operator. Numerical results presented there shown that a faster convergence rate can also be obtained if the parameters were selected appropriately. The aim of this paper is to study the convergence property of generalized additive and multiplicative Schwarz algorithms for solving the obstacle problems with a T-monotone operator. We will also give some numerical results.

The paper is organized as follows: In Sections 2 and 3, we introduce the generalized Schwarz algorithms with two subdomains, and establish the convergence theorems. In Section 4, we discuss the generalized Schwarz algorithms with more than two subdomains. Finally, in Section 5, we give some preliminary numerical results to verify the efficiency of the algorithms.

2 Obstacle Problems with T-Monotone Operator

Definition 2.1. *[See [13, 17]] Let V be a reflexive Banach space of real functional defined on a domain $\Omega \subset R^d$ ($d = 1$ or 2), V^* its dual space, K a subset of V, and A an operator from K to V^*. Let every element v of K be expressed by $v = v^+ + v^-$ with $v^+ = \max\{v, 0\}$ and $v^- = \min\{v, 0\}$. Operator A is called T-monotone over K if*

$$\langle Av - Aw, (v-w)^+\rangle \geqslant 0, \qquad \forall v, w \in K, \tag{2.1}$$

where $\langle \cdot, \cdot\rangle$ denotes a dual product. Moreover, if for all v and $w \in K$, $\langle Av - Aw, (v-w)^+\rangle = 0$ is equivalent to $(v-w)^+ = 0$, then A is called strictly T-monotone over K.

Definition 2.2. *[See [23]] Let K and V be defined as in Definition 2.1. Operator A is called monotone over K if*

$$\langle Av - Aw, v-w\rangle \geqslant 0, \qquad \forall v, w \in K. \tag{2.2}$$

A is called strictly monotone over K if strict inequality holds in (2.2) whenever $v \neq w$, and A is called strongly monotone over K if there is a constant $\alpha > 0$ such that

$$\langle Av - Aw, v-w\rangle \geqslant \alpha\|v-w\|_V^2, \qquad \forall v, w \in K, \tag{2.3}$$

where $\|\cdot\|_V$ is a norm defined on V.

Remark 2.1. Let

$$\mathcal{K} = \{v \in H_0^1(\Omega) |\ v \geqslant \varphi \text{ a.e. in } \Omega\}, \tag{2.4}$$

$$\langle \mathcal{A}w, v\rangle = \int_\Omega \nabla w \nabla v \mathrm{d}x + \int_\Omega f(x, w) v \mathrm{d}x, \tag{2.5}$$

where, Ω is a polygonal domain in R^2, φ belongs to $W^{2,s}$ for some $s > 2$ satisfying $\varphi|_{\partial\Omega} \leqslant 0$, and $f(x, v)$ is twice continuously differentiable that satisfies $\dfrac{\partial f}{\partial v}|_{\bar{\Omega}\times\{v\geqslant\varphi\}} \geqslant 0$. Since for any v and $w \in \mathcal{K}$, we have

$$
\begin{aligned}
&\langle \mathcal{A}v - \mathcal{A}w, (v-w)^+\rangle \\
&= \int_\Omega \nabla(v-w)\nabla(v-w)^+ \mathrm{d}x + \int_\Omega (f(x,v)-f(x,w))(v-w)^+\mathrm{d}x \\
&= \int_\Omega \nabla(v-w)^+\nabla(v-w)^+ \mathrm{d}x + \int_\Omega\int_0^1 \frac{\partial f}{\partial v}(x, v+t(w-v))(v-w)(v-w)^+\mathrm{d}t\mathrm{d}x \\
&= \int_\Omega (\nabla(v-w)^+)^2 \mathrm{d}x + \int_\Omega\int_0^1 \frac{\partial f}{\partial v}(x, v+t(w-v))(v-w)^+(v-w)^+\mathrm{d}t\mathrm{d}x \\
&\geqslant \int_\Omega (\nabla(v-w)^+)^2 \mathrm{d}x,
\end{aligned}
$$

it follows from Definition 2.1 that $\mathcal{A}$ is strictly T-monotone over $\mathcal{K}$. It is also easy to see that $\mathcal{A}$ is strongly monotone over $\mathcal{K}$.

Remark 2.2 Let $\mathcal{K}$ and $\mathcal{A}$ be defined as in Remark 2.1, and V be the conforming linear element space in $H_0^1(\Omega)$. If A is the corresponding discrete operator of $\mathcal{A}$, where the lump region is introduced to deal with the nonlinear term $f(x,v)$ (see, e.g., [18, 20]), and $K=\{v\in V|\ v\geqslant \phi\}$, where $\phi=\varphi_I$ is the interpolating of φ, then it is easy to prove that A is strictly T-monotone and strongly monotone over K.

The obstacle problem with the operator A specified in Remark 2.1 arises from a mathematical model of the diffusion problem involving Michaelis-Menten or second-order irreversible reactions (see, e.g., [15, 16]).

Lemma 2.3. *If A is (strictly) T-monotone over K, then A is (strictly) monotone over K.*

Proof. If A is T-monotone over K, then we have

$$
\langle Av - Aw, (v-w)^+\rangle \geqslant 0, \quad \forall v, w\in K.
$$

Therefore,

$$
\begin{aligned}
\langle Av - Aw, v-w\rangle &= \langle Av - Aw, (v-w)^+\rangle + \langle Av - Aw, (v-w)^-\rangle \\
&\geqslant \langle Av - Aw, (v-w)^-\rangle \\
&= \langle Aw - Av, (w-v)^+\rangle \\
&\geqslant 0.
\end{aligned} \tag{2.6}
$$

This means that A is monotone over K. If A is strictly T-monotone over K, then for any $v\neq w\in K$, we have either $(v-w)^+>0$ or $(v-w)^+>0$, which together with (2.6) implies that $\langle Av-Aw, v-w\rangle>0$. That is, A is strictly monotone over K. □

Let Ω be polygonal domain in R^d ($d=1$ or 2), and $V\subset H_0^1(\Omega)$ the conforming linear finite element space. We consider the following finite-dimensional obstacle problem of finding a $u\in K$, such that

$$
\langle Au, v-u\rangle \geqslant 0, \qquad \forall v\in K, \tag{2.7}
$$

where $K = \{v \in V|\ v \geqslant \phi\}$, and the operator A, satisfying $Av = DJ(v)$ for some functional $J(v)$, is coercive, continuous, and strictly T-monotone over K.

It is well-known that problem (2.7) has a unique solution (see e.g., [24]). As was pointed out in Remarks 2.1 and 2.2, problem (2.7) can be considered as the discretization of the obstacle problem with some V-elliptic differential operator.

Let domain Ω be decomposed into two overlapped subdomains Ω_1 and Ω_2 such that $\Omega = \Omega_1 \bigcup \Omega_2$ and $\Omega_1 \bigcap \Omega_2 \neq \emptyset$. We assume that any finite element does not across the boundaries of Ω_1 and Ω_2. Denote by $\Omega_h = \{x_1, x_2, \cdots, x_N\}$ the inner node-set of the triangulation. For $i = 1, 2$, we use $I_i = \{x \in \Omega_h|\ x \in \Omega_i\}$ to stand for the inner node-set over Ω_i, $\Gamma_i = \{x \in \Omega_h|\ x \in \partial\Omega_i \backslash \partial\Omega\}$ to stand for the node-set over the inner boundary of Ω_i. Let $I_i' = \Omega_h \backslash (I_i \bigcup \Gamma_i)$.

In order to pose the Robin condition on Γ_i, we define

$$g_{i,\theta}(v(x)) = \theta_i v(x) + (1-\theta_i)\frac{\partial v(x)}{\partial n_i}, \qquad x \in \Gamma_i, \tag{2.8}$$

where n_i is the outer normal direction of $\partial\Omega_i$ at x, $\theta = (\theta_1, \theta_2) \in (0,1]^2$. When $d = 1$ and $x \in \Gamma_i$, the directional derivative $\partial v(x)/\partial n_i$ is indeed a difference quotient. When $d = 2$ and $x \in \Gamma_i$, there are two edges containing vertex x and locating on $\partial\Omega_i \backslash \partial\Omega$. Denote the two edges by s_1 and s_2. Let n_{ij} be the outer normal direction of s_j on $\partial\Omega_i$. Then, we can define for $v \in V$ and $x \in \Gamma_i$ that

$$\frac{\partial v(x)}{\partial n_i} = \frac{1}{2}\sum_{j=1}^{2}\frac{\partial v(x)}{\partial n_{ij}}.$$

As a result, $g_{i,\theta}(v)$ given in (2.8) is well defined for the case of $d = 2$ (Note that since $v \in V$ is piece-wisely linear, the outer normal direction at $x \in \Gamma_i$ does not exist in ordinary sense).

In [22], we presented generalized additive and multiplicative Schwarz algorithms for linear complementarity problems. Numerical results show that the algorithms are competitive to the classical Schwarz algorithms. In this paper, we try to extend the algorithms to solve the obstacle problems with T-monotone operator. First we give generalized additive and multiplicative Schwarz algorithms with two subdomains as follows.

Generalized Additive Schwarz Algorithm (GASA1).

Step 1. Given $\epsilon > 0$, vector $\theta = (\theta_1, \theta_2)$ with θ_1 and θ_2 in $(0,1]$, and positive weights ω_1, ω_2 with $\omega_1 + \omega_2 = 1$. Choose initial point $u^0 \in K$. Set $n := 0$.

Step 2. For $i = 1, 2$, solve the subproblems of finding a $u^{n,i} \in K_{i,\theta}^n$ such that

$$\langle Au^{n,i}, v - u^{n,i}\rangle \geqslant 0, \quad \forall v \in K_{i,\theta}^n, \tag{2.9}$$

where

$$K_{i,\theta}^n = \{v \in K|\ v = u^n \text{ in } I_i', \quad g_{i,\theta}(v) = g_{i,\theta}(u^n) \text{ on } \Gamma_i\}. \tag{2.10}$$

Step 3. Let $u^{n+1} = \sum\limits_{i=1}^{2}\omega_i u^{n,i}$. If $\|u^{n+1} - u^n\| < \epsilon$, stop. Otherwise, set $n := n+1$ and go to Step 2.

Remark 2.3 Usually, we choose the weights ω_1, ω_2 by letting $\omega_1 = \omega_2 = \omega = 1/2$.

Generalized Multiplicative Schwarz Algorithm (GMSA1).

Step 1. Given $\epsilon > 0$ and vector $\theta = (\theta_1, \theta_2)$ with θ_1 and θ_2 in $(0, 1]$. Choose initial point $u^0 \in K$. Set $n := 0$.

Step 2. Solve the subproblem over Ω_1 of finding a $u^{n+1/2} \in K^n_{1,\theta}$ such that

$$\langle Au^{n+1/2}, v - u^{n+1/2}\rangle \geqslant 0, \quad \forall v \in K^n_{1,\theta}. \tag{2.11}$$

Step 3. Solve the subproblem over Ω_2 of finding a $u^{n+1} \in K^{n+1/2}_{2,\theta}$ such that

$$\langle Au^{n+1}, v - u^{n+1}\rangle \geqslant 0, \quad \forall v \in K^{n+1/2}_{2,\theta}, \tag{2.12}$$

where

$$K^{n+1/2}_{2,\theta} = \{v \in K |\ v = u^{n+1/2} \text{ in } \Gamma'_2, \quad g_{2,\theta}(v) = g_{2,\theta}(u^{n+1/2}) \text{ on } \Gamma_2\}. \tag{2.13}$$

Step 4. If $\|u^{n+1} - u^n\| < \epsilon$, stop. Otherwise, set $n := n + 1$ and go to Step 2.

Remark 2.4. If we choose $\theta_1 = \theta_2 = 1$, then the above algorithms correspond to classical additive and multiplicative Schwarz algorithms respectively. see e.g., [18, 19, 20] for details.

3 Convergence of the Algorithms

To prove the convergence of the algorithms we proposed, we first show that the algorithms are well-defined.

Lemma 3.1. *For given $u^n \in V$, $K^n_{i,\theta}$ is a nonempty, closed and convex subset of V. Therefore, if A is coercive, continuous and strictly T-monotone, problem (2.9) has a unique solution.*

Proof. Since $u^n \in K^n_{i,\theta}$, $K^n_{i,\theta}$ is nonempty. For any v_1, $v_2 \in K^n_{i,\theta}$ and $0 \leqslant \lambda \leqslant 1$, it is easy to see that $g_{i,\theta}(\lambda v_1 + (1-\lambda)v_2) = \lambda g_{i,\theta}(v_1) + (1-\lambda) g_{i,\theta}(v_2) = g_{i,\theta}(u^n)$ and then $\lambda v_1 + (1-\lambda)v_2 \in K^n_{i,\theta}$. So $K^n_{i,\theta}$ is convex. Obviously, $K^n_{i,\theta}$ is a closed subset of V. Then we complete the proof. □

Lemma 3.2. *Assume for some functional $J(v)$, $Av = DJ(v)$. Then problems (2.7) and (2.9) are equivalent to the following two functional minimization problems, respectively:*

$$u \in K, \qquad J(u) = \inf_{v \in K} J(v), \tag{3.1}$$

$$u^{n,i} \in K^n_{i,\theta}, \qquad J(u^{n,i}) = \inf_{v \in K^n_{i,\theta}} J(v). \tag{3.2}$$

Proof. Let u be the solution of (2.7). For any $v \in K$, denote $I(s) = J(u + s(v - u))$. we have

$$I'(0) = \langle Au, v - u\rangle \geqslant 0.$$

Therefore, by the use of Lemma 2.3, we have

$$\begin{aligned} J(v)-J(u) &= J(v)-J(u)-\langle Au, v-u\rangle+\langle Au, v-u\rangle \\ &\geqslant \langle Av_t - Au, v-u\rangle \\ &= \langle Av_t - Au, v_t-u\rangle/t \\ &\geqslant 0. \end{aligned}$$

where $v_t = u+t(v-u)$ for some scale $t \in (0,1)$. Then, u is the solution of (3.1). On the other hand, let u be the solution of (3.1). Since for any $v \in K$, $u+s(v-u) \in K$ for $s \in [0,1]$, we have that $J(u+s(v-u)) \geqslant J(u)$, which implies that $I(s) \geqslant I(0)$. Consequently, $I(0)$ is a minimization of functional $I(s)$ on $[0,1]$. Thereby $I'(0) = \langle Au, v-u\rangle \geqslant 0$. That is, u is the solution of (2.7). The equivalence between problems (2.9) and (3.2) can be proved in a similar way. □

In order to prove the convergence of the proposed algorithms, we need the so-called condition A, which was introduced in [22]. We say that nodes x_k and $x_l \in \Omega_h$ are adjacent if they belong to the same element. Let

$$\Gamma_i' = \{x \in I_i |\ x \text{ is adjacent to a node on } \Gamma_i\}.$$

Condition A $(\Gamma_1' \bigcup \Gamma_1) \bigcap (\Gamma_2' \bigcup \Gamma_2) = \emptyset$.

Remark 3.1. Let $\delta = \text{dist}(\partial\Omega_1 \bigcap \Omega, \partial\Omega_2 \bigcap \Omega)$. This means that δ is the overlapping size. $\delta > 0$ means that the decomposition $\Omega = \Omega_1 \bigcup \Omega_2$ is a uniform overlapping decomposition (see, e.g., [4]). Let h be the finite element mesh-size. It is obvious that condition A holds if $h < \delta/3$. So condition A is natural in the case of uniform overlapping decomposition.

Lemma 3.3. *Assume condition A holds. Then (2.7) is equivalent to the following problem of finding a $u^* \in K$ such that*

$$\langle Au^*, v-2u^*\rangle \geqslant 0, \qquad \forall v \in K_{1,\theta}^* + K_{2,\theta}^*, \tag{3.3}$$

where

$$K_{i,\theta}^* = \{v \in K |\ v = u^* \text{ in } I_i', \quad g_{i,\theta}(v) = g_{i,\theta}(u^*) \text{ on } \Gamma_i\}. \tag{3.4}$$

Proof. Assume u is the solution of (2.7). Let $u^* = u$. Since $K_i^* \subset K$, we get from (2.7) that for $i = 1, 2$,

$$\langle Au^*, v_i - u^*\rangle \geqslant 0, \qquad \forall v_i \in K_{i,\theta}^*. \tag{3.5}$$

Summing (3.5) for $i = 1, 2$ we obtain

$$\langle Au^*, v_1+v_2-2u^*\rangle \geqslant 0, \qquad \forall v_1+v_2 \in K_{1,\theta}^* + K_{2,\theta}^*, \tag{3.6}$$

which means that u is the solution of (3.3).

Conversely, suppose that u^* is the solution of (3.3). We are going to prove that u^* is the solution of (2.7). For any $v \in K$, take $v_1 \in V$ such that

$$v_1 = \begin{cases} u^*, & \text{in } (\Omega_h \backslash I_1) \bigcup \Gamma_1', \\ v, & \text{in } (\Omega_h \backslash I_2) \bigcup \Gamma_2', \\ v, & \text{elsewhere (maybe empty)}. \end{cases} \tag{3.7}$$

From condition A, we know that v_1 is well defined by (3.7). Obviously $v_1 \in K$, since $v, u^* \in K$. From (3.7), we have

$$g_{1,\theta}(v_1) = g_{1,\theta}(u^*) \qquad \text{on } \Gamma_1, \tag{3.8}$$

$$g_{2,\theta}(v_1) = g_{2,\theta}(v) \qquad \text{on } \Gamma_2. \tag{3.9}$$

It then follows from (3.7) and (3.8) that $v_1 \in K_{1,\theta}^*$.

Let $v_2 = v + u^* - v_1$. Then it is easy to see that $v_2 \in K$ and $v_2 = u^*$ in I_2'. By (3.9) we have for $x \in \Gamma_2$,

$$\begin{aligned} g_{2,\theta}(v_2(x)) &= g_{2,\theta}(v(x) + u^*(x) - v_1(x)) \\ &= g_{2,\theta}(v(x)) + g_{2,\theta}(u^*(x)) - g_{2,\theta}(v_1(x)) \\ &= g_{2,\theta}(u^*(x)). \end{aligned}$$

Hence $v_2 \in K_{2,\theta}^*$. So, for any $v \in K$, it follows from (3.3) that

$$\langle Au^*, v - u^* \rangle = \langle Au^*, v_1 + v_2 - 2u^* \rangle \geqslant 0,$$

which means that u^* is the solution of (2.7). The proof is then completed. □

Now we prove the following theorem.

Theorem 3.4. *Assume that A, satisfying $Av = DJ(v)$ for some functional $J(v)$, is coercive, continuous, and strictly T-monotone over K. Assume further that condition A holds. Then, any sequence $\{u^n\}$ generated by the algorithm GASA1 converges to the solution u of problem (2.7).*

Proof. By Lemma 2.3, for any $v \neq w \in K$, we have

$$\langle DJ(v), v - w \rangle - \langle DJ(w), v - w \rangle = \langle Av - Aw, v - w \rangle > 0.$$

Then, J is a strictly convex functional over K. Therefore, by Lemma 3.2 and the third step of GASA1, we get

$$J(u) \leqslant J(u^{n+1}) \leqslant \omega_1 J(u^{n,1}) + \omega_2 J(u^{n,2}) \leqslant \omega_1 J(u^n) + \omega_2 J(u^n) = J(u^n). \tag{3.10}$$

Hence

$$\lim_{n \to +\infty} J(u^n) = \lim_{n \to +\infty} J(u^{n,1}) = \lim_{n \to +\infty} J(u^{n,2}) = l. \tag{3.11}$$

Since A is coercive, we have

$$\lim_{v \in K, \|v\| \to +\infty} J(v) = +\infty. \tag{3.12}$$

Therefore, by (3.11), sequences $\{u^n\}$, $\{u^{n,1}\}$ and $\{u^{n,2}\}$ are bounded. So they have convergent subsequences, denoted still by $\{u^n\}$, $\{u^{n,1}\}$ and $\{u^{n,2}\}$ converging to u^*, u_1^* and u_2^*, respectively. It is obvious that

$$J(u^*) = J(u_1^*) = J(u_2^*), \qquad \text{and} \qquad u^* = \sum_{i=1}^{2} \omega_i u_i^*. \tag{3.13}$$

This together with the strict convexity of $J(v)$ implies $u^* = u_1^* = u_2^*$. Therefore, by the second step of GASA1, we obtain

$$\langle Au^*, v - u^* \rangle \geqslant 0, \qquad \forall v \in K_{i,\theta}^*, \quad i = 1, 2, \tag{3.14}$$

where $K_{i,\theta}^*$ $(i = 1, 2)$ are defined by (3.4). Summing (3.14) for $i = 1, 2$ we have

$$\langle Au^*, v - 2u^* \rangle \geqslant 0, \qquad \forall v \in K_{1,\theta}^* + K_{2,\theta}^*, \tag{3.15}$$

which is (3.3). By Lemma 3.3, we get $u^* = u$. The proof is completed. □

For algorithm GMSA1, if (2.3) holds, we have that

$$\begin{aligned}
&J(u^n) - J(u^{n+1/2}) \\
&= J(u^n) - J(u^{n+1/2}) - \langle Au^{n+1/2}, u^n - u^{n+1/2} \rangle + \langle Au^{n+1/2}, u^n - u^{n+1/2} \rangle \\
&= \int_0^1 \langle Au_t^n - Au^{n+1/2}, u_t^n - u^{n+1/2} \rangle \frac{dt}{t} + \langle Au^{n+1/2}, u^n - u^{n+1/2} \rangle \\
&\geqslant \alpha \int_0^1 t \|u^n - u^{n+1/2}\|_V^2 dt + \langle Au^{n+1/2}, u^n - u^{n+1/2} \rangle \\
&\geqslant \frac{\alpha}{2} \|u^n - u^{n+1/2}\|_V^2,
\end{aligned} \tag{3.16}$$

where $u_t^n = u^{n+1/2} + t(u^n - u^{n+1/2})$, the first inequality follows from (2.3) and the second inequality follows from (2.11).

Similar to (3.10) and (3.11), we have that

$$J(u) \leqslant J(u^{n+1}) \leqslant J(u^{n+1/2}) \leqslant J(u^n), \tag{3.17}$$

and hence,

$$\lim_{n\to\infty} J(u^n) = \lim_{n\to\infty} J(u^{n+1/2}) = l. \tag{3.18}$$

By (3.16) and (3.18), we have

$$0 \leftarrow J(u^n) - J(u^{n+1/2}) \geqslant \frac{\alpha}{2} \|u^n - u^{n+1/2}\|_V^2.$$

Consequently,

$$u^n - u^{n+1/2} \to 0. \tag{3.19}$$

Similarly, we have that

$$0 \leftarrow J(u^{n+1/2}) - J(u^{n+1}) \geqslant \frac{\alpha}{2} \|u^{n+1/2} - u^{n+1}\|_V^2$$

and then

$$u^{n+1/2} - u^{n+1} \to 0. \tag{3.20}$$

By (3.12) and (3.18), sequences $\{u^n\}$ and $\{u^{n+1/2}\}$ are bounded. Let u^* be a limit of a subsequence $\{u^{n_k}\}$ of $\{u^n\}$. By (3.19) and (3.20), we get

$$\lim_{k\to\infty} u^{n_k+1/2} = \lim_{k\to\infty} u^{n_k+1} = \lim_{k\to\infty} u^{n_k} = u^*.$$

This together with (2.11) and (2.12) implies

$$u^* \in K^*_{1,\theta}, \qquad \text{and} \qquad \langle Au^*, v - u^*\rangle \geqslant 0, \qquad \forall v \in K^*_{1,\theta}$$

and

$$u^* \in K^*_{2,\theta}, \qquad \text{and} \qquad \langle Au^*, v - u^*\rangle \geqslant 0, \qquad \forall v \in K^*_{2,\theta},$$

where $K^*_{1,\theta}$ and $K^*_{2,\theta}$ are defined by (3.4). Therefore, we have the following theorem.

Theorem 3.5. *Assume that A, satisfying $Av = DJ(v)$ for some functional $J(v)$, is coercive, continuous, strictly T-monotone, and strongly monotone over K. Assume further that condition A holds. Then, any sequence $\{u^n\}$ generated by the algorithm GMSA1 converges to the solution u of problem (2.7).*

4 Generalized Schwarz Algorithms with More Than Two Subdomains

In this section, we consider generalized Schwarz algorithms with more than two subdomains. For simplicity, we assume that the subdomains $\Omega_1, \Omega_2, \cdots, \Omega_m$ satisfy $\Omega = \bigcup_{i=1}^{m} \Omega_i$ and that for any $i \in \{1, 2, \cdots, m\}$, there exists $j \in \{1, 2, \cdots, m\}, j \neq i$, such that $\Omega_i \cap \Omega_j \neq \emptyset$. We further assume that any finite element does not across the boundaries of Ω_i $(i \in \{1, 2, \cdots, m\})$. Define Ω_h, Γ_i, Γ'_i, I_i, $g_{i,\theta}$, $\dfrac{\partial v}{\partial n_i}$, $K^*_{i,\theta}$, $K^n_{i,\theta}$ and $K^{n+i/m}_{i,\theta}$ in a similar way. The corresponding algorithms are stated as follows:

Algorithm GASA2.

Step 1. Given $\epsilon > 0$, $\theta = (\theta_1, \theta_2, \cdots, \theta_m)$ with $\theta_i (i = 1, 2, \cdots, m)$ in $(0, 1]$, and positive weights $\omega_1, \omega_2, \cdots, \omega_m$ with $\sum_{i=1}^{m} \omega_i = 1$. Choose initial point $u^0 \in K$. Set $n := 0$.

Step 2. For $i = 1, 2, \cdots, m$, solve the subproblems of finding a $u^{n,i} \in K^n_{i,\theta}$ such that

$$\langle Au^{n,i}, v - u^{n,i}\rangle \geqslant 0, \quad \forall v \in K^n_{i,\theta}.$$

Step 3. Let $u^{n+1} = \sum_{i=1}^{m} \omega_i u^{n,i}$. If $\|u^{n+1} - u^n\| < \epsilon$, stop. Otherwise, set $n := n + 1$ and go to Step 2.

Remark 4.1. Similarly, we may choose average weights $\omega_1 = \omega_2 = \cdots = \omega_m = \omega = 1/m$.

Algorithm GMSA2.

Step 1. Given $\epsilon > 0$ and $\theta = (\theta_1, \theta_2, \cdots, \theta_m)$ with $\theta_i (i = 1, 2, \cdots, m)$ in $(0, 1]$. Choose initial point $u^0 \in K$. Set $n := 0$.

Step 2. For $i = 1, 2, \cdots, m$, solve the following subproblems of finding a $u^{n+i/m} \in K_{i,\theta}^{n+(i-1)/m}$ such that

$$\langle Au^{n+i/m}, v - u^{n+i/m} \rangle \geqslant 0, \quad \forall v \in K_{i,\theta}^{n+(i-1)/m}.$$

Step 3. If $\|u^{n+1} - u^n\| < \epsilon$, stop. Otherwise, set $n := n + 1$ and go to Step 2.

To establish the convergence of above two algorithms, instead of Condition A, we introduce the following condition.

Condition B If $\Omega_i \bigcap \Omega_j \neq \emptyset$ and $i \neq j$, then $(\Gamma_i' \bigcup \Gamma_i) \bigcap (\Gamma_j' \bigcup \Gamma_j) = \emptyset$.

Similarly to Lemma 3.3, we can prove the following lemma.

Lemma 4.1. *Assume condition B holds. Then problem (2.7) is equivalent to the following problem of finding a* $u^* \in K$ *such that*

$$\langle Au^*, v - mu^* \rangle \geqslant 0, \quad \forall v \in K_1^* + K_2^* + \cdots K_m^*. \tag{4.1}$$

Using the above lemma, it is not difficult to derive the following conclusions.

Theorem 4.2. *Assume that* A*, satisfying* $Av = DJ(v)$ *for some functional* $J(v)$*, is coercive, continuous and strictly* T*-monotone over* K*. Assume further that condition B holds. Then, any sequence* $\{u^n\}$ *generated by algorithm GASA2 converges to the solution* u *of problem (2.7).*

Theorem 4.3. *Assume that* A*, satisfying* $Av = DJ(v)$ *for some functional* $J(v)$*, is coercive, continuous, strictly* T*-monotone and strongly monotone over* K*. Assume further that condition B holds. Then, any sequence* $\{u^n\}$ *generated by algorithm GMSA2 converges to the solution* u *of problem (2.7).*

5 Numerical Examples

In this section we give some numerical experiments to investigate the behavior of the algorithms presented in this paper. In the tests, we let $\Omega = (0, 1)$, $\mathcal{AV} = -\mathcal{V}'' + f(x, \mathcal{V})$, where $f(x, \mathcal{V}) = \mathcal{V}^2 + g(x) = \mathcal{V}^2 - [\sin(x - 0.1) - x(\sin(0.1) + \sin(0.9)) + \sin(0.1)]^2 - \sin(x - 0.1)$. We let the mesh-size be $h = \frac{1}{N+1} = 1/1000$ and use the Lagrange linear finite element space as V. Here, as mentioned in Remark 2.2, the lump region is introduced to deal with the nonlinear term $f(x, \mathcal{V})$. Hence, the operator A in problem (2.7) can be written as $Av = A_1 v + A_2 v$, where

$$A_1 v = \begin{pmatrix} 2 & -1 & & & \\ -1 & 2 & -1 & & \\ & \ddots & \ddots & \ddots & \\ & & -1 & 2 & -1 \\ & & & -1 & 2 \end{pmatrix}_{N \times N} \quad \text{and} \quad A_2 v = (h^2 f(x_i, v_i))_{i=1}^N.$$

Obviously, $Av = DJ(v)$, where

$$J(v) = \frac{1}{2}v^T A_1 v + h^2 \sum_{i=1}^{N} (\frac{1}{3}v_i^3 + g(x_i)v_i).$$

We choose the initial point $u^0 = 0$. The stopping criterion is that the maximum norm of the difference between the successive iterative solutions is less than $\epsilon = 10^{-6}$. The subproblems are solved by linearized PSOR algorithm (c.f. [14]), where the stopping criterion is that the maximal norm of the difference between the successive iterative solutions is less than 10^{-8}.

We decompose Ω into two subdomains $\Omega_1 = (0, 0.5 + \rho_1)$ and $\Omega_2 = (0.5 - \rho_2, 1)$. Consider the following two cases:

Case I: $\rho_1 = \rho_2 = h$, where the intersects $I_{12} \stackrel{\text{def}}{=} I_1 \bigcap I_2 = \{0.5\}$ is a singleton and $I'_{12} \stackrel{\text{def}}{=} (\Gamma_1 \bigcup \Gamma'_1) \bigcap (\Gamma_2 \bigcup \Gamma'_2) = \{0.5\} \neq \varnothing$;

Case II: $\rho_1 = 2h$, $\rho_2 = h$, where $I_{12} = \{0.5, 0.55\}$ and $I'_{12} = \varnothing$.

To compare the proposed algorithms with the conventional linearized PSOR algorithm, we use linearized PSOR algorithm to solve (2.7) with the same meshsize. Numerical results are listed in Table 1. From the table, we see that the optimal relaxation factor is about $\omega = 1.99$. Therefore, when we apply generalized Schwarz algorithms to solve the problem, the subproblems are solved by linearized PSOR algorithm, in which the relaxation factor is chosen as 1.99. Also, when we solve problem (2.7) by the generalized Schwarz algorithms (GASA1 and GMSA1) for above two cases, we select $\omega_1 = \omega_2 = \omega = 1/2$ and $\theta_1 = \theta_2 = \theta$. The numerical results are listed in Table 2. In the tables, we us "iter." to denote the iterative number, and "cpu" to denote the CPU time used.

For three subdomains, we also solve the problem for two different cases. Case I: The intersects of the subdomains are singletons. Case II: The intersects of the subdomains have two points. The numerical results are list in Table 3.

Table 1 Results of linearized PSOR algorithm

ω	iter.	cpu
1.00	391 699	338
1.05	363 518	328
1.15	312 029	278
1.25	265 904	212
1.35	224 059	201
1.45	185 634	158
1.55	149 911	139
1.65	116 248	109
1.75	83 990	79
1.85	52 310	47
1.90	19 510	20
1.99	4 339	4

Table 2 Results of two subdomains

Case	Case I				Case II			
algorithm	GASA1		GMSA1		GASA1		GMSA1	
θ	iter.	cpu	iter.	cpu	iter.	cpu	iter.	cpu
0.15	93	102	29	27	94	104	29	28
0.25	56	52	19	20	57	67	19	22
0.35	40	32	14	13	41	44	14	13
0.45	30	25	9	9	30	30	9	8
0.55	23	19	6	6	23	24	6	6
0.65	21	15	5	5	20	21	5	4
0.75	24	16	5	4	23	20	5	4
0.85	35	27	9	6	34	29	9	6
0.95	83	50	25	14	83	63	24	11
1.00	2272	864	1308	484	1310	443	528	161

Table 3 Results of three subdomains

Case	Case I				Case II			
algorithm	GASA2		GMSA2		GASA2		GMSA2	
θ	iter.	cpu	iter.	cpu	iter.	cpu	iter.	cpu
0.25	119	228	51	134	119	226	48	132
0.35	86	119	27	49	82	132	27	53
0.45	62	70	18	23	62	72	18	26
0.55	47	45	12	11	47	47	12	13
0.65	37	33	14	7	37	37	8	6
0.75	36	28	6	4	35	30	6	4
0.85	57	39	13	7	57	39	11	6
0.95	149	82	31	14	146	78	31	14
1.00	3663	2897	921	426	1454	421	518	225

Figures 1 and 2 reveal the relationships among the iteration errors, denoted by "pre.", CPU times and the parameter θ for the generalized additive and multiplicative Schwarz algorithms respectively. In the figures, GASA11 stands for generalized additive Schwarz algorithm for Case I, while GASA12 stands for generalized additive Schwarz algorithm for Case II. Similarly, GMSA11 and GMSA12 stand for generalized multiplicative Schwarz algorithm for Cases I and II respectively. Figures 3 and 4 show the variation of CPU time with different relaxation factors of generalized Schwarz algorithms with $\theta = 0.65$ and linearized PSOR. Figure 5 compares the performances among the generalized Schwarz algorithm with $\theta = 0.65$, classical Schwarz algorithm (that is the case of generalized Schwarz algorithm with $\theta = \theta_1 = \theta_2 = 1$, where A-Schwarz and M-Schwarz stand for additive and multiplicative Schwarz algorithm respectively) and linearized PSOR algorithm.

From the tables and figures we may see that

(1) With an appropriate choice of θ, the performances of the generalized Schwarz algorithms are much faster than the performance of the classical Schwarz algorithms.

(2) In the Case I, condition A does not hold. However, the generalized Schwarz algorithms still work well.

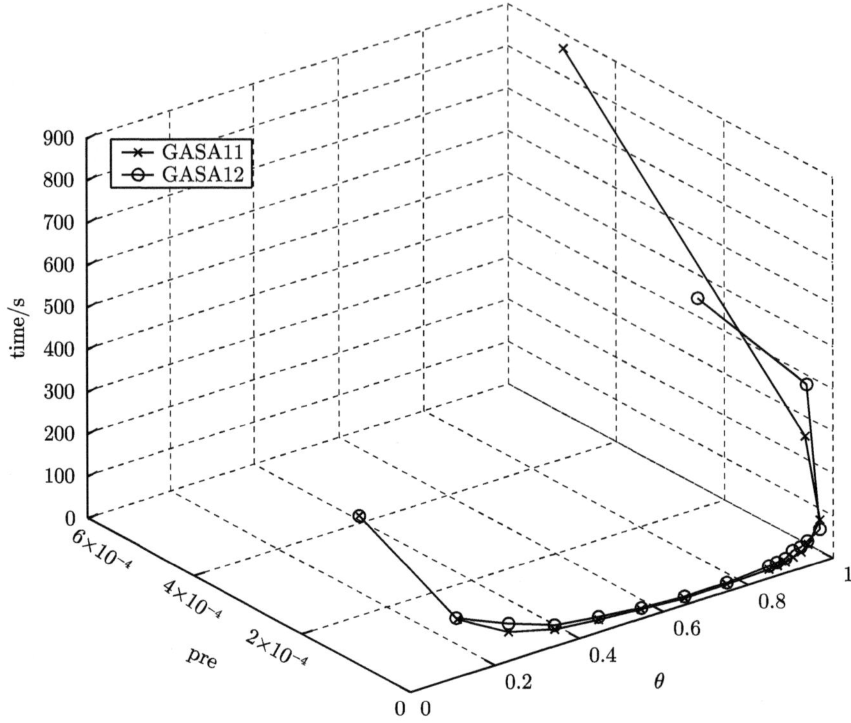

Figure 1　CPU times and iterative errors of GASA1 with different weights θ

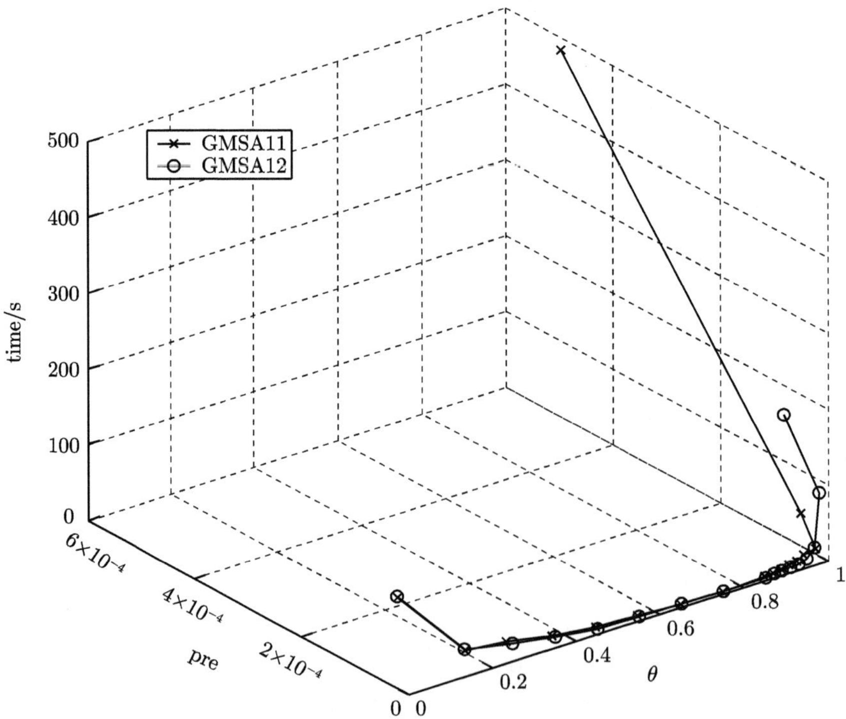

Figure 2　CPU times and iterative errors of GASA1 with different weights θ

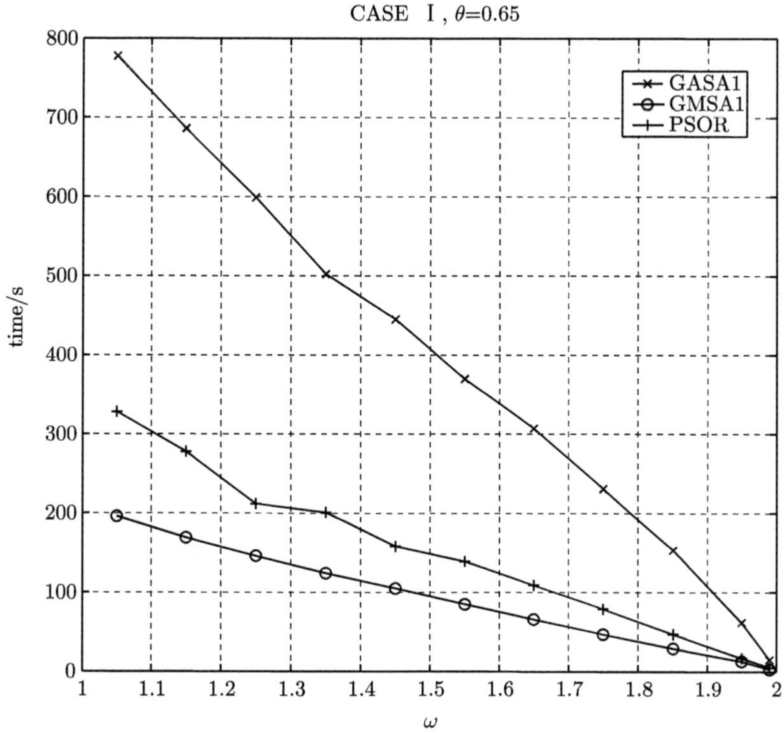

Figure 3　CPU times of GSA for case Ⅰ with different relaxation factors ω

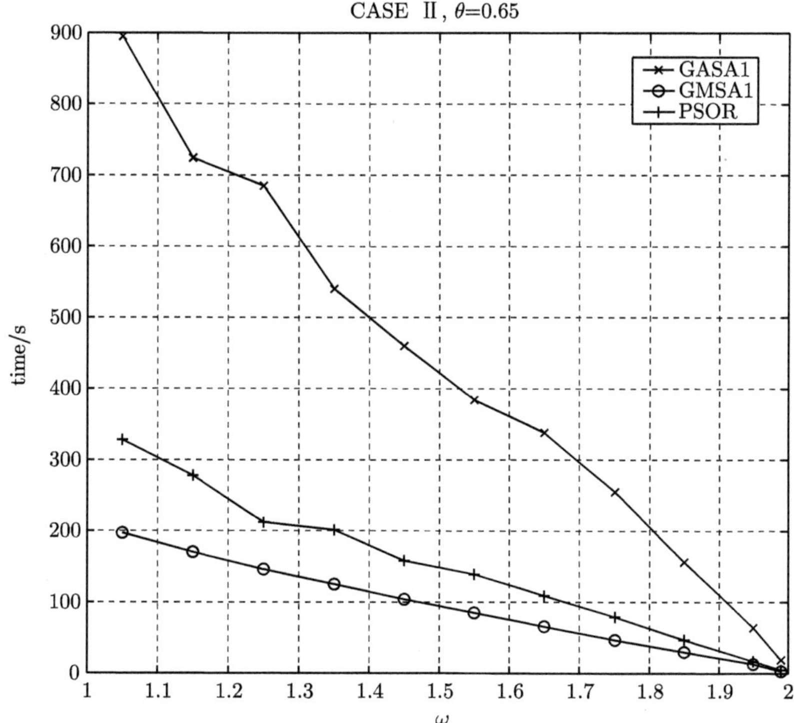

Figure 4　CPU times of GSA for case Ⅱ with different relaxation factors ω

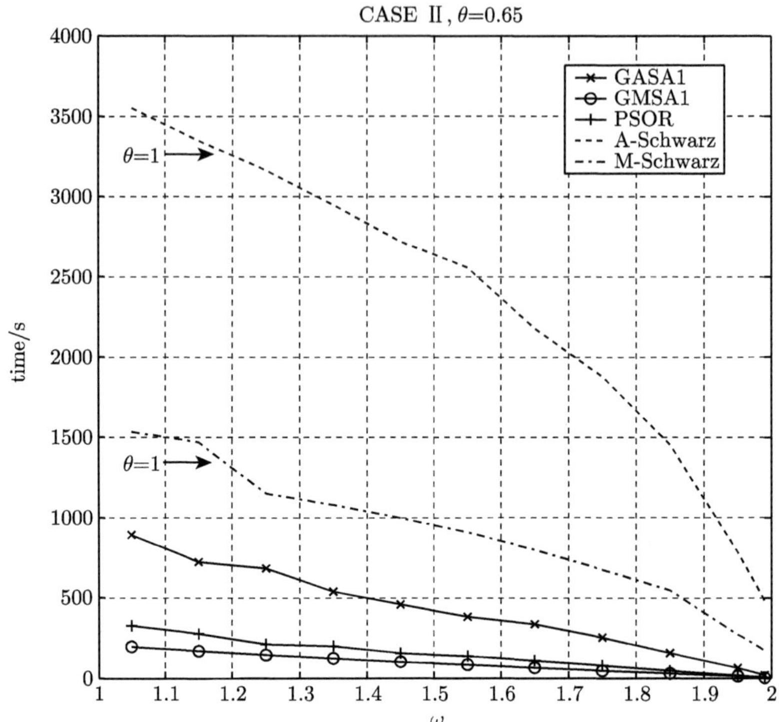

Figure 5 CPU times of generalized and classical Schwarz algorithms with different relaxation factors ω

(3) It takes much more CPU time for classical multiplicative and additive Schwarz algorithms than the conventional PSOR algorithm. Generalized Schwarz algorithms with an appropriate choice of the parameter becomes competitive to PSOR algorithm. However, it seems that the linearized PSOR algorithm works better than additive algorithm. As a result, it is necessary to compute in parallel.

We refer the reader to [22] for more numerical results.

Acknowledgments We thank the anonymous referee for his helpful comments and suggestions, which improved the paper.

References

[1] T. Lü, T. M. Shih and C. B. Liem, *Domain Decomposition Methods—New Numerical Technique for Solving PDE*, Science Press, Beijing, 1992.

[2] B. Smith, P. Bjørstad and W. Gropp, *Domain Decomposition*, Cambridge University Press, Cambridge, 1996.

[3] L. Badea, On the Schwarz alternating method with more than two subdomains for nonlinear monotone problems, *SIAM Journal on Numerical Analysis*, 28 (1991), 179-204.

[4] P. L. Lions, On the Schwarz alternating method I, in: R. Glowinski, G. H. Golub, G. A. Meurant, J. Périaux (Eds.), *Proceedings of Domain Decomposition Method*, SIAM, 1988, pp.1-40.

[5] P. L. Lions, On the Schwarz alternating method II, in: T. F. Chan, R. Glowinski, J. Périaux, O. B. Widlund (Eds.), *Proceedings of Domain Decomposition Method*, SIAM, 1989, pp.47-70.

[6] T. Lü, T. M. Shih and C. B. Liem, Parallel algorithms for variational inequalities based on domain decomposition, *Journal of Computational Mathematics*, 9(1991), 340-349.

[7] F. Scarpini, The alternative Schwarz method applied to some biharmonic variational inequalities, *Calcolo*, 27 (1990), 57-72.

[8] L. Badea and J. Wang, An additive Schwarz method for variational inequalities, *Mathematics of Computation*, 69 (1999), 1341-1354.

[9] Y. Kuznetsov, P. Neittaanmäki and P. Tarvainen, Schwarz methods for obstacle problems with convection-diffusion operators, in: D. E. Keyes and J. C. Xu (Eds.), *Proceedings of Domain Decomposition Methods in Scientifical and Engineering Computing*, AMS, 1995, pp.251-256.

[10] X.C. Tai, *Convergence rate analysis of domain decomposition methods for obstacle problems*, Report no. 124, 1999, University of Bergen.

[11] P. Tarvainen, On the convergence of block relaxation methods for algebraic obstacle problems with M-matrices, *East-West Journal of Numerical Mathematics*, 4 (1996) 69-82.

[12] J. P. Zeng, and S. Z. Zhou, On monotone and geometric convergence of Schwarz methods for two-side obstacle problems, *SIAM Journal on Numerical Analysis*, 35 (1998), 600-616.

[13] J.P. Zeng and S. Z. Zhou, A domain decomposition method for a kind of optimization problems, *Journal of Computational and Applied Mathematics*, 146 (2002), 127-139.

[14] R. Glowinski, J. L. Lions and R. Tremolieres, *Numerical Analysis of Variational Inequalities*, North-Holland, Amsterdam, 1981.

[15] C.M. Elliott and J. R. Ockendon, *Weak and variational methods for moving boundary problems*, Research Notes in Mathematics, no. 59, Pitman, London, 1982.

[16] G.H. Meyer, Free boundary problems with nonlinear source terms, *Numerische Mathematik*, 43 (1984), 463-482.

[17] R.H. Hoppe, Multigrid algorithms for variational inequalities, *SIAM Journal on Numerical Analysis*, 24 (1987), 1046-1065.

[18] K.H. Hoffmann and J. Zou, Parallel solution of variational inequality problems with nonlinear source terms, *IMA Journal of Numerical Analysis*, 16 (1996), 31-45.

[19] S.Z. Zhou and J.P. Zeng, Schwarz algorithm for obstacle problems with a type of nonlinear operators, *Acta Mathematicae Applicatae Sinica*, 20 (1997), 521-530.

[20] J.P. Zeng and S.Z. Zhou, Schwarz algorithm for the solution of variational inequalities with nonlinear source terms, *Applied Mathematics and Computation*, 97 (1998), 23-35.

[21] W. Tang, Generalized Schwarz splittings, *SIAM Journal on Scientific and Statistical Computing*, 13 (1992), 573-595.

[22] S.Z. Zhou, J.P. Zeng and X.M. Tang, Generalized Schwarz algorithm for obstacle problems, *Computers and Mathematics with Applications*, 38 (1999), 263-271.

[23] R.T. Rockafellar and J-B Wets, *Variational Analysis*, Springer, Berlin, 1998.

[24] U. Mosco, *An Introduction to Approximate Solution of Variational Inequalities*, Cremonese, Roma, 1973.

A New Minimization Protocol for Solving Nonlinear Poisson-Boltzmann Mortar Finite Element Equation*

D.X. Xie (谢德宣)† S.Z. Zhou (周叔子)‡

Abstract The nonlinear Poisson-Boltzmann equation (PBE) is a widely-used implicit solvent model in biomolecular simulations. This paper formulates a new PBE nonlinear algebraic system from a mortar finite element approximation, and proposes a new minimization protocol to solve it efficiently. In particular, the PBE mortar nonlinear algebraic system is proved to have a unique solution, and is equivalent to a unconstrained minimization problem. It is then solved as the unconstrained minimization problem by the subspace trust region Newton method. Numerical results show that the new minimization protocol is more efficient than the traditional merit least squares approach in solving the nonlinear system. At least 80 percent of the total CPU time was saved for a PBE model problem.

Keywords Poisson-Boltzmann equation, mortar finite element, nonlinear system, unconstrained minimization, biomolecular simulations.

1 Introduction

The Poisson-Boltzmann equation (PBE) is one widely-used implicit solvent model for electrostatic interactions of a biomolecular system in an ionic aqueous environment. It has been applied to many important applications in biology and chemistry [1–8]. In the Poisson-Boltzmann approach, all the solute atoms are described explicitly with point partial charges at atomic positions while the solute and solvent domains are modelled as the continuum regions with low and high dielectric constants, respectively. Several program packages such as DelPhi [9], UHBD [10] and APBS [11] have been developed for solving PBE using different numerical techniques (such as finite difference, and finite element methods) and typical linear and nonlinear iterative solvers (such as the successive over-relaxation method, the conjugate gradient method, the

* 发表于: BIT Numerical Mathematics, Vol. 47, 2007, pp. 853-871.

† Department of Mathematical Sciences, University of Wisconsin, Milwaukee, WI 53211, USA.

‡ Department of Applied Mathematics, Hunan University, Changsha 410082, China.

inexact-Newton method, and the multigrid method) [2, 11–14].

In this paper, we focus on the numerical solution of the PBE nonlinear algebraic system that arises from a mortar finite element discretization. The mortar finite element technique has been shown to be effective to treat the difficulties caused by discontinuous diffusion coefficients, corner point singularities, and singular source terms [15–18]. In the case of PBE, the domain is naturally decomposed into two subdomains—the solute and solvent domains, and the solute domain is surrounded completely by the solvent domain. The interface between them is the boundary of the solute domain, which may be very irregular since it consists of biomolecular surfaces. Also, PBE contains discontinuous coefficients and singular source terms. Hence, it is of particularly interest to discretize PBE by mortar finite element techniques. A mortar finite element application to the linear PBE problem was analyzed in [19]. In this paper, we intend to develop a new minimization protocol to solve the nonlinear PBE mortar finite element equation efficiently and effectively. Its error estimates will be studied in the future.

Currently, a mortar finite element equation can be formulated in either the constrained mortar space approach [20] or the saddle-point approach based on the unconstrained product space and a Lagrange multiplier space [16]. In this paper, we follow the constrained mortar space approach since it can lead to a nonlinear algebraic system with a positive definite Jacobian matrix, which is required in our new minimization protocol. According to the mortar finite element theory (see [20], for example), we set the solute and solvent regions as the slave and master subdomains, respectively, define the interface finite element space as the restriction of the finite element space of the solute region to the interface, and use a finer mesh size in the solute domain. In this paper, we obtain a mortar finite element equation to the nonlinear PBE in both a variational form and an algebraic form. We then prove that the nonlinear PBE mortar finite element equation has the unique solution and is equivalent to a unconstrained minimization problem. Consequently, the nonlinear PBE mortar finite element equation can be solved naturally as a unconstrained minimization problem.

In order to numerically confirm our theoretical results and demonstrate the performance of the new minimization protocol, we developed a MATLAB program package for solving a PBE model problem. The package was written based on the finite element program PUFFIN [21] and the PDE and optimization toolboxes of MATLAB [22]. While there exist several effective unconstrained optimization algorithms [23], we selected the subspace trust-region Newton method [24, 25]) for solving the PBE unconstrained minimization problem.The subspace trust-region Newton method is a widely-used global convergence algorithm, which is available in the MATLAB unconstrained minimization solver function *fminunc* as the algorithm for the large scale case.

As comparison, we also solved the nonlinear PBE mortar finite element equation by the MATLAB nonlinear system solver function *fsolve*. In fact, the algorithm in *fsolve* for the large scale case uses the same subspace trust-region Newton method as the one in *fminunc* except that it is applied to solving the merit least squares minimization problem: $\min\{\mathbf{f}(U) \mid U \in R^n\}$, where $\mathbf{f}(U) = \dfrac{1}{2}\sum_{i=1}^{n}[F_i(U)]^2$ is a widely-used merit function, and the nonlinear system to be solved is in the form $\mathcal{F}(U) = 0$ with $\mathcal{F}(U) = (F_1(U), F_2(U), \cdots, F_n(U))^t$. The merit function $\mathbf{f}(U)$ is also used in another traditional nonlinear solver—the damped inexact Newton method, which was well studied in [13] for solving the PBE nonlinear algebraic system that arises from an adaptive finite element approximation.

The main role of the merit function $\mathbf{f}(U)$ is to make globalization strategies (such as line search and trust region techniques) applicable in solving the nonlinear system such that a numerical solution can be generated from a Newton or inexact Newton method with any initial guess [23]. Without using a globalization strategy, the Newton or inexact Newton method may converge only if an initial guess is "close" enough to the solution. However, since a local minimizer of $\mathbf{f}(U)$ is not necessarily a root of $\mathcal{F}(U)$, some global convergence behaviors of a global convergent Newton or inexact Newton method may be degraded [23]. Moreover, in the merit least squares approach, the gradient vector g_k and Hessian matrix H_k have the forms: $g_k = J_k^t\mathcal{F}(U^{(k)})$ and $H_k = J_k^t J_k$, where J_k denotes the Jacobian matrix of $\mathcal{F}(U)$ at the kth iterate $U^{(k)}$ of the subspace trust-region Newton method. Hence, H_k may become dense even though J_k is sparse.Thus, it becomes expansive to solve the trust region subproblem (see (4.1) for details) by the preconditioned conjugate gradient method (PCG) [26].

In contrast, with our new minimization protocol, we have that $H_k = J_k$ and $g_k = \mathcal{F}(U^{(k)})$; thus, the trust region subproblem can be solved efficiently by PCG, and a numerical minimum point is guaranteed to be a numerical solution of the nonlinear system. Consequently, it can be claimed that the new minimization protocol is more effective and efficient than the traditional merit least squares approach. We confirmed this claim via our MATLAB package for solving the model problem. Numcrical results show that the new minimization protocol could save at least 80% of the total CPU time in solving the nonlinear system of PBE compared to the merit least squares approach. We also made numerical experiments using different initial guesses (many of them were selected randomly), confirming that the PBE mortar nonlinear system has a unique solution.

The remainder of the paper is organized as follows. Section 2 introduces and analyzes the PBE mortar finite element approximation and its equivalent functional minimization problem. Section 3 formulates their algebraic forms. Section 4 gives the new minimization protocol. Finally, numerical results are presented in Section 5.

2 PBE Mortar Finite Element Approximation

Let Ω be a bounded domain in R^3 with boundary $\partial\Omega$. It consists of three non-overlapping subdomains Ω_1, Ω_e, and Ω_I satisfying that $\Omega = \Omega_1 \cup \Omega_e \cup \Omega_I$, Ω_1 is surrounded by Ω_e, and Ω_e by Ω_I. We set $\Omega_2 = \Omega_e \cup \Omega_I$ and define PBE by

$$\begin{cases} -\nabla\cdot(\epsilon(x)\nabla u)+\kappa(x)\sinh u=f(x), & \text{in }\ \Omega, \\ u=g, & \text{on }\ \partial\Omega, \end{cases} \tag{2.1}$$

where g is a given function, $\epsilon(x)$ and $\kappa(x)$ are the two piecewise constant functions as given below:

$$\epsilon(x)=\begin{cases} \epsilon_1, & \text{for }\ x\in\ \Omega_1, \\ \epsilon_2, & \text{for }\ x\in\ \Omega_2, \end{cases} \qquad \kappa(x)=\begin{cases} 0, & \text{for }\ x\in\Omega_1\cup\Omega_e, \\ \overline{\kappa}, & \text{for }\ x\in\Omega_I, \end{cases} \tag{2.2}$$

and the right hand side function $f(x)$ is given by

$$f(x)=\begin{cases} \bar{c}\sum_{i=1}^{n_c} q_i\delta(x-x^i), & \text{for }\ x\in\Omega_1, \\ 0, & \text{for }\ x\in\Omega_2. \end{cases} \tag{2.3}$$

Here $x^i\in\Omega_1$ is the position vector of atom i of a biomolecule (e.g., a protein) in Ω_1, q_i is the charge at x^i, n_c is the total number of point charges of the protein, $\bar{c}=4\pi e_c/(k_BT)$ with e_c as an electron charge constant, k_B Boltzmann's constant, and T the absolute temperature, and $\delta(x)$ denotes the Delta-function, which is defined as 1 at the origin and zero otherwise. According to the Debye-Hückel theory of continuum molecular electrostatics [27], a solution u of PBE gives an electrostatic potential in the region of $\Omega, \epsilon(x)$ indicates the permittivity in both the solute region Ω_1 and the solvent region Ω_2, and κ describes the distribution of ions in region Ω. Here Ω_e is the ionic exclusive region while Ω_I the ionic region. For simplicity, we set $g=0$ in this paper. The case of nonzero g can be easily treated in the way as shown in [28].

Let $v|_S$ denote the restriction of v onto region S, and Γ the interface between Ω_1 and Ω_2.To discretize PBE by a mortar technique, we define two independent conforming finite element spaces, V_{Ω_1} and V_{Ω_2}, based on two independent conforming triangulations $\mathcal{T}_{1,h_1}$ and $\mathcal{T}_{2,h_2}$ of Ω_1 and Ω_2, respectively, where h_i denotes the mesh size of Ω_i for $i=1$ and 2. The finite element space on the interface, denoted by Λ_{h_1}, is defined as the restriction of V_{Ω_1} to the interface Γ. The mortar condition (or a week matching condition across the interface) is assumed as below:

$$\int_\Gamma (v|_{\Omega_1}-v|_{\Omega_2})w\mathrm{d}s=0, \quad \forall w\in\Lambda_{h_1}. \tag{2.4}$$

According to the mortar finite element theory (see [20], for example), we set $h_1<h_2$ and refer to Ω_1 and Ω_2 as the slave and master subdomain, respectively, since the coefficient ϵ_2 on Ω_2 is larger than the coefficient ϵ_1 on Ω_1.

Let V_h be the product function space of V_{Ω_1} and V_{Ω_2}. The constrained mortar finite element space, denoted by $\tilde{V}_h$, is defined as a subspace of V_h of functions that satisfy the mortar condition (2.4). Here $h=\max\{h_1,h_2\}$. We then can obtain the mortar finite element equation of PBE as follows:

Find $u_h \in \tilde{V}_h$ such that

$$a(u_h, v) + \int_\Omega \kappa(x) v \sinh u_h \mathrm{d}x = \int_\Omega f v \mathrm{d}x, \quad \forall v \in \tilde{V}_h, \tag{2.5}$$

where $a(u, v)$ is a symmetric bilinear functional defined by

$$a(u, v) = \int_{\Omega_1} \epsilon_1 \nabla u \cdot \nabla v \mathrm{d}x + \int_{\Omega_2} \epsilon_2 \nabla u \cdot \nabla v \mathrm{d}x. \tag{2.6}$$

From [20] it can imply that there exists a positive constant α such that

$$a(v, v) \geqslant \alpha ||v||^2_{\tilde{V}_h}, \quad \forall v \in \tilde{V}_h, \tag{2.7}$$

where $||\cdot||^2_{\tilde{V}_h}$ denotes a norm in V_h. Hence, the corresponding stiffness matrix obtained from a set of basis functions of $\tilde{V}_h$ is symmetric positive definite. In particular, we obtain a new proof to directly prove the following theorem with respect to the case of PBE.

Theorem 2.1. *Let $a(u, v)$ be the bilinear functional defined in (2.6). Then*

$$a(v, v) > 0, \quad \forall v \in \tilde{V}_h \text{ and } v \neq 0.$$

Proof. Suppose that there exists a nonzero $v^0 \in \tilde{V}_h$ such that $a(v^0, v^0) = 0$. For clarity, we denote by $v^{0,1}$ and $v^{0,2}$ the restrictions of v^0 onto the closures of Ω_1 and Ω_2, respectively. Since $v^0 \in \tilde{V}_h$, we have that $v^0 = 0$ on $\partial\Omega$, and the mortar condition (2.4) holds, i.e.,

$$\int_\Gamma (v^{0,1} - v^{0,2}) w \mathrm{d}s = 0, \quad \forall w \in \Lambda_{h_1}. \tag{2.8}$$

Since the two terms of $a(v^0, v^0)$ in (2.6) are nonnegative, they must be zero when $a(v^0, v^0) = 0$. That is,

$$\int_{\Omega_1} \epsilon_1 \nabla v^{0,1} \cdot \nabla v^{0,1} \mathrm{d}x = 0, \quad \text{and} \quad \int_{\Omega_2} \epsilon_2 \nabla v^{0,2} \cdot \nabla v^{0,2} \mathrm{d}x = 0. \tag{2.9}$$

Since $v^{0,2} \in H^1(\Omega_2)$, and $v^{0,2} = 0$ on $\partial\Omega$, from [28] we know that $v^{0,2}$ satisfies the inequality

$$\int_{\Omega_2} \epsilon_2 \nabla v^{0,2} \cdot \nabla v^{0,2} \mathrm{d}x \geqslant \gamma ||v^{0,2}||^2_{H^1(\Omega_2)}, \tag{2.10}$$

where γ is a positive constant. Hence, applying the second expression of (2.9) to the above inequality yields $v^{0,2} = 0$.

As a result, the mortar condition (2.8) becomes

$$\int_\Gamma v^{0,1} w \mathrm{d}s = 0, \quad \forall w \in \Lambda_{h_1},$$

and from which we get that $v^{0,1} = 0$ on Γ. Note that Γ is exactly the boundary $\partial\Omega_1$ of Ω_1. Hence, $v^{0,1} \in H_0^1(\Omega_1)$. Thus, from [28] we can know that there exists a positive constant, β, such that

$$\int_{\Omega_1} \epsilon_1 \nabla v^{0,1} \cdot \nabla v^{0,1} \mathrm{d}x \geqslant \beta ||v^{0,1}||^2_{H^1(\Omega_1)}. \tag{2.11}$$

Therefore, applying the first expression of (2.9) to the above inequality immediately yields $v^{0,1} = 0$.

Since both $v^{0,1}$ and $v^{0,2}$ are zero, we get that $v^0 = 0$. This is a contradiction to the assumption that $v^0 \neq 0$. Therefore, $a(v, v)$ must be positive for all nonzero $v \in \tilde{V}_h$. □

In general, for a nonlinear problem given in the form

$$\text{Find } u \in H \text{ such that } \quad (B(u), v) = 0, \quad \forall v \in H,$$

where H is a Hilbert space, it is difficult to find a functional, $\mathcal{J}(v)$, such that its G-derivative $\mathcal{J}'(u)$ satisfies

$$(\mathcal{J}'(u), v) = (B(u), v), \quad \forall v \in H.$$

However, in the case of PBE, we find such a functional and obtain the following theorem.

Theorem 2.2. *Let $\mathcal{J}(v)$ be a functional defined by*

$$\mathcal{J}(v) = \frac{1}{2}a(v, v) + \int_\Omega \kappa(x) \cosh v \mathrm{d}x - \int_\Omega f v \mathrm{d}x. \tag{2.12}$$

Then, the minimization problem

$$\textit{Find } u \in \tilde{V}_h \textit{ such that } \quad \mathcal{J}(u) = \min\{\mathcal{J}(v) \mid v \in \tilde{V}_h\} \tag{2.13}$$

has a unique solution. Moreover, it is equivalent to the PBE mortar finite element equation (2.5).

Proof. Clearly, $\mathcal{J}(v)$ is Gâteaux differentiable. Its first and second G-derivatives at u, $\mathcal{J}'(u)$ and $\mathcal{J}''(u)$, can be found as below:

$$(\mathcal{J}'(u), v) = a(u, v) + \int_\Omega \kappa(x) v \sinh u \mathrm{d}x - \int_\Omega f v \mathrm{d}x, \quad \forall v \in V_h, \tag{2.14}$$

and

$$(\mathcal{J}''(u)v, v) = a(v, v) + \int_\Omega \kappa(x) v^2 \cosh u \mathrm{d}x, \quad \forall v \in V_h. \tag{2.15}$$

From the above expression, together with Theorem 2.1 and the definition of $\kappa(x)$ in (2.2), it is easy to see that $(\mathcal{J}''(u)v, v) > 0$ for all nonzero $v \in \tilde{V}_h$. Thus, $\mathcal{J}(v)$ is a strictly convex functional in $\tilde{V}_h$.

Furthermore, since the delta function $\delta(x - x^i)$ is a linear functional in the sense that $\int_\Omega \delta(x - x^i) v \mathrm{d}x = v(x^i)$ for all $v \in V_h$, the function f defined in (2.3) can be regarded as a linear functional of $\tilde{V}_h$, and satisfy that

$$|(f, v)| \leqslant ||f||_{\tilde{V}_h^*} ||v||_{\tilde{V}_h}, \quad \forall v \in \tilde{V}_h, \tag{2.16}$$

where $\tilde{V}_h^*$ denotes the dual space of $\tilde{V}_h$. Here $||f||_{\tilde{V}_h^*}$ is bounded since $\tilde{V}_h$ is finite dimensional. Hence, a combination of (2.16) and (2.7) with the nonnegativity of $\kappa(x)$ immediately gives that

$$\mathcal{J}(v) \geqslant a(v,v) - (f,v) \geqslant \alpha||v||_{\tilde{V}_h}^2 - ||f||_{\tilde{V}_h^*}||v||_{\tilde{V}_h}, \quad \forall v \in \tilde{V}_h.$$

From the above inequality it implies that $\mathcal{J}(v) \to \infty$ as $||v||_{\tilde{V}_h} \to \infty$. Thus, $\mathcal{J}(v)$ is coercive. Therefore, according to [29, Proposition 1.2], we claim that the unconstrained minimization problem (2.13) has a unique solution.

Finally, the equivalence between (2.5) and (2.13) is followed from (2.14) [30]. □

3 Formulations of Nonlinear Algebraic Problems

In this section, we formulate the algebraic expressions of the PBE mortar finite element equation (2.5), the minimization problem (2.13), and the mortar condition (2.4). In addition, some details on computing the source and nonlinear terms of the PBE mortar nonlinear algebraic system are discussed.

Let $\{\varphi_j\}_{j=1}^N$ be a set of basis functions of the product space V_h based on the N mesh nodes, x_j for $j = 1, 2, \cdots, N$, of the triangulations $\mathcal{T}_{1,h_1}$ and $\mathcal{T}_{2,h_2}$ (except of the mesh points lying on the boundary $\partial\Omega$). Then, each function u of V_h can be expressed in the form $u = \sum_{j=1}^N u_j\varphi_j$ such that the finite element equation of (2.5) within V_h can be expressed as a system of nonlinear algebraic equations:

$$\sum_{i=1}^N a(\varphi_i, \varphi_j)u_i + s_j(u_1, u_2, \cdots, u_N) = f_j, \quad j = 1, 2, \cdots, N, \tag{3.1}$$

where u_j is a numerical approximation of $u(x_j)$, $s_j(u_1, u_2, \cdots, u_N)$ is a nonlinear function of the unknowns $u_1, u_2, \cdots$, and u_N, which is defined by

$$s_j(u_1, u_2, \cdots, u_N) = \int_\Omega \kappa(x)\varphi_j \sinh\left(\sum_{i=1}^N u_i\varphi_i\right) \mathrm{d}x,$$

and $f_j = \displaystyle\int_\Omega f\varphi_j \mathrm{d}x$. Clearly, the nonlinear function s_j can be simplified as

$$s_j = \int_{\tau^j} \kappa(x)\varphi_j \sinh\left(\sum_{i=1}^N u_i\varphi_i\right) \mathrm{d}x,$$

where τ^j denotes the support set of φ_j, and can be evaluated approximately by a numerical quadrature. For example, $s_j \approx \kappa(x_j)|\tau^j|\sinh(u_j)$, where $|\tau^j|$ denotes the size of τ^j. From the definition of delta function we can get the following formula for evaluating f_j:

$$f_j = \begin{cases} \bar{c}\displaystyle\sum_{i=1}^{n_c} q_i\varphi_j(x^i), & \text{if } \varphi_j \in V_{\Omega_1}; \\ 0, & \text{otherwise}, \end{cases}$$

where $j = 1, 2, \ldots, N$, and x^i is the position vector of atom i of the protein.

To write the nonlinear system (3.1) in a 4-block matrix form, we assume that the N mesh nodes are labelled globally in the following ordering: first on the nodes in Ω_1, then on Γ_{Ω_1}, next in $\tilde{\Omega}_2$, and finally on Γ_{Ω_2}. We then construct four subspaces of V_h as follows:

$$\mathcal{V}_{\Omega_1} = \text{Span}\ \{\varphi_j | x_j \in \Omega_{1,h_1}\}, \quad \mathcal{V}_{\Gamma_{\Omega_1}} = \text{Span}\ \{\varphi_j | x_j \in \Gamma_{\Omega_1,h_1}\}, \tag{3.2}$$

$$\mathcal{V}_{\Omega_2} = \text{Span}\ \{\varphi_j | x_j \in \Omega_{2,h_2}\}, \quad \text{and} \quad \mathcal{V}_{\Gamma_{\Omega_2}} = \text{Span}\ \{\varphi_j | x_j \in \Gamma_{\Omega_2,h_2}\}. \tag{3.3}$$

For convenience, we rewrite their basis functions as $\{\varphi_j\}_{j=1}^{n_1}, \{\tilde{\varphi}_j\}_{j=1}^{n_2}, \{\hat{\varphi}_j\}_{j=1}^{l}$, and $\{\overline{\varphi}_j\}_{j=1}^{m}$, where n_1, n_2, l and m are the dimensions of $\mathcal{V}_{\Omega_1}, \mathcal{V}_{\Omega_2}, \mathcal{V}_{\Gamma_{\Omega_1}}$ and $\mathcal{V}_{\Gamma_{\Omega_2}}$ respectively.Thus, the restrictions of a function v of V_h onto these four subspaces, $v|_{\Omega_1}, v|_{\Omega_2}, v|_{\Gamma_{\Omega_1}}$, and $v|_{\Gamma_{\Omega_2}}$ can be expressed as

$$v|_{\Omega_1} = \sum_{j=1}^{n_1} v_j \varphi_j,\ v|_{\Omega_2} = \sum_{j=1}^{n_2} \tilde{v}_j \tilde{\varphi}_j,\ v|_{\Gamma_{\Omega_1}} = \sum_{j=1}^{l} \hat{v}_j \hat{\varphi}_j,\ v|_{\Gamma_{\Omega_2}} = \sum_{j=1}^{m} \overline{v}_j \overline{\varphi}_j. \tag{3.4}$$

Hence, the corresponding vector V of function v can have the 4-block form

$$V = \begin{pmatrix} v_{\Omega_1} \\ v_{\Gamma_{\Omega_1}} \\ v_{\Omega_2} \\ v_{\Gamma_{\Omega_2}} \end{pmatrix},$$

where

$$v_{\Omega_1} = \begin{pmatrix} v_1 \\ v_2 \\ \vdots \\ v_{n_1} \end{pmatrix}, \quad v_{\Gamma_{\Omega_1}} = \begin{pmatrix} \hat{v}_1 \\ \hat{v}_2 \\ \vdots \\ \hat{v}_l \end{pmatrix}, \quad v_{\Omega_2} = \begin{pmatrix} \tilde{v}_1 \\ \tilde{v}_2 \\ \vdots \\ \tilde{v}_{n_2} \end{pmatrix}, \quad v_{\Gamma_{\Omega_2}} = \begin{pmatrix} \overline{v}_1 \\ \overline{v}_2 \\ \vdots \\ \overline{v}_m \end{pmatrix}.$$

Using the above block vector partition, we can express the nonlinear system (3.1) in the 4-block form

$$\mathcal{F}(U) = 0 \quad \text{with} \quad \mathcal{F}(U) = AU + S(U) - F, \tag{3.5}$$

where $U, S(U)$, and F are the three column vectors with u_j, s_j, and f_j as the jth entries, respectively, for $j = 1, 2, \ldots, N$, A is a symmetric matrix of $N \times N$ with the (i, j)th entry being $a(\varphi_i, \varphi_j)$ for $i, j = 1, 2, \ldots, N$, and their block forms are given by

$$U = \begin{pmatrix} U_{\Omega_1} \\ U_{\Gamma_{\Omega_1}} \\ U_{\Omega_2} \\ U_{\Gamma_{\Omega_2}} \end{pmatrix}, \quad F = \begin{pmatrix} f_{\Omega_1} \\ f_{\Gamma_{\Omega_1}} \\ f_{\Omega_2} \\ f_{\Gamma_{\Omega_2}} \end{pmatrix}, \quad S(V) = \begin{pmatrix} S_{\Omega_1} \\ S_{\Gamma_{\Omega_1}} \\ S_{\Omega_2} \\ S_{\Gamma_{\Omega_2}} \end{pmatrix},$$

$$A = \begin{pmatrix} A_{\Omega_1} & A_{\Omega_1,\Gamma_{\Omega_1}} & 0 & 0 \\ A^t_{\Omega_1,\Gamma_{\Omega_1}} & A_{\Gamma_{\Omega_1}} & 0 & 0 \\ 0 & 0 & A_{\Omega_2} & A_{\Omega_2,\Gamma_{\Omega_2}} \\ 0 & 0 & A^t_{\Omega_2,\Gamma_{\Omega_2}} & A_{\Gamma_{\Omega_2}} \end{pmatrix}.$$

The following lemma gives the algebraic form of the mortar condition (2.4).

Lemma 3.1. *Let $\{\psi_j\}_{j=1}^l$ be a set of basis functions of the interface finite element space Λ_{h_1}. Then, the mortar condition (2.4) can be expressed in the matrix form*

$$Mv_{\Gamma_{\Omega_1}} - Wv_{\Gamma_{\Omega_2}} = 0, \tag{3.6}$$

where M and W are the two matrices of $l \times l$ and $l \times m$, respectively, with entries m_{ji} and w_{jk} being defined by

$$m_{ij} = \int_\Gamma \psi_i \hat{\varphi}_j \mathrm{d}s \quad \text{and} \quad w_{ik} = \int_\Gamma \psi_i \overline{\varphi}_k \mathrm{d}s \tag{3.7}$$

for $i, j = 1, 2, \cdots, l$ and $k = 1, 2, \cdots, m$. Moreover, M is nonsingular.

Proof. By using the expressions of $v|_{\Gamma_{\Omega_1}}$ and $v|_{\Gamma_{\Omega_2}}$ in (3.4), the bilinear functional in (2.4) with $v \in V_h$ can be written as

$$\int_\Gamma (v|_{\Gamma_{\Omega_1}} - v|_{\Gamma_{\Omega_2}}) w \mathrm{d}s = \int_\Gamma \left(\sum_{i=1}^l \hat{v}_i \hat{\varphi}_i - \sum_{k=l}^m \overline{v}_k \overline{\varphi}_k \right) w \mathrm{d}s, \quad \forall w \in \Lambda_{h_1}.$$

Selecting $w = \psi_j$ and using the expressions of m_{ij} and w_{kj} in (3.7) give that

$$\begin{aligned} &\sum_{i=1}^l \hat{v}_i \int_\Gamma \hat{\varphi}_i \psi_j \mathrm{d}s - \sum_{k=l}^m \overline{v}_k \int_\Gamma \overline{\varphi}_k \psi_j \mathrm{d}s \\ =& \sum_{i=1}^l m_{ji} \hat{v}_i - \sum_{k=l}^m w_{jk} \overline{v}_k \end{aligned}$$

for $j = 1, 2, \cdots, l$. Clearly, the above expressions can be written as the matrix form $Mv_{\Gamma_{\Omega_1}} - Wv_{\Gamma_{\Omega_2}}$. Thus, the mortar condition (2.4) is expressed in the matrix form (3.6).

We next prove the non-singularity of matrix M. Let M_j denote the jth row of M. We consider their linear combination of zero:$\sum_{j=1}^l c_j M_j = 0$, which is equivalent to $\sum_{j=1}^l c_j m_{ji} = 0$ for $i = 1, 2, \cdots, l$, or equivalently,

$$\int_\Gamma \hat{\varphi}_i \sum_{j=1}^l c_j \psi_j \mathrm{d}s = 0 \quad \text{for } i = 1, 2, \cdots, l.$$

Here the definition of w_{ji} in (3.7) has been used. Multiplying $\hat{v}_i$ to the both sides of the above equations and adding them together for all i immediately give

$$\int_\Gamma v \sum_{j=1}^l c_j \psi_j \mathrm{d}s = 0, \quad \forall v \in V_h,$$

where the fact $v|_{\Gamma_{\Omega_1}} = \sum_{i=1}^{l} \hat{v}_i \hat{\varphi}_i$ has been used. Hence, from the above equation it implies $\sum_{j=1}^{l} c_j \psi_j \mathrm{d}s = 0$. Since $\{\psi_j\}_{j=1}^{l}$ is a set of basis functions, all c_j must be zero. This shows that the matrix M has a rank of l. Hence, M is nonsingular. □

Corresponding to Theorem 2.2, we obtain the algebraic forms of (2.5) and (2.13) and their equivalence in the following theorem.

Theorem 3.2. *Let R^n be the Euclidean vector space with $n = N - l$, and $T = M^{-1}W$ with the matrices M and W being defined in (3.7). Then, the PBE mortar finite element equation (2.5) and the corresponding variational problem (2.13) can be formulated, respectively, as the nonlinear algebraic system (3.8) and the unconstrained minimization problem (3.9) as given below:*

$$\tilde{\mathcal{F}}(\tilde{U}) = 0 \quad with \quad \tilde{\mathcal{F}}(\tilde{U}) = \tilde{A}\tilde{U} + \tilde{S}(\tilde{U}) - \tilde{F} \in R^n, \tag{3.8}$$

and

$$Find \quad \tilde{U} \in R^n \quad such\ that \quad \tilde{J}(\tilde{U}) = \min\{\tilde{J}(\tilde{V}) \mid \tilde{V} \in R^n\}, \tag{3.9}$$

where $\tilde{V} = (\tilde{v}_1, \tilde{v}_2, \cdots, \tilde{v}_n)^t$, $\tilde{J}(\tilde{V})$ is defined by

$$\tilde{J}(\tilde{V}) = \frac{1}{2}\tilde{V}^t\tilde{A}\tilde{V} + \int_{\Omega_1} \kappa(x)\cosh\left(\sum_{x_i\in\Omega_I} \tilde{v}_i\varphi_i\right)\mathrm{d}x - \tilde{F}^t\tilde{V}, \tag{3.10}$$

and $\tilde{U}, \tilde{A}, \tilde{F}$, and $\tilde{S}(\tilde{U})$ are defined by

$$\tilde{U} = \begin{pmatrix} u_{\Omega_1} \\ u_{\Omega_2} \\ u_{\Gamma_{\Omega_2}} \end{pmatrix}, \quad \tilde{A} = \begin{pmatrix} A_{\Omega_1} & 0 & A_{\Omega_1,\Gamma_{\Omega_1}}T \\ 0 & A_{\Omega_2} & A_{\Omega_2,\Gamma_{\Omega_2}} \\ T^tA^t_{\Omega_1,\Gamma_{\Omega_1}} & A^t_{\Omega_2,\Gamma_{\Omega_2}} & T^tA_{\Gamma_{\Omega_1}}T + A_{\Gamma_{\Omega_2}} \end{pmatrix}, \tag{3.11}$$

$$\tilde{F} = \begin{pmatrix} f_{\Omega_1} \\ 0 \\ T^tf_{\Gamma_{\Omega_1}} + f_{\Gamma_{\Omega_2}} \end{pmatrix}, \quad and \quad \tilde{S}(\tilde{U}) = \begin{pmatrix} 0 \\ S_{\Omega_2}(\bar{B}\tilde{U}) \\ 0 \end{pmatrix}. \tag{3.12}$$

Moreover, both (3.8) and (3.9) have the unique solution and are equivalent.

Proof. Since the inverse M^{-1} exists, equation (3.6) can be rewritten as

$$u_{\Gamma_{\Omega_1}} = Tu_{\Gamma_{\Omega_2}} \quad with \quad T = M^{-1}W. \tag{3.13}$$

Thus, we can set the transform matrix $\overline{B}$ such that $U = \overline{B}\tilde{U}$, where

$$U = \begin{pmatrix} u_{\Omega_1} \\ u_{\Gamma_{\Omega_1}} \\ u_{\Omega_2} \\ u_{\Gamma_{\Omega_2}} \end{pmatrix}, \quad \text{and} \quad \overline{B} = \begin{pmatrix} I & 0 & 0 \\ 0 & 0 & T \\ 0 & I & 0 \\ 0 & 0 & I \end{pmatrix}.$$

Multiplying the transpose of $\bar{B}, \bar{B}^t$, on the both sides of (3.5) and replacing U by $\overline{B}\tilde{U}$ gives the expression of $\tilde{\mathcal{F}}(\tilde{U})$ with $\tilde{A} = \bar{B}^tA\overline{B}, \tilde{S}(\tilde{U}) = \bar{B}^tS(\overline{B}\tilde{U})$, and $\tilde{F} = \bar{B}^tF$.By matrix-vector operations, we then immediately obtain the block form of $\tilde{A}$ in (3.11), and the block forms of $\tilde{F}$ and $\tilde{S}(\tilde{U})$ as below:

$$\tilde{F} = \begin{pmatrix} f_{\Omega_1} \\ f_{\Omega_2} \\ T^t f_{\Gamma_{\Omega_1}} + f_{\Gamma_{\Omega_2}} \end{pmatrix}, \quad \text{and} \quad \tilde{S}(\tilde{U}) = \begin{pmatrix} S_{\Omega_1}(\overline{B}\tilde{U}) \\ S_{\Omega_2}(\overline{B}\tilde{U}) \\ T^t S_{\Gamma_{\Omega_1}}(\overline{B}\tilde{U}) + S_{\Gamma_{\Omega_2}}(\overline{B}\tilde{U}) \end{pmatrix}.$$

The above forms are then simplified to the forms of (3.12) since from the definitions of $\kappa(x)$ and $f(x)$ in (2.2) and (2.3) it is easy to see that $f_{\Omega_2} = 0$ and $S_{\Omega_1}(U) = S_{\Gamma_{\Omega_1}}(U) = S_{\Gamma_{\Omega_2}}(U) = 0$.

Similarly, we can obtain (3.9).Clearly, $\tilde{J}(\tilde{V})$ is a twice continuously differentiable function of n real variables $\tilde{v}_j$ for $j = 1, 2, \cdots, n$. Its gradient vector $\nabla \tilde{J}$ and Hessian matrix $\nabla^2 \tilde{J}$ can be easily found as below:

$$\nabla \tilde{J}(\tilde{V}) = \tilde{\mathcal{F}}(\tilde{V}), \quad \text{and} \quad \nabla^2 \tilde{J}(\tilde{V}) = \nabla \tilde{\mathcal{F}}(\tilde{V}), \tag{3.14}$$

where $\tilde{\mathcal{F}}(\tilde{V})$ is given in (3.8), and $\nabla \tilde{\mathcal{F}}(\tilde{V})$ is the Jacobian matrix of $\tilde{\mathcal{F}}(\tilde{V})$. Hence, from the above identities it follows the equivalence between (3.8) and (3.9). □

The formulation of the PBE mortar nonlinear system (3.8) depends on the transform matrix T of (3.13). To reduce the computing cost, the basis functions can be selected based on the dual Lagrange multiplier approach [18] to satisfy the bi-orthogonality relation $m_{ii} = \int_{\Gamma} \hat{\varphi}_i \mathrm{d}s$ and $m_{ij} = 0$ if $i \neq j$, where $i, j = 1, 2, \cdots, l$. In this way, T becomes a sparse scaled mass matrix.

The Jacobian matrix $\nabla \tilde{\mathcal{F}}(\tilde{V})$ can be found as below:

$$\nabla \tilde{\mathcal{F}}(\tilde{V}) = \tilde{A} + \nabla \tilde{S}(\tilde{V}), \tag{3.15}$$

where $\nabla \tilde{S}(\tilde{V})$ is a symmetric matrix of $N \times N$ with entry s_{ij} being defined by

$$s_{ij} = \int_{\Omega_I} \kappa(x) \varphi_i \varphi_j \cosh\left(\sum_{x_i \in \Omega_I} \tilde{v}_k \varphi_k \right) \mathrm{d}x \quad \text{for } i, j = 1, 2, \cdots, N. \tag{3.16}$$

From the definition of $\kappa(x)$ in (2.2) we see that the above s_{ij} is nonzero only if the corresponding mesh nodes x_i and x_j of basis functions φ_i and φ_j are located in the ionic region Ω_I.In particular, if s_j is approximated by the numerical quadrature, $s_j \approx \kappa(x_j)|\tau^j| \sinh(\tilde{v}_j)$, $\nabla \tilde{S}(\tilde{V})$ is reduced to the diagonal matrix with the nonzero diagonal entries $s_{jj} = \kappa(x_j)|\tau^j| \cosh(\tilde{v}_j)$ for $j = 1, 2, \cdots, n$.

4 New Minimization Protocol for Solving PBE

Based on Theorem 3.2, we propose to solve the nonlinear system (3.8) through solving the unconstrained minimization problem (3.9) by a widely-used global convergent optimization algorithm, the subspace trust-region Newton method [24, 25, 31–33]. By this new minimization protocol, a numerical solution of (3.8) can be obtained efficiently and effectively with any initial guess.

The subspace trust region Newton iterates $\{\tilde{U}^{(k)}\}$ for solving (3.9) is defined in the form

$$\tilde{U}^{(k+1)} = \tilde{U}^{(k)} + \mathbf{s}^k, \quad k = 0, 1, 2, \cdots,$$

where $\tilde{U}^{(0)}$ is a given initial guess, and $\mathbf{s}^k$ denotes the kth trial step vector, which satisfies that $\tilde{J}(\tilde{U}^{(k)} + \mathbf{s}^k) < \tilde{J}(\tilde{U}^{(k)})$, and is selected by PCG for solving the trust region subproblem:

$$\min\left\{\frac{1}{2}\mathbf{s}^t H_k \mathbf{s} + \mathbf{s}^t \mathbf{g}_k + \mathbf{f}_k \mid \mathbf{s} \in \mathcal{N}\right\}, \tag{4.1}$$

where $\mathbf{f}_k = \tilde{J}(\tilde{U}^{(k)})$, $\mathbf{g}_k = \nabla \tilde{J}(\tilde{U}^{(k)})$, $H_k = \nabla^2 \tilde{J}(\tilde{U}^{(k)})$, and $\mathcal{N}$ is a trust region. In the subspace trust region Newton method, $\mathcal{N}$ is defined as below:

$$\mathcal{N} = \{\mathbf{s} \in \text{Span}\ \{\mathbf{g}_k, \mathbf{u}_k\} \mid ||\mathbf{s}|| \leqslant \delta_k\},$$

where $\delta_k > 0$ is the trust region radius, $||\cdot||$ denotes the Euclidean norm, and $\mathbf{u}_k$ is the jth iterate $s^{(j)}$ of PCG if it is well defined and satisfies the termination rule $||\mathbf{g}_k + H_k s^{(j)}|| < \text{TolPCG}||\mathbf{g}_k||$; otherwise, $\mathbf{u}_k$ is a negative curvature direction $\mathbf{s}$ produced from PCG that satisfies $s^t H_k s < 0$. Here TolPCG is a given tolerance for controlling the accuracy of PCG iterates. Clearly, the dominant work is to determine the two-dimensional subspace Span$\{\mathbf{g}_k, \mathbf{u}_k\}$, which is spanned linearly by two vectors $\mathbf{g}_k$ and $\mathbf{u}_k$. Once this subspace is computed, solving (4.1) becomes trivial. See [31, 32] for the details of the subspace trust region Newton method.

We denote by J_k as the Jacobian matrix $\nabla\tilde{\mathcal{F}}(\tilde{U}^{(k)})$. Because of (3.14), we see that $H_k = J_k$ and $\mathbf{g}_k = \tilde{\mathcal{F}}(\tilde{U}^{(k)})$. Thus, the trust region subproblem (4.1) can be solved efficiently by PCG since J_k is a sparse symmetric positive definite matrix. Due to that $\mathbf{g}_k = \tilde{\mathcal{F}}(\tilde{U}^{(k)})$, a numerical minimizer of $\tilde{J}(\tilde{V})$ satisfying that $||\mathbf{g}_k|| < \epsilon$ (e.g., $\epsilon = 10^{-6}$) is also a numerical solution of the nonlinear algebraic system (3.8) with the error $||\tilde{\mathcal{F}}(\tilde{U}^{(k)})||$ less than ϵ.

As comparison, we also consider the traditional merit least squares approach: Find a numerical solution of the nonlinear system (3.8) through solving the following merit least squares minimization problem:

$$\text{Find}\ \ U \in R^n \ \ \text{such that} \quad \mathbf{f}(U) = \min\{\mathbf{f}(V) \mid V \in R^n\}, \tag{4.2}$$

where $\mathbf{f}(V) = \frac{1}{2}||\tilde{\mathcal{F}}(V)||_2^2$ with the vector function $\tilde{\mathcal{F}}$ being given in (3.8). When the subspace trust region Newton method is used to solve (4.2), the trust region subproblem (4.1) is defined with $\mathbf{f}_k = \frac{1}{2}[\tilde{\mathcal{F}}(\tilde{U}^{(k)})]^t\tilde{\mathcal{F}}(\tilde{U}^{(k)})$, $\mathbf{g}_k = J_k^t\tilde{\mathcal{F}}(\tilde{U}^{(k)})$, and $H_k = J_k^t J_k$. Clearly, H_k may become dense even though J_k is sparse, and may have a larger condition number than J_k. Thus, the cost for solving (4.1) by PCG may increase significantly. Hence, our new minimization protocol is expected to be more effective and efficient than the traditional merit least squares approach in solving the PBE nonlinear algebraic system (3.8).

5 Numerical Results

To illustrate our minimization protocol, we developed a MATLAB program package for solving a two-dimensional (2D) model problem of PBE with domains $\Omega = (0,1)\times(0,1), \Omega_1 = (1/4,3/4)\times(1/4,3/4)$, and $\Omega_2 = \Omega - \Omega_1$. The values of $\epsilon(x)$ and $\kappa(x)$ were set as $\epsilon_1 = 2, \epsilon_2 = 78$, and $\bar{\kappa} = 60$.The n charge positions $\{x^i | i = 1, 2, \cdots, n\}$ of the right hand side function (2.3) were set as the centers of the n triangles selected randomly from the triangulation $\mathcal{T}_{1,h_1}$.The charge q_i at position x^i was set as a random value between -1 and 1 to mimic an atomic charge of a protein. In addition, we set the electron charge constant $e_c = 1$ and the temperature $T = 298K$.

The MATLAB program package was written based on the finite element MATLAB program PUFFIN [21] and the PDE toolbox of MATLAB [22]. The PBE mortar nonlinear system (3.8) and its corresponding unconstrained minimization problem (3.9) were formulated based on two independent linear finite element spaces V_{Ω_1} and V_{Ω_2}. The interface finite element space Λ_{h_1} was set as a linear finite element space inherited from V_{Ω_1}. For test purpose, all the involved matrices including the Hessian matrix and the Jacobian matrix were simply stored in 2D matrix arrays.

Note that the subspace trust region Newton method described in the previous section has been implemented in the MATLAB library, and used in the MATLAB unconstrained minimization solver function *fminunc* for solving a large scale unconstrained minimization problem and the MATLAB nonlinear solver function *fsolve* for solving a large scale nonlinear system based on the merit least square approach. Hence, we simply called *fminunc* and *fsolve* to solve the minimization problem (3.9) and the nonlinear system (3.8), respectively, in the package.

In the tests, we provided both *fminunc* and *fsolve* with the same calculation routines on $\tilde{\mathcal{F}}(\tilde{U}^{(k)})$ and $\nabla\tilde{\mathcal{F}}(\tilde{U}^{(k)})$. For simplicity, we used the same parameters for all the tests, which included the parameter options of *fsolve* and *fminunc* (e.g., diagonal preconditioning), the termination tolerance on unknowns TolX$= 10^{-8}$, the termination tolerance on the function value TolFun$= 10^{-12}$, and the termination tolerance on the PCG inner loop TolPCG$= 10^{-3}$.To test the effect of an accuracy of the PCG solution on the performance of the subspace trust region Newton method, some tests were repeated with TolPCG$= 10^{-6}$.

The tests with $h_1 = 1/32, h_2 = 1/16$, and the total charge number $n_c = 470$ were done on a PC computer (Dell Latitude D610 with 1.86 GHz Pentium 4 and 1 GB of RAM) at the University of Wisconsin-Milwaukee. The results are reported in Figures 1 to 8 and Tables 1 and 2.

Figure 1 displays the mortar triangulation mesh of domain Ω generated by our package. The solution of the corresponding PBE model problem is visualized in Figure 2. From the figure we see that the electrostatic potential energy in Ω_1 dominated over the whole region Ω. Although the potential in the solvent region Ω_2 is relatively small, it does nonlinearly change toward the interface Γ as shown in Figure 3.

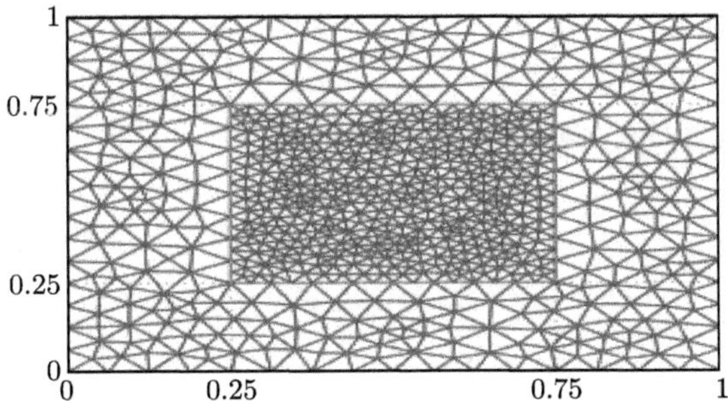

Figure 1　The mortar triangular mesh of domain Ω with grid sizes $h_1 = 1/32$ and $h_2 = 1/16$

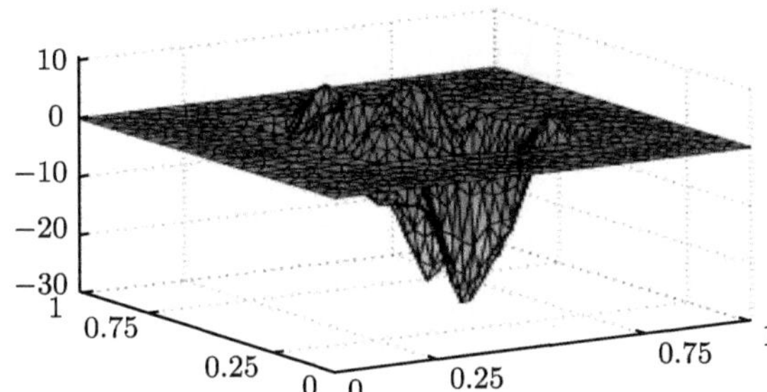

Figure 2　The numerical solution of the PBE model problem

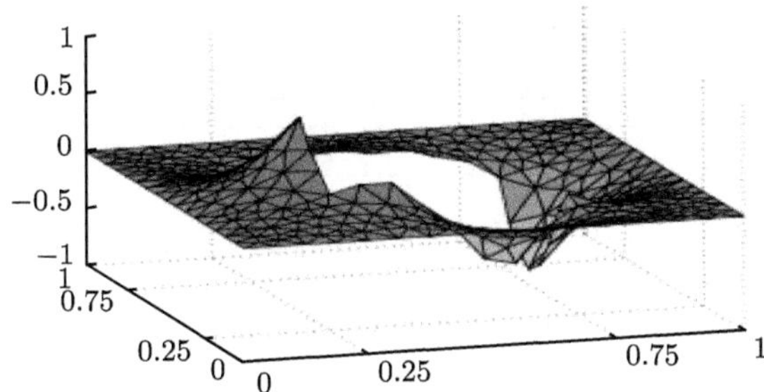

Figure 3　The part of the PBE numerical solution in the solvent region Ω_2

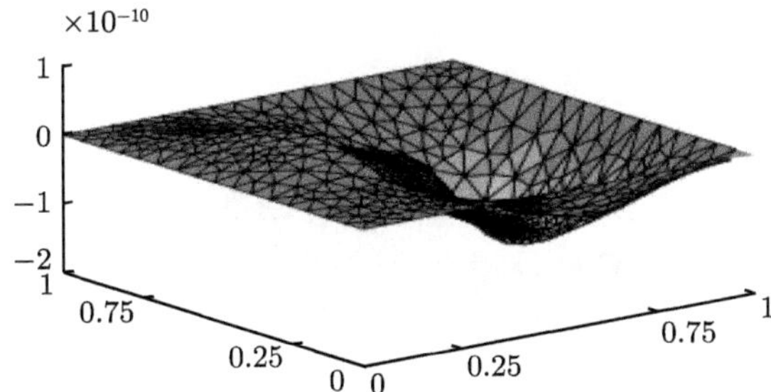

Figure 4　The errors between the two numerical solutions produced by solving (3.9) and (4.2) with $\tilde{U}^{(0)} = \tilde{A}^{-1}\tilde{F}$

Tables 1 and 2 compare the performance of the subspace trust region Newton method for solving the PBE minimization problem (3.9) with that for solving the merit least squares problem (4.2). From the tables we see that our PBE minimization protocol has significantly speeded up the searching process of a PBE solution: it took only at most 20% of the total CPU time spent by solving the traditional merit least squares problem. Most of the CPU time savings came from the inner PCG loop for solving the trust region subproblem (4.1). From the tables we also see that using $\tilde{U}^{(0)} = \tilde{A}^{-1}\tilde{F}$ is a good choice, with which 96% of the total CPU time was saved compared the case with an initial guess of zero.

Table 1 Performance of the subspace trust region Newton method for solving the PBE minimization problem (3.9). Here TolPCG is the PCG termination tolerance, ITE denotes the total number of iterations, s is a trial step vector, and #PCG denotes the total number of PCG iterations

With the initial guess $\tilde{U}^{(0)} = \tilde{A}^{-1}\tilde{F}$ (CPU in seconds)						
TolPCG	ITE	$\tilde{J}$	$\|\|\nabla\tilde{J}\|\|$	$\|\|\mathbf{s}\|\|$	#PCG	CPU
10^{-6}	6	−7 173.55	7.25×10^{-11}	1.6×10^{-9}	393	5.39
10^{-3}	6	−7 173.55	7.25×10^{-11}	1.6×10^{-9}	245	4.82
With the initial guess $\tilde{U}^{(0)} = 0$ (CPU in minutes)						
TolPCG	ITE	$\tilde{J}$	$\|\|\nabla\tilde{J}\|\|$	$\|\|\mathbf{s}\|\|$	#PCG	CPU
10^{-6}	224	−7 173.55	1.16×10^{-3}	9.66×10^{-9}	12 537	2.79
10^{-3}	224	−7 173.55	6.80×10^{-4}	6.14×10^{-9}	7 560	2.54

Table 2 Performance of the subspace trust region Newton method for solving the merit least squares problem (4.2). Here f is the merit function and g its gradient vector

With the initial guess $\tilde{U}^{(0)} = \tilde{A}^{-1}\tilde{F}$ (CPU in seconds)						
TolPCG	ITE	f	$\|\|\mathbf{g}\|\|$	$\|\|\mathbf{s}\|\|$	#PCG	CPU
10^{-6}	5	8.6×10^{-17}	2.7×10^{-7}	5.3×10^{-8}	5716	36.70
10^{-3}	5	8.5×10^{-17}	2.08×10^{-6}	5.3×10^{-8}	3434	23.62
With the initial guess $\tilde{U}^{(0)} = 0$ (CPU in minutes)						
TolPCG	ITE	f	$\|\|\mathbf{g}\|\|$	$\|\|\mathbf{s}\|\|$	# PCG	CPU
10^{-6}	215	5.22×10^{-18}	7.47×10^{-8}	1.43×10^{-8}	199 488	20.88
10^{-3}	216	5.22×10^{-18}	1.57×10^{-7}	1.67×10^{-8}	79 115	10.14

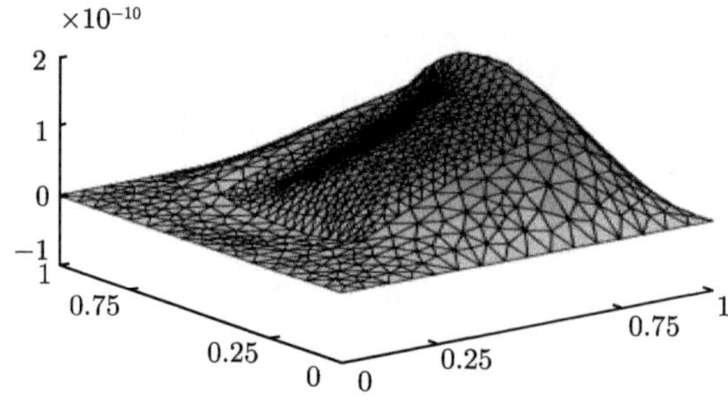

Figure 5 The errors between the two numerical solutions produced by solving (3.9) with $\tilde{U}^{(0)} =$ and $\tilde{U}^{(0)} = \tilde{A}^{-1}\tilde{F}$

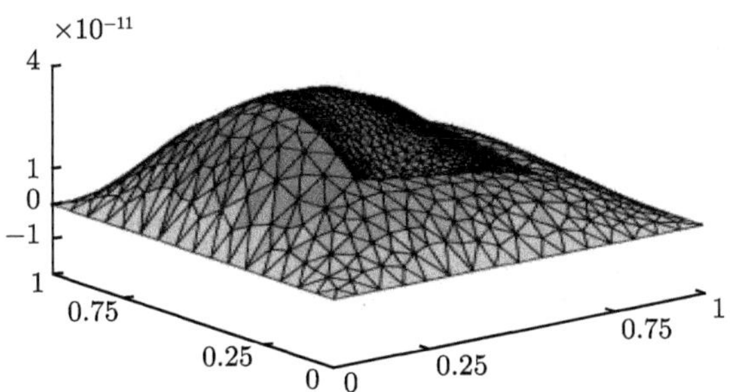

Figure 6　The errors between the two numerical solutions produced by solving (3.9) with $\tilde{U}^{(0)} = \tilde{A}^{-1}\tilde{F}$ and a randomly selected initial guess

Figures 7 and 8 compare the global convergence behaviors (in terms of gradient norms) of the subspace trust region Newton methods for solving (3.9) and (4.2). Here the initial guess was respectively set as zero and a randomly selected initial guess. Each component of the randomly selected initial guess was a random number between 0 and 10. The starting and ending values of the gradient norm are marked out in the figures for easy to compare. From the figures we see that the subspace trust region Newton method has a superlinear convergence rate at the last few iterations. We noted that the gradient norm was not small enough in the case of solving (3.9) with the initial guess of zero due to the iteration satisfying the termination rule:$||\tilde{U}^{(k)} - \tilde{U}^{(k-1)}|| = ||\mathbf{s}^{k-1}|| <$ TolX, where we set TolX$= 10^{-8}$ for all the tests.This rule is one of the practical termination rules in *fminunc* since it guarantees the solution $\tilde{U}^{(k)}$ to have an enough accuracy.

To confirm that the PBE problem (3.8) has the unique solution, we made many tests using different initial guesses (most of them selected randomly). It was found that the differences between any two computed solutions were less than $O(10^{-10})$. As demonstrations, three of them were displayed in Figures 4 − 6. These test results confirm what is claimed in Theorem 3.2.

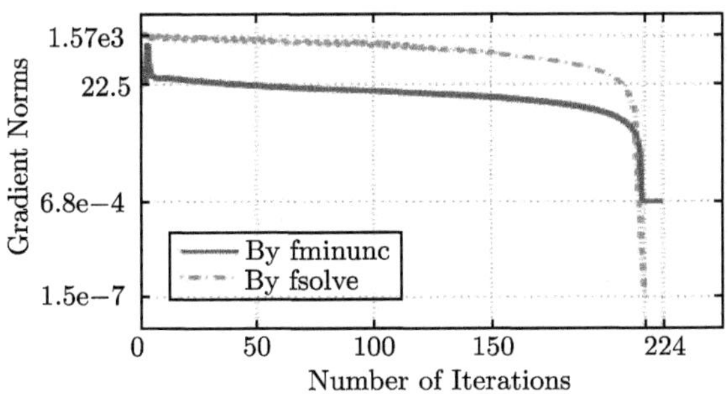

Figure 7　Comparison of the gradient norms of the subspace trust region Newton method for solving the PBE minimization problems (3.9) with that for the merit least squares problem (4.2). Here $\tilde{U}^{(0)} = 0$

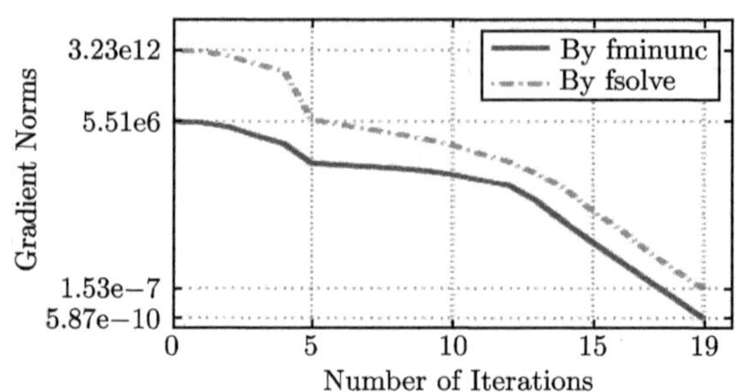

Figure 8 Comparison of the gradient norms of the subspace trust region Newton method for solving (3.9) with that for (4.2). Here $\tilde{U}^{(0)}$ is a randomly selected initial guess

Table 3 Comparison of the performance of the new minimization protocol (*fminunc*) with that of the traditional least squares approach (*fsolve*). Here the model problem was set with $h_1 = 1/64$, $h_2 = 1/32$, $n_c = 1500$, and $n = 3042$

Solve PBE minimization problem (3.9) by *fminunc* (CPU in minutes)						
$\tilde{U}^{(0)}$	ITE	$\tilde{J}$	$\|\|\nabla\tilde{J}\|\|$	$\|\|\mathbf{s}\|\|$	# PCG	CPU
$\tilde{A}^{-1}\tilde{F}$	7	−24 930.3	1.09×10^{-8}	2.39×10^{-10}	601	0.713
0	618	−24 930.3	3.36×10^{-9}	1.53×10^{-3}	9365	53.18
Solve PBE nonlinear system (3.8) by *fsolve*						
$\tilde{U}^{(0)}$	ITE	**f**	$\|\|\mathbf{g}\|\|$	$\|\|\mathbf{s}\|\|$	# PCG	CPU
$\tilde{A}^{-1}\tilde{F}$	6	2.76×10^{-17}	2.10×10^{-7}	4.90×10^{-7}	41 026	9.93
0	618	1.61×10^{-15}	6.27×10^{-7}	1.12×10^{-6}	396 180	469

Finally, we made tests with $h_1 = 1/64, h_2 = 1/32$, and $n_c = 1500$ to demonstrate the performance of our new minimization protocol on a finer mesh. In this case, the nonlinear algebraic system consisted of 3042 equations and 3042 unknowns, which resulted in a 3042×3042 Jacobian matrix. To speed up the calculations, we made these tests on a powerful Linux workstation (a Dell Precision 490 with dual core Intel Xeon processor 2GHz and 4GB main memory). The results were given in Table 3. From it we see that using our new minimization protocol saved more CPU time for a larger nonlinear system. For example, in the case of using the initial guess $\tilde{U}^{(0)} = \tilde{A}^{-1}\tilde{F}$, our new minimization protocol speeded up by 14 times the performance of the traditional merit lease-squares approach. In the future, we plan to develop a new program package with sparse matrix techniques to further improve the performance of our PBE minimization protocol in the large scale case.

Acknowledgments We are indebted to the two anonymous referees and the reviewing editor for valuable comments and helpful suggestions.

References

[1] N.A. Baker, Improving implicit solvent simulations: a Poisson-centric view, *Current Opinion in Structural Biology*, 15 (2005), 137-143.

[2] N.A. Baker, D. Sept, S. Joseph, M. Holst and J.A. MoCammon, Electrostatics of nanosystems: Application to microtubules and the ribosome, *Proceedings of the National Academy of Sciences of the United States of America*, 98 (2001), 10037-10041.

[3] K.E. Forsten, R.E. Kozack, D.A. Lauffenburger and S. Subramaniam, Numerical solution of the nonlinear Poisson-Boltzmann equation for a membrane-electrolyte system, *The Journal of Physical Chemistry*, 98 (1994), 5580-5586.

[4] B. Honig and A. Nicholls, Classical electrostatics in biology and chemistry, *Science*, 268(1995), 1144-1149.

[5] B. Al-Lazikani, J. Jung, Z. Xiang and B. Honig, Protein structure prediction, *Current Option in Chemical Biology*, 5 (2001), 51-56.

[6] B. Roux and T. Simonson, Implicit solvent models, *Biophysical Chemistry*, 78 (1999), 1-20.

[7] Y. N. Vorobjev and H.A. Scheraga, A fast adaptive multigrid boundary element method for macromolecular electrostatic computations in a solvent, *Journal of Computational Chemistry*, 18 (1997), 569-583.

[8] J. Wagoner and N.A. Baker, Solvatin forces on biomolecular structures: A comparison of explicit solvent and Poisson-Boltzmann models, *Journal of Computational Chemistry*, 25 (2004), 1623-1629.

[9] W. Rocchia, E. Alexov and B. Hoing, Extending the applicability of the nonlinear Poisson-Boltzmann equation: Multiple dielectric constants and multivalent ions, *Journal of Physical Chemistry B*, 105 (2001), 6507-6514.

[10] M. E. Davis, J.D. Madura, B.A. Luty and J.A. McCammon, Electrostatics and diffusion of molecules in solution: Simulations with the University of Houston Brownian dynamics program, *Computer Physics Communications*, 62 (1991), 187-197.

[11] N. Baker, D. Sept, M. Holst and J.A. McCammon, The adaptive multilevel finite element solution of the Poisson-Boltzmann equation on massively parallel computers, *IBM Journal of Research & Development*, 45 (2001), 427-438.

[12] R. Bank and M. Holst, A new paradigm for parallel adaptive meshing algorithms, *SIAM Review*, 45 (2003), 291-323.

[13] M. Holst and F. Saied, Numerical solution of the nonlinear Poisson-Boltzmann equation: Developing more robust and efficient methods, *Journal of Computational Chemistry*, 16 (1995), 337-364.

[14] M. Holst, N. Baker and F. Wang, Adaptive multilevel finite element solution of the Poisson-Boltzmann equation; I: Algorithms and examples, *Journal of Computational Chemistry*, 21 (2000), 1319-1342.

[15] Y. Achdou, Y. Maday and O. Widlund, Iterative substructuring preconditioners for mortar element methods in two dimensions, *SIAM Journal on Numerical Analysis*, 36 (1999), 551-580.

[16] F.B. Belgacem, The mortar finite element method with Lagrange multipliers, *Numerische Mathematik*, 84 (1999), 173-197.

[17] B.P. Lamichhane and B.I. Wohlmuth, Mortar finite elements for interface problems, *Computing*, 72 (2004), 333-348.

[18] B.I. Wohlmuth, A V-cycle multigrid approach for mortar finite elements, *SIAM Journal on Numerical Analysis*, 42 (2005), 2476-2405.

[19] W. Chen, Y. Shen and Q. Xia, A mortar finite element approximation for the linear Poisson-Boltzmann equation, *Applied Mathematics and Computation*, 164 (2005), 11-23.

[20] C. Bernardi, Y. Maday and A. Patera, A new nonconforming approach to domain decomposition: The mortar element method, in: H. Brezis et al., eds., *Nonlinear Partial Differential Equations and Their Applications, vol. XI*, Pitman Res. Notes. Math. Ser., 299, Longman Sci. Tech., Harlow, 1994, pp. 13-51.

[21] J. Hoffman and A. Logg, PUFFIN Web page: http://www.fenics.org/puffin/, 2004.

[22] The MathWorks, Inc., MATLAB Partial Differential Equation Toolbox 1.0.8 Web page: http://www.mathworks.com/products/pde/, 2005.

[23] J. Nocedal and S.J. Wright, *Numerical Optimization*, Springer, New York, 2000.

[24] T.F. Coleman and Y. Li, An interior, trust region approach for nonlinear minimization subject to bounds, *SIAM Journal on Optimization*, 6 (1996), 418-445.

[25] T.F. Coleman and Y. Li, On the convergence of reflective Newton methods for large-scale nonlinear minimization subject to bounds, *Mathematical Programming*, 67 (1994), 189-224.

[26] G.H. Golub and C.F. van Loan, *Matrix Computations*, 3rd edn., John Hopkins University Press, Baltimore, MD, 1996.

[27] C. Tanford, *Physical Chemistry of Macromolecules*, John Wiley & Sons, New York, 1961.

[28] S.C. Brenner and L.R. Scott, *The Mathematical Theory of Finite Element Methods*, 2nd edn., Springer, New York, 2002.

[29] I. Ekeland and R. Témam, *Convex Analysis and Variational Problems*, SIAM, Philadelphia, 1999.

[30] L. Debnath and P. Mikusinnski, *Introduction to Hilbert Spaces with Applications*, Academic Press, New York, 1990.

[31] M.A. Branch, T.F. Coleman and Y. Li, A subspace, interior, and conjugate gradient method for large-scale bound-constrained minimization problems, *SIAM Journal on Scientific Computing*, 21 (1999), 1-23.

[32] R.H. Byrd, R.B. Schnabel and G.A. Shultz, *Approximate solution of the trust region problem by minimization over two-dimensional subspaces*, Mathematical Programming, 40 (1988), 247-263.

[33] J.J. Moré and D.C. Sorensen, Computing a Trust Region Step, *SIAM Journal on Scientific and Statistical Computing*, 3 (1983), 553-572.

Mathematical Analysis of EEP Method for One-Dimensional Finite Element Postprocessing*

Q.H. Zhao (赵庆华)† S.Z. Zhou (周叔子)† Q.D. Zhu (朱起定)‡

Abstract For a class of two-point boundary value problems, by virtue of one-dimensional projection interpolation, it is proved that the nodal recovery derivative obtained by Yuan's element energy projection(EEP) method has the accuracy $O(h^{\min\{k+4,2k\}})$. The theoretical analysis coincides the reported numerical results.

Keywords Superconvergence stress, element energy projection method, finite element, two-point boundary value problems, projection interpolation.

Introduction

In the finite element history, there have been many investigations on derivative recovery technique. For one-dimensional problems, some recent developments include the derivative patch interpolating recovery technique by Zhang [1], global ultraconvergence corretion scheme by Zhu and Zhao [2], ultraconvergence based on the new error expansion by Chen et al. [3].

In 2004, Yuan and Wang [4] proposed the so called "element energy projection method" based on mechanical interpretation. The nodal derivative recovery formula is simple. A surprising observation from numerical tests is that the accuracy of the nodal recovery derivative is $O(h^{2k})$ when the finite element degree $k \leqslant 4$ and $p = 1$. in (1.1). We shall prove for a class of two-point boundary value problems that the nodal recovery derivatives have convergence rate of $O(h^{\min\{2k,k+4\}})$ for $k \geqslant 1$, which is just $O(h^{2k})$ for $k \leqslant 4$.

We shall use C to denote a generic constant which may take different values at different occurrences. We shall also use the standard notations for the Sobolev spaces and norms as in Ref.[5].

* 发表于: Applied Mathematics and Mechanics (English Edition), Vol. 28, 2007, pp. 441-445.

† College of Mathematics and Econometrics, Hunan University, China.

‡ College of Mathematics and Computer Science, Hunan Normal University, China.

1 Problem and Theorem

Consider the following two-point boundary value problem:

$$Lu = -(pu')' + qu = f, \quad \text{in} \quad I = [0,1], \tag{1.1}$$

$$u(0) = u(1) = 0, \tag{1.2}$$

where $p(x) \geqslant c > 0, q(x) > 0$, p, q and f are sufficiently smooth.

Suppose that J_h is a quasi-uniform partition over the interval $I = [0,1]$: $0 = x_0 < x_1 < ... < x_N = 1$. denote $h_i = x_{i+1} - x_i$, $h = \max h_i$. According to Ref. [4], let e be any element in J_h. Let $\overline{x}_1$, $\overline{x}_2$ denote the two endpoints of the element e.

The weak formulation of (1.1) and (1.2) is to find $u \in H_0^1\ (I)$ such that

$$a(u,v) = (f,v), \quad \forall\, v \in H_0^1(I), \tag{1.3}$$

where

$$a(u,v) = \int_0^1 (pu'v' + quv)\mathrm{d}x.$$

The finite element solution of (1.1) and (1.2) is to find $u_h \in S_h$ such that

$$a(u - u_h, v_h) = 0, \quad \forall\, v_h \in S_h,$$

where $S_h \subset H_0^1$ is the kth-order Lagrange type finite element space associated with the partition J_h.

The following nodal derivative recovery formula is obtained by Ref. [4]:

$$u'^*(\overline{x}_i) = u_h'(\overline{x}_i) + (-1)^{i-1}\frac{1}{p(\overline{x}_i)}\int_{\overline{x}_1}^{\overline{x}_2}(f - Lu_h)N_i\mathrm{d}x, \quad i = 1,2, \tag{1.4}$$

where $N_i(x)$ is the linear element basis function satisfying

$$N_i(\overline{x}_j) = \delta_{ij}, \quad i,j = 1,2.$$

Our main theorem is as follows:

Theorem 1.1. *Suppose $u \in W^{k+1,\infty}(I)$, the postprocessing scheme (1.4) for equation (1.1) and (1.2) with $p \equiv 1$ have the accuracy of $O(h^{2k})$ for $k \leqslant 4$ and $O(h^{k+4})$ for $k \geqslant 5$, that is,*

$$|u'(\overline{x}_i) - u'^*(\overline{x}_i)| \leqslant Ch^{\min\{2k,k+4\}}\|u\|_{k+1,\infty}, \quad k \geqslant 1, \quad i = 1,2. \tag{1.5}$$

2 One-Dimensional Projection Interpolation

For any element $e = (\overline{x}_1, \overline{x}_2)$, let $x_e = (\overline{x}_1 + \overline{x}_2)/2, h_e = (\overline{x}_2 - \overline{x}_1), \tilde{h}_e = \frac{h_e}{2}$, then

$$e = (x_e - \tilde{h}_e, x_e + \tilde{h}_e).$$

Now we can use one-dimensional projection interpolation theory to e (see Ref. [5]).

Consider the normalized system of orthogonal polynomials in $L^2(e)$,

$$\begin{cases} L_0(x)=\sqrt{\dfrac{1}{2}}\tilde{h}_e^{-\frac{1}{2}}, \\ L_i(x)=\sigma_i\left(\dfrac{\mathrm{d}}{\mathrm{d}x}\right)^i [A(x)]^i, \quad i\geqslant 1, \\ \sigma_i=\sqrt{\dfrac{2i+1}{2}}\dfrac{1}{i!}\tilde{h}_e^{-i-\frac{1}{2}}=O(\tilde{h}_e^{-i-\frac{1}{2}}), \end{cases} \tag{2.1}$$

where $A(x)=\dfrac{1}{2}[(x-x_e)^2-\tilde{h}_e^2]$, and the scaled version of the Lobatto polynomials,

$$\begin{cases} \omega_0=1, \\ \omega_{i+1}=\tilde{h}_e^{-\frac{1}{2}}\displaystyle\int_{I_x} L_i(x)\mathrm{d}x=\tilde{h}_e^{-\frac{1}{2}}\sigma_i\left(\dfrac{\mathrm{d}}{\mathrm{d}x}\right)^{i-1}[A(x)]^i, \quad i\geqslant 1 \end{cases} \tag{2.2}$$

where $I_x=(x_e-\tilde{h}_e,x)$.

Now suppose $\nu\in H^1(e)$. Then $\nu'\in L^2(e)$, and we have the Fourier expansion [5],

$$\nu'=\alpha_0+\alpha_1L_1(x)+\alpha_2L_2(x)+\cdots+\alpha_kL_k(x)+\cdots, \tag{2.3}$$

where $\alpha_k=(\nu',L_k)_e$. It follows that

$$\nu(x)=\sum_{j=0}^{\infty}\beta_j\omega_j(x), \quad \forall\,\nu\in H^1(e) \tag{2.4}$$

where

$$\begin{cases} \beta_0=\nu(x_e-\tilde{h}_e), \\ \beta_1=\sqrt{\dfrac{1}{2}}(\nu(x_e+\tilde{h}_e)-\nu(x_e-\tilde{h}_e)), \\ \beta_j=\tilde{h}_e^{\frac{1}{2}}\alpha_{j-1}, \quad j\geqslant 2 \end{cases} \tag{2.5}$$

Define the kth-degree interpolation operator,

$$i_k: H^1(I)\to P_k(e), i_k\nu(x)=\sum_{j=0}^{k}\beta_j\omega_j(x), \quad \forall\,x\in e. \tag{2.6}$$

The operator i_k has the following properties [5]:

$$\|\nu-i_k\nu\|_{0,\infty,e}\leqslant Ch^{k+1}|\nu|_{k+1,\infty}, \quad \forall\,\nu\in W^{k+1,\infty}(e); \tag{2.7}$$

$$\|\nu-i_k\nu\|_{0,1,e}\leqslant Ch^{k+1}|\nu|_{k+1,1}, \quad \forall\,\nu\in W^{k+1,1}(e); \tag{2.8}$$

$$(\nu-i_k\nu,p)_e=0, \quad \forall\,p\in P_{k-2}(e), \ \ \nu\in H^1(e). \tag{2.9}$$

Let u_h be kth-degree finite element solution to two-point boundary value problems (1.1) and (1.2) with $p=1$. Then we have ultra-approximation,

$$\|u_h-i_ku\|_{0,\infty}\leqslant Ch^{k+3}|u|_{k+1,\infty}, \ \ k\geqslant 3. \tag{2.10}$$

3 Proof of Theorem 1.1

We need the following existing results to prove Theorem 1.1.

First we have the standard L_∞ error estimation,

$$\|u-u_h\|_{0,\infty,e} \leqslant Ch^{k+1}\|u\|_{k+1,\infty}, \quad k \geqslant 1. \tag{3.1}$$

Next,we have the following high accuracy result for $e_h = u - u_h$ (see Ref.[6]):

$$\max_j |e_h(x_j) - e_h(x_{j-1})| \leqslant Ch^{2k+1}\|u\|_{k+1,\infty}, \quad k \geqslant 1. \tag{3.2}$$

Proof of Theorem 1.1. We only prove the case $i = 1$. The argument for $i = 2$ is similar. It is easy to see that, for $x \in e$, there holds

$$f - Lu_h = Le_h. \tag{3.3}$$

Hence we have

$$\begin{aligned} u'^*(\overline{x}_1) &= u_h'(\overline{x}_1) + \frac{1}{p(\tilde{x}_i)}\int_{\overline{x}_1}^{\overline{x}_2}(f - Lu_h)N_1 \mathrm{d}x \\ &= u_h'(\overline{x}_1) + \frac{1}{p(\tilde{x}_i)}(Le_h, N_1)_e. \end{aligned}$$

Therefore, it follows from $p = 1$ that

$$\begin{aligned} u'^*(\overline{x}_1) &= u_h'(\overline{x}_1) + [(LN_1, e_h)_e - e_h' N_1|_{\overline{x}_1}^{\overline{x}_2} + e_h N_1'|_{\overline{x}_1}^{\overline{x}_2}] \\ &= u_h'(\overline{x}_1) + (qN_1, e_h)_e + e_h'(\overline{x}_1) - \frac{1}{\overline{x}_2 - \overline{x}_1}[e_h(\overline{x}_2) - e_h(\overline{x}_1)] \\ &= u'(\overline{x}_1) + (e_h, qN_1)_e - \frac{1}{\overline{x}_2 - \overline{x}_1}[e_h(\overline{x}_2) - e_h(\overline{x}_1)]. \end{aligned} \tag{3.4}$$

It follows from (3.2) and the quasi-uniform property of partition that

$$\left|\frac{1}{\overline{x}_2 - \overline{x}_1}[e_h(\overline{x}_2) - e_h(\overline{x}_1)\right| \leqslant h_e^{-1} \cdot Ch^{2k+1}\|u\|_{k+1,\infty} \leqslant Ch^{2k}\|u\|_{k+1,\infty}. \tag{3.5}$$

Now we estimate $(e_h, qN_1)_e$. By (3.1) we know

$$\begin{aligned} |\ (e_h, qN_1)_e\ | &\leqslant \|e_h\|_{0,\infty,e} \cdot \|qN_1\|_{0,1,e} \leqslant Ch^{k+1}\|u\|_{k+1,\infty} \cdot Ch \\ &\leqslant Ch^{k+2}\|u\|_{k+1,\infty}, \quad k \geqslant 1. \end{aligned} \tag{3.6}$$

When $k \geqslant 3$, we have

$$\begin{aligned} |\ (e_h, qN_1)_e\ | &\leqslant |\ (u - i_k u, qN_1)_e\ | + |\ (i_k u - u_h, qN_1)_e\ | \\ &\leqslant |\ (u - i_k u, qN_1 - i_{k-2}(qN_1))_e\ | + |\ (i_k u - u_h, qN_1)_e\ |. \end{aligned}$$

It follows from Hölder inequality, (2.7) and (2.8) that

$$\begin{aligned}|(u-i_ku,qN_1-i_{k-2}(qN_1))_e| &\leqslant \|u-i_ku\|_{0,\infty,e}\cdot\|qN_1-i_{k-2}(qN_1)\|_{0,1,e}\\ &\leqslant Ch^{k+1}\|u\|_{k+1,\infty}\cdot h^{k-1}\|qN_1\|_{k-1,1,e}\\ &\leqslant Ch^{2k}\|u\|_{k+1,\infty}.\end{aligned}$$

On the other hand we know by (2.10) that

$$\begin{aligned}|(i_ku-u_h,qN_1)_e| &\leqslant \|i_ku-u_h\|_{0,\infty,e}\cdot\|qN_1\|_{0,1,e}\\ &\leqslant Ch^{k+3}\|u\|_{k+1,\infty}\cdot h\\ &\leqslant Ch^{k+4}\|u\|_{k+1,\infty}.\end{aligned}$$

Hence we have for $k\geqslant 3$ that

$$|(e_h,qN_1)_e|\leqslant Ch^{2k}\|u\|_{k+1,\infty}+Ch^{k+4}\|u\|_{k+1,\infty}=Ch^{\min\{2k,k+4\}}\|u\|_{k+1,\infty}. \tag{3.7}$$

From (3.6) and (3.7) we have

$$|(e_h,qN_1)_e|\leqslant Ch^{\min\{k+4,2k\}}\|u\|_{k+1,\infty},\quad k\geqslant 1. \tag{3.8}$$

Finally,by (3.4), (3.5) and (3.8) we obtain

$$|u'(\overline{x}_1)-u'^*(\overline{x}_1)|\leqslant Ch^{\min\{k+4,2k\}}\|u\|_{k+1,\infty},\quad k\geqslant 1.$$

The proof is complete.

4 Conclusion

For a class of two-point boundary value problems,we have proved that the nodal recovery derivative in Ref. [4] based on EEP method have the accuracy of $O(h^{\min\{k+4,2k\}})$. Our theoretical justification satisfactorily interpret the numerical results in Ref. [4].

Remark 4.1. The method and result in this article is also applicable to mixed boundary problems.

See complementary numerical test in the next section.

5 Complementary Numerical Test

Consider the following two-point boundary value problems:

$$\begin{cases}-u''+u=1, & x\in I=(0,1).\\ u(0)=u'(1)=0.\end{cases} \tag{5.1}$$

The exact solution is $u = \frac{1}{1+e^2}(e^x + e^{2-x}) + 1$. We solved (5.1) on uniform grids with $h = \frac{1}{N}$, $N = 2, 4, ..., 32$ and $k = 4, 5$. The results, presented in Table 1, show that error $e'^*(0) = u'(0) - u'^*(0)$ still keep accuracy $O(h^{2k})$ for $k = 5$.

Table 1 $e'^*(0)$ for $k = 4,5$

N	$e'^*(0)$ $(k = 4)$	Convergence rate	$e'^*(0)$ $(k = 5)$	Convergence rate
2	−2.4128e−11	7.60	−5.6194e−14	9.92
4	−1.0048e−13	7.91	−5.5721e−17	9.98
8	−3.9872e−16	7.98	−5.4627e−20	9.99
16	−1.5636e−18	7.99	−5.3399e−23	10.00
32	−6.1138e−21	8.00	−5.2160e−26	10.00

References

[1] T. Zhang, The derivative patch interpolating recovery technique and superconvergence, *Chinese Journal of Numerical Mathematics and Applications*, 23 (2001), 1-10.

[2] Q. Zhu and Q. Zhao, SPR technique and finite element correction, *Numerische Mathematik*, 96 (2003), 185-196.

[3] C. Chen, Z. Xie and J. Liu, New error expansion for one dimensional finite element and ultraconvergence, *Numerical Mathematics A journal of Chinese Universities*, 14 (2005), 296-304.

[4] S. Yuan and M. Wang, An element energy projection method for post-computation of superconvergent solution in one-dimensional FEM, *Engineering Mechanics*, 21 (2004), 1-9 (in Chinese).

[5] Q. Lin and Q. Zhu, *The preprocessing and postprocessing for the finite element method*, Shanghai: Shanghai Science and Technology Publishers, 1994 (in Chinese).

[6] J. Douglas and T. Dupont, Galerkin approximations for the two point boundary problems using continuous, piecewise polynomial spaces, *Numerische Mathematik*, 22 (1974), 99-109.

A New Iterative Method for Discrete HJB Equations*

S.Z. Zhou (周叔子)† Z.Y. Zou (邹战勇)†

Abstract A successive relaxation iterative algorithm for discrete HJB equations is proposed. Monotone convergence has been proved for the algorithm.

Keywords Discrete HJB equations, iterative method, monotone convergence.

1 Introduction

Consider the following Hamilton-Jacobi-Bellman (HJB) equation:

$$\begin{cases} \max\limits_{1\leqslant i\leqslant k}\{L^i u - f^i\} = 0, & \text{in } \Omega, \\ u = 0, & \text{on } \partial\Omega, \end{cases} \tag{1.1}$$

where Ω is a bounded domain in R^d, $L^i, i = 1, \cdots, k$, are elliptic operators of second order. Equation (1.1) is arising in stochastic control problems. See [1] and the references therein.

Equation(1.1) can be discretized by finite difference method or finite element method. See [2–4] and the references therein. Then we obtain the following discrete HJB equation:

$$\max_{1\leqslant j\leqslant k}\{A^j U - F^j\} = 0, \tag{1.2}$$

where $A^j \in R^{n\times n}, F^j \in R^n, j = 1, \cdots, k$. Equation (1.2) is a system of nonsmooth nonlinear equations. Many numerical algorithms for solving (1.3) have been proposed, see [3–9] and the references therein.

Lions and Mercier [3] has given two iterative algorithms for solving (1.2). At each iteration, a linear complementarity subproblem or a linear equation system subproblem is solved. Hoppe [5] has proposed a multigrid method for solving (1.2). It uses the schemes of Lions and Mercier [3] and adopts multigrid method to solve the subproblems. Huang et al. [6] and Sun [7] have studied alternative direction algorithms for (1.2). Sun [4] has constructed a domain

* 发表于: Numerische Mathematik, Vol. 111, 2008, pp. 159-167.

† Department of Applied Mathematics, Hunan University, Changsha 410082, China.

decomposition method based on the algorithms of Lions and Mercier [3], and Zhou and Zhan [9] has proposed another domain decomposition method based on a quasivariational inequality formulation equivalent to (1.2).

All the iterative algorithms mentioned above need to solve a subproblem at each iteration, which is a linear equation system or a linear complementarity problem.

Zhou and Chen [8] has given a new iterative algorithm for (1.2). It does not solve a subproblem but carry out arithmetic operations at each iteration.

We call $V \in R^n$ a subsolution for (1.2) if

$$\max_{1\leqslant j\leqslant k}\{A^jV - F^j\} \leqslant 0. \tag{1.3}$$

The set of all the subsolutions for (1.2) is denoted by S_0. We call $V \in R^n$ a supersolution for (1.2) if

$$\max_{1\leqslant j\leqslant k}\{A^jV - F^j\} \geqslant 0. \tag{1.4}$$

The set of all the subsolutions for (1.2) is denoted by S_1.

The algorithm proposed in [8] is as follows.

Algorithm A

Step 1. Given $\varepsilon > 0$, $U^0 \in S_0$, $m := 0$.

Step 2. Compute

$$R^m = \max_{1\leqslant j\leqslant k}\{A^jU^m - F^j\}.$$

Step 3. Compute

$$\delta^m = -R^m/a, \tag{1.5}$$

where

$$a = \max_{1\leqslant j\leqslant k}\max_{1\leqslant s\leqslant n} a^j_{ss}, \quad A^j = (a^j_{sl}). \tag{1.6}$$

Step 4. Compute

$$U^{m+1} = U^m + \delta^m. \tag{1.7}$$

Step 5. If $\|U^{m+1} - U^m\| < \varepsilon$ then stop. Otherwise, let $m := m + 1$ and go to Step 2.

Obviously, Algorithm A is a simultaneous iterative scheme, similar to the Jacobi iteration for linear system.

In this paper we propose a successive iterative scheme, similar to the Gauss-Seidel iteration for linear system. Numerical tests show that it is faster than Algorithm A. It may be a new choice for solving the subproblems of domain decomposition methods for (1.2) . It is also a possible nonlinear smoother if a new multigrid method for (1.2) is constructed.

In Section 2 we give the new iterative scheme and the convergence theorem. We discuss the choice of initial value in Section 3. Some theoretical discussions are given in Section 4.

2 Algorithm and Its Convergence

Now we give the new iterative method for solving (1.2).

Algorithm 1

Step 1. Given $\varepsilon > 0$, $U^0 \in S_1$, $\omega \in (0,1]$, $m := 0$, $l := 1$.

Step 2. Compute

$$\alpha_l^m = \max_{1\leqslant j\leqslant k}\left\{\sum_{s=1}^{l-1} a_{ls}^j U_s^{m+1} + \sum_{s=l}^{n} a_{ls}^j U_s^m - F_l^j\right\}.$$

Step 3. Compute $\delta_l^m = -\alpha_l^m/a_l$, where

$$a_l = \max_{1\leqslant j\leqslant k} a_{ll}^j.$$

Step 4. Compute $V_l^{m+1} = U_l^m + \delta_l^m$, and

$$U_l^{m+1} = (1-\omega)U_l^m + \omega V_l^{m+1}.$$

Step 5. If $l = n$ then go to Step 6. Otherwise let $l := l+1$ and go to Step 2.

Step 6. If $\|U^{m+1} - U^m\| < \varepsilon$ then stop. Otherwise let $m := m+1$, $l := 1$ and go to Step 2.

Algorithm 2

Step 1. Given $\varepsilon > 0$, $U^0 \in S_0$, $\omega \in (0,1]$, $m := 0$, $l := 1$.

Steps 2–6. The same as those in Algorithm 1.

We denote the residual of U^m by R^m:

$$R^m = \max_{1\leqslant j\leqslant k}\{A^j U^m - F^j\}. \tag{2.1}$$

Then we have the following lemma.

Lemma 2.1. *Assume $A^j, j = 1, \cdots, k$, are L-matrices. Then for Algorithm 1 we have*

$$R^m \geqslant 0, \quad U^{m+1} \leqslant U^m, \quad m = 0, 1, 2, \cdots. \tag{2.2}$$

Proof. Since $U^0 \in S_1$ we have

$$R^0 \geqslant 0. \tag{2.3}$$

Hence we know by Step 1 of Algorithm 1 that

$$\alpha_1^0 = \max_{1\leqslant j\leqslant k}\left\{(A^j U^0)_1 - F_1^i\right\} = R_1^0 \geqslant 0. \tag{2.4}$$

$A^j, j = 1, \cdots, k$, are L-matrices means that (See [10])

$$a_{ss}^j > 0, \quad a_{ls}^j \leqslant 0 \quad \text{for} \quad l \neq s, \quad j = 1, \cdots, k. \tag{2.5}$$

It follows from Step 4 of Algorithm 4 that

$$U_1^1 = U_1^0 + \omega\delta_1^0 = U_1^0 - \omega\alpha_1^0/a_1 \leqslant U_1^0. \tag{2.6}$$

We derive from (2.3), (2.5) and (2.6) that

$$\alpha_2^0 = \max_{1\leqslant j\leqslant k}\left\{a_{21}^j(U_1^1 - U_1^0) + \sum_{s=1}^n a_{2s}^j U_s^0 - F_2^j\right\} \geqslant R_2^0 \geqslant 0. \tag{2.7}$$

By the similar way and induction we obtain

$$\alpha_l^0 \geqslant 0, \quad U_l^1 \leqslant U_l^0, \quad l = 1, \cdots, n. \tag{2.8}$$

Now we prove that

$$R^1 \geqslant 0. \tag{2.9}$$

Denote

$$\beta_l^{m,p} = \max_{1\leqslant j\leqslant k}\left\{\sum_{s=1}^p a_{ls}^j U_s^m + \sum_{s=p+1}^n a_{ls}^j U_s^{m-1} - F_l^j\right\}. \tag{2.10}$$

Then we have by (2.6), (2.4), $0 < a_{11}^j \leqslant a_1$ and $\omega \in (0,1]$ that

$$\begin{aligned}
\beta_1^{1,1} &= \max_{1\leqslant j\leqslant k}\left\{a_{11}^j U_1^1 + \sum_{s=2}^n a_{1s}^j U_s^0 - F_1^j\right\} \\
&= \max_{1\leqslant j\leqslant k}\left\{a_{11}^j \omega\delta_1^0 + \sum_{s=1}^n a_{1s}^j U_s^0 - F_1^j\right\} \\
&= \max_{1\leqslant j\leqslant k}\left\{-a_1^{-1}a_{11}^j\omega\alpha_1^0 + \sum_{s=1}^n a_{1s}^j U_s^0 - F_1^j\right\} \\
&\geqslant \max_{1\leqslant j\leqslant k}\left\{-\omega R_1^0 + \sum_{s=1}^n a_{1s}^j U_s^0 - F_1^j\right\} \\
&= (1-\omega)R_1^0 \geqslant 0.
\end{aligned} \tag{2.11}$$

Therefore, we obtain by (2.6) and (2.11) that

$$\begin{aligned}
\beta_1^{1,2} &= \max_{1\leqslant j\leqslant k}\left\{a_{12}^j(U_2^1 - U_2^0) + a_{11}^j U_1^1 + \sum_{s=2}^n a_{1s}^j U_s^0 - F_1^j\right\} \\
&\geqslant \max_{1\leqslant j\leqslant k}\left\{a_{11}^j U_1^1 + \sum_{s=2}^n a_{1s}^j U_s^0 - F_1^j\right\} = \beta_1^{1,1} \geqslant 0.
\end{aligned}$$

Similarly, we derive

$$\beta_1^{1,3} \geqslant 0, \quad \cdots, \quad \beta_1^{1,n} \geqslant 0.$$

Hence we have

$$R_1^1 = \beta_1^{1,n} \geqslant 0. \tag{2.12}$$

By the way similar to that for (2.12) we obtain

$$R_2^1 \geqslant 0, \quad \cdots, \quad R_n^1 \geqslant 0,$$

which and (2.12) means (2.9) holds.

Finally, the similar argument and induction lead to (2.2). The proof is completed. □

Similarly, we may prove the following lemma.

Lemma 2.2. *Assume $A^j, j = 1, \cdots, k$, are L-matrices. Then for Algorithm 2 we have*

$$R^m \leqslant 0, \quad U^{m+1} \geqslant U^m, \quad m = 0, 1, 2, \cdots. \tag{2.13}$$

Now we can prove the convergence of the algorithms.

Theorem 2.3. *Assume that $A^j, j = 1, \cdots, k$, are L-matrices, that at least one A^j is nonsingular, and that the set S_1 has lower bound. Then $\{U^m\}$ produced by Algorithm 1 is monotonically decreasing and convergent to a solution of (1.2) .*

Proof. Without loss of generality, we may assume A^1 is nonsingular. Then equation

$$A^1\tilde{U} = F^1 \tag{2.14}$$

has a unique solution $\tilde{U}$. Obviously, $\tilde{U} \in S_1$ and we may take $U^0 = \tilde{U}$. Hence Algorithm 1 is well-defined in this case. By lemma 2.1 and assumption we know $\{U^m\}$ is a monotonically decreasing supersolution sequence and has lower bound. So $\{U^m\}$ converges. Denote

$$U^* = \lim_{m\to\infty} U^m.$$

Then we have

$$\lim_{m\to\infty} V^{m+1} = \omega^{-1}\left(\lim_{m\to\infty} U^{m+1} - \lim(1-\omega)U^m\right) = U^*,$$

which implies for $l = 1, \cdots, n$ holds

$$\lim_{m\to\infty} \alpha_l^m = -a_l \lim_{m\to\infty} \delta_l^m = -a_l \lim_{m\to\infty} (V_l^{m+1} - U_l^m) = 0. \tag{2.15}$$

Finally, it follows from the expression of α_l^m and (2.15) that

$$\max_{1\leqslant j\leqslant k}\{(A^jU^* - F^j)_l\} = \lim_{m\to\infty} \alpha_l^m = 0, \quad l = 1, \cdots, n,$$

which means U^* is a solution of (1.2) . The proof is completed. □

Theorem 2.4. *Assume that $A^j, j = 1, \cdots, k$, are L-matrices, that at least one A^j is an M-matrix, and that S_0 is nonempty. Then $\{U^m\}$ produced by Algorithm 2 is monotonically increasing and convergent to a solution of (1.2) .*

Proof. Without loss of generality, we may assume A^1 is an M-matrix. Denote $\tilde{U} = (A^1)^{-1}F^1$. By Lemma 2.2 we have $R^m \leqslant 0$. Hence

$$A^1U^m - F^1 \leqslant 0,$$

which and the fact that A^1 is an M-matrix implies

$$U^m \leqslant \tilde{U}, \quad m = 0, 1, 2, \cdots.$$

Therefore, $\{U^m\}$ has a limit denoted by U^*. The rest part of the proof is almost the same as that for Theorem 2.3. □

3 Choise of U^0

At first, let us formulate conditions introduced in [10].

Condition A. $A^j, j = 1, \cdots, k$, are L-matrices, and have strict diagonal dominance.

Condition B. $A^j, j = 1, \cdots, k$, are L-matrices, and all the matrices $A(p_1, \cdots, p_n) = (a_{ls}^{p_l}), p_l = 1, \cdots, k, l = 1, \cdots, n$, are irreducible, weakly diagonally dominant.

Proposition 3.1([8]). *If condition A or Condition B holds then S_0 is nonempty.*

Zhou and Chen [8] has given a method for producing $U^0 \in S_0$ in Algorithm 2 provided that Condition A or Condition B holds.

For Algorithm 1, if A^1 is nonsingular, just as in the proof of Theorem 2.3, we may choose $\tilde{U}$ in (2.14) as U^0. If Condition A holds then we may take $U^0 = \theta e$, where $e = (1, \cdots, 1)^T$ and

$$\theta = \max_{1\leqslant j\leqslant k} \max_{1\leqslant l\leqslant n} \left(\sum_{s=1}^{n} a_{ls}^j \right)^{-1} F_l^j.$$

The first choice above needs to solve a linear system for producing U^0, and the second choice does not need to. But, the first choice can be applied in more general case. In addition, numerical test shows that the first choice needs less iteration number even in the case that Condition A holds.

4 Discussion on "S_1 Has Lower Bound"

Now we discuss a condition in Theorem 2.3: S_1 has lower bound.

Proposition 4.1. *If all the matrices $A(p_1, \cdots, p_n), p_l = 1, \cdots, k, l = 1, \cdots, n$ are M-matrices, and (1.2) has at least one solution U, then*

$$V \geqslant U, \qquad \forall V \in S_1. \tag{4.1}$$

Remark 4.1. (4.1) Shows that U is a lower bound of S_1 in this case.

Proof. For any $V \in S_1$ there exist indices $p_1, \cdots, p_n$ such that

$$A(p_1, \cdots, p_n)V - F(p_1, \cdots, p_n) = \max_{1 \leqslant j \leqslant k} \{A^j V - F^j\},$$

where $F(p_1, \cdots, p_n) = (F_l^{p_l})$. Hence we have

$$A(p_1, \cdots, p_n)V - F(p_1, \cdots, p_n) \geqslant 0. \tag{4.2}$$

On the other hand, since U is a solution of (1.2) , we have

$$A(p_1, \cdots, p_n)U - F(p_1, \cdots, p_n) \leqslant 0,$$

which combining with (4.2) and the fact that $A(p_1, \cdots, p_n)$ is an M-matrix implies (4.1). The proof is completed. □

Corollary 4.2. *If Condition A or Condition B holds then S_1 has lower bound.*

Proof. In these cases, (1.2) has a unique solution (see [9]), and it is easy to see all the matrices $A(p_1, \cdots, p_n)$, $p_l = 1, \cdots, k, l = 1, \cdots, n$, are M-matrices. Hence (4.1) holds. The proof is completed. □

Corollary 4.3. *Under the assumptions of Proposition 4.1, holds*

$$V \geqslant U \geqslant W, \qquad \forall V \in S_1, \quad W \in S_0.$$

Corollary 4.4. *If Condition A or Condition B holds then both Algorithms 1 and 2 are convergent monotonically.*

Now we consider a condition weaker than both Condition A and Condition B.

Condition C. $A^j, j = 1, \cdots, k$, are irreducible, weakly diagonally dominant M-matrices.

The following example shows Condition C does not imply that S_1 has lower bound.

Example 1. $k = n = 2$, and

$$A^1 = \begin{pmatrix} 1 & -1 \\ -1 & 2 \end{pmatrix}, \quad A^2 = \begin{pmatrix} 2 & -1 \\ -1 & 1 \end{pmatrix}, \quad F^1 = F^2 = \begin{pmatrix} 1 \\ -1 \end{pmatrix}.$$

In this case,

$$U = (\alpha + 1, \alpha)^T, \quad -\infty < \alpha \leqslant -1 \tag{4.3}$$

satisfy (1.2) . Set (4.3) does not have lower bound but is contained in S_1. Hence S_1 does not have lower bound in this example.

The example above shows also that Condition C does not imply the uniqueness of the solution of (1.2). The following example shows that it does not imply the existence of the solution.

Example 2. $k = n = 2, A^1, A^2$ are as above, and

$$F^1 = F^2 = \begin{pmatrix} 1 \\ -2 \end{pmatrix}.$$

It is not difficult to show by direct calculation that (1.2) does not have any solution in this case.

References

[1] A. Bensoussan and J.L. Lions, *Applications of Variational Inequalities in Stochastic Control*, North-Holland, Amsterdam, 1982

[2] M. Boulbrachene and M. Haiour, The finite element approximation of Hamilton-Jacobi- Bellman equations, *Computer and Mathematics with Applications*, 14 (2001), 993-1007.

[3] P.L. Lions and B. Mercier, Approximation numerique des equations de Hamilton-Jacobi-Bellman, *RAIRO-Analyse numérique*, 14 (1980), 369-393.

[4] M. Sun, Domain decomposition method for solving HJB equations, *Numerical Functional Analysia and Optimizaiton*, 14 (1993), 145-166.

[5] R.H.W. Hoppe, Multigrid methods for Hamilton-Jacobi-Belman equations, *Numerishe Mathematik*, 49 (1986), 239-254.

[6] C.S. Huang, S. Wang and K.S. Teo, On application of an alternating direction method to HJB equations, *Journal of Computational and Applied Mathematics*, 166 (2004), 153-166.

[7] M. Sun, Alternating direction algorithms for solving HJB equations, *Appled Mathematics and Optimizaiton*, 34 (1996), 267-277.

[8] S.Z. Zhou and G.H. Chen, A monotone iterative algorithm for a discrete HJB equation. *Mathematica Applicata* 18, 639-643, 2005 (in Chinese).

[9] S.Z. Zhou and W.F. Zhan, A new domain decomposition method for an HJB equatoin. *Journal of Computational and Applied Mathematics*, 159 (2003), 195-204.

[10] D. Young, *Iterative solution of Large Linear Systems*, AP, New York, 1971

Domain Decomposition Method for a System of Quasivariational Inequalities*

S.Z. Zhou (周叔子)† Z.Y. Zou (邹战勇)‡ G.H. Chen (陈光华)†

Abstract We propose a domain decomposition method for a system of quasivariational inequalities related to HJB equation. The monotone convergence of the algorithm is also established.

Keywords Domain decomposition method, system of quasivariational inequalities, HJB equation, monotone convergence.

1 Introduction

In order to study the HJB equation, Evans and Friedman, Lions and Mendali introduced an approximate problem (see [1, 2] and the references therein): find $(u^1, \cdots, u^M) \in (H_0^1(\Omega))^M$ such that $u^i \in K^i$ and

$$a^i(u^i, v - u^i) \geqslant (f^i, v - u^i), \quad \forall v \in K^i,\ i = 1, \cdots, M, \tag{1.1}$$

where

$$K^i = \{v \in H_0^1(\Omega) : v \leqslant k + u^{i+1}\},$$

$u^{M+1} = u^1$, k is a positive number small enough,

$$a^i(u, v) = \int\limits_\Omega \Big(\sum_{j,k=1}^d a_{jk}^i \frac{\partial u}{\partial x_j} \frac{\partial v}{\partial x_k} + \sum_{k=1}^d a_k^i \frac{\partial u}{\partial x_k} + a_0^i uv \Big) \mathrm{d}x,$$

$$\sum_{j,k=1}^d a_{jk}^i(x) \xi_j \xi_k \geqslant \alpha |\xi|^2, \quad \forall x \in \bar{\Omega},\ \xi \in R^N,$$

$$a_{jk}^i = a_{kj}^i, \quad a_0^i(x) \geqslant c_0 \geqslant 0, \quad \forall x \in \bar{\Omega},$$

* 发表于：Acta Mathematicae Applicatae Sinica, Vol. 25, 2009, pp. 75-82.

† Department of Applied Mathematics, Hunan University, Changsha, China.

‡ School of Mathematics & Computer Science, Guangdong University of Business Studies, Guangzhou, China.

$\alpha > 0$, a^i_{jk}, a^i_k, a^i_0, f^i are smooth, Ω is a bounded domain in R^d ($d = 1, 2, 3$). We assume there exists $\delta > 0$ such that

$$a^i(v, v) \geqslant \delta \|v\|^2_{H^1(\Omega)}, \quad \forall v \in H^1_0(\Omega).$$

The discrete problem of (1.1) is stated as follows: find $U^i \in C^i$ such that

$$(L^i U^i - F^i, V - U^i) \geqslant 0, \quad \forall V \in C^i,\ i = 1, \cdots, M, \tag{1.2}$$

where $C^i = \{V \in R^N : V \leqslant U^{i+1} + ke\}$, $e = (1, \cdots, 1)^T$, $L^i \in R^{N\times N}$, $F^i \in R^N$.

In [1] an iterative algorithm was constructed for solving (1.2). If $\tilde{U}^n = (U^{1,n}, \cdots, U^{M,n})$ is given then compute $\tilde{U}^{n+1} = (U^{1,n+1}, \cdots, U^{M,n+1})$ by solving the following independent variational (not quasivariational!) inequalities separately:

$$\begin{cases} U^{i,n+1} \in C^{i,n+1}, \\ (L^i U^{i,n+1} - F^i, V - U^{i,n+1}) \geqslant 0, \quad \forall V \in C^{i,n+1},\ i = 1, \cdots, M, \end{cases} \tag{1.3}$$

where $C^{i,n+1} = \{V \in R^N : V \leqslant U^{i+1,n} + ke\}$, $U^{M+1,n} = U^{1,n}$.

In [1] the convergence of algorithm (1.3) was proved under the conditions:

(a) $F^i \geqslant 0$, $i = 1, \cdots, M$.

(b) $\tilde{U}^0 = (0, \cdots, 0)$ or $\tilde{U}^0 = (U^{1,0}, \cdots, U^{M,0})$ satisfies $L^i U^{i,0} = F^i$, $i = 1, \cdots, M$.

In this paper we propose a domain decomposition method for solving (1.2), which is based on (1.3) and is easy to be implemented parallelly. We use an argument approach, which we proposed in [3], to prove the convergence of our method without conditions (a) and (b).

2 Supersolution and Subsolution

If we use appropriate finite element triangulation, then it holds that (see [4, 5])

$$L^i,\ i = 1, \cdots, M, \text{ are M-matrixes.} \tag{2.1}$$

From now on we assume (2.1) is true. Hence (1.2) has a unique solution (see [1] and the references therein).

It is well-known that (1.2) is equivalent to the following system:

$$\max(L^i U^i - F^i,\ U^i - U^{i+1} - ke) = 0, \quad i = 1, \cdots, M. \tag{2.2}$$

We call $\tilde{V} \in (V^1, \cdots, V^M) \in (R^N)^M$ a supersolution of (1.2) if

$$\max(L^i V^i - F^i,\ V^i - V^{i+1} - ke) \geqslant 0, \quad i = 1, \cdots, M. \tag{2.3}$$

The set of all supersolutions of (1.2) is denoted by S_1. We call $\tilde{V} \in (R^N)^M$ a subsolution of (1.2) if

$$\max(L^i V^i - F^i, V^i - V^{i+1} - ke) \leqslant 0, \quad i = 1, \cdots, M. \tag{2.4}$$

The set of all subsolutions of (1.2) is denoted by S_0.

Obviously,the unique solution of (1.2) is both a subsolution and a supersolution. If $F^i \geqslant 0, i = 1,\cdots,M$, then $\tilde{V} = 0 \in (R^N)^M$ is a subsolution. If $F^i \leqslant 0$, $i = 1,\cdots,M$, then $\tilde{V} = 0 \in (R^N)^M$ is a supersolution. Now we give more examples.

Example 1. Assume U_a^i is the solution of the following system:

$$L^i U_a^i = |F^i|, \quad i = 1,\cdots,M, \tag{2.5}$$

where $|F^i| = (|F_1^i|,\cdots,|F_N^i|)^T$. Then $L^i U_a^i - F^i \geqslant 0$, and $\tilde{U}_a = (U_a^1,\cdots,U_a^M) \in S_1$.

Example 2. Assume U_b^i is the solution of the following system:

$$L^i U_b^i = \max_i F^i, \quad i = 1,\cdots,M, \tag{2.6}$$

where $\max\limits_i F^i = (\max\limits_i F_1^i,\cdots,\max F_N^i)$. Then $\tilde{U}_b \in S_1$.

Example 3. Let $L^i = (l_{kj}^i)_{k,j=1,N}$, $F^i = (f_j^i)_{j=1,\cdots,N}$. If L^i, $i = 1,\cdots,M$, have strict diagonal dominance then $V_a = \theta e$ satisfies (see [3]):

$$L^i V_a - F^i \leqslant 0, \quad i = 1,\cdots,M, \tag{2.7}$$

where

$$\theta \leqslant \min_{i=1,\cdots,M,k=1,\cdots,N} \left(l_{kk}^i + \sum_{k\neq j} l_{kj}^i \right)^{-1} f_k^i.$$

Let $\tilde{V}_a = (V_a,\cdots,V_a) \in (R^N)^M$. Then $\tilde{V}_a \in S_0$.

Example 4. If L^i, $i = 1,\cdots,M$, are irreducible and have weak diagonal dominance then under mild assumption we may construct $V_b \in R^N$ satisfying (2.7) and $\tilde{V}_b \in S_0$ (see [6]).

Now we prove the set S_0 has an upper bound.

Lemma 2.1. *For any $\tilde{V} \in S_0$ there holds*

$$\tilde{V} \leqslant \tilde{U}_a, \quad \tilde{V} \leqslant \tilde{U}_b,$$

where $\tilde{U}_a$ and $\tilde{U}_b$ are defined by (2.5) and (2.6) respectively.

Proof. We have by definition:

$$L^i V^i - F^i \leqslant 0, \quad i = 1,\cdots,M.$$

Subtracting the above inequalities from (2.5) we obtain:

$$L^i(U_a^i - V^i) \geqslant |F^i| - F^i \geqslant 0, \quad i = 1,\cdots,M. \tag{2.8}$$

Since L^i, $i = 1,\cdots,M$, are M-matrixes, (2.8) means that $U_a^i \geqslant V^i$, $i = 1,\cdots,M$. Similarly, we derive $U_b^i \geqslant V^i$, $i = 1,\cdots,M$. The proof is complete. □

3 Domain Decomposition Method

Let $\tilde{N}=\{1,\cdots,N\}$. Decompose $\tilde{N}=N_1\bigcup N_2\bigcup\cdots\bigcup N_m$. Define domain decomposition method as follows.

Algorithm DDM

Step 1. Given $\varepsilon>0$, $\tilde{U}^0\in S_0$, $n:=0$, $i:=1$.

Step 2. Solve the following subproblems for $j=1,\cdots,m$ parallely: find $U^{i,n+1,j}\in C^{i,n+1,j}$ such that

$$(L^iU^{i,n+1,j}-F^i,V-U^{i,n+1,j})\geqslant 0,\quad \forall V\in C^{i,n+1,j},\tag{3.1}$$

where

$$\begin{aligned}&C^{i,n+1,j}=\{V\in R^N: V_s\leqslant U_s^{i+1,n}+k\ \text{ if }\ s\in N_j,\ V_s=U_s^{i,n}\ \text{ if }\ s\in\tilde{N}\backslash N_j\},\\&U^{M+1,n}=U^{1,n}.\end{aligned}$$

Step 3. $U^{i,n+1}=\max\limits_j\{U^{i,n+1,j}\}$, if $i=M$ then go to Step 4 otherwise $i:=i+1$ and go to Step 2.

Step 4. If $\|\tilde{U}^{n+1}-\tilde{U}^n\|<\varepsilon$ then output $\tilde{U}^{n+1}$ otherwise $n:=n+1, i:=1$ and go to Step 2.

Remark 3.1. Subproblem (3.1) is equivalent to the following system:

$$U_s^{i,n+1,j}=U_s^{i,n}\quad\text{for}\quad s\in\tilde{N}\backslash N_j,\tag{3.2}$$

$$\max((L^iU^{i,n+1,j}-F^i)_s,U_s^{i,n+1,j}-U_s^{i+1,n}-k)=0\quad\text{for}\quad s\in N_j.\tag{3.3}$$

Remark 3.2. We may take $\tilde{U}^0\in S_1$ in Step 1. Then we compute $U^{i,n+1}$ in Step 3 as follows:

$$U^{i,n+1}=\min_j\{U^{i,n+1,j}\}.$$

Now we prove the convergence of the above algorithm.

Theorem 3.1. *Assume $\{\tilde{U}^n\}$ is produced by Algorithm DDM. Then*

(i) If $\tilde{U}^0\in S_0$ then $\{\tilde{U}^n\}$ is monotonely increasing, $\{\tilde{U}^n\}\subset S_0$, and converges to the solution of (1.2);

(ii) If $\tilde{U}^0\in S_1$ then $\{\tilde{U}^n\}$ is monotonely decreasing, $\{\tilde{U}^n\}\subset S_1$, either $\{\tilde{U}^n\}$ converges to the solution of (1.2), or $\|\tilde{U}^n\|\to\infty$ $(n\to\infty)$.

Proof. (i) Since $\tilde{U}^0\in S_0$, we have

$$\max(L^iU^{i,0}-F^i,\ U^{i,0}-U^{i+1,0}-ke)\leqslant 0,\quad i=1,\cdots,M.\tag{3.4}$$

At first we prove

$$U^{1,1,1}\geqslant U^{1,0}.\tag{3.5}$$

By algorithm we know (see (3.2), (3.3)):

$$U_s^{1,1,1} = U_s^{1,0} \quad \text{for} \quad s \in \tilde{N}\backslash N_1, \tag{3.6}$$

and

$$\max((L^1U^{1,1,1} - F^1)_s, U_s^{1,1,1} - U_s^{2,0} - k) = 0 \quad \text{for} \quad s \in N_1. \tag{3.7}$$

By (3.4) we know

$$\max((L^1U^{1,0} - F^1)_s,\ U_s^{1,0} - U_s^{2,0} - k) \leqslant 0 \quad \text{for} \quad s \in N_1. \tag{3.8}$$

Let $I = \{s \in N_1 : (L^1U^{1,0} - F^1)_s(U_s^{1,0} - U_s^{2,0} - k) > 0\}$. Then

$$(L^1U^{1,0} - F^1)_s(U_s^{1,0} - U_s^{2,0} - k) = 0 \qquad \text{for} \qquad s \in N_1\backslash I, \tag{3.9}$$

$$U_s^{1,0} - U_s^{2,0} - k < 0 \qquad \text{for} \qquad s \in I. \tag{3.10}$$

Choose $\delta_s > 0$ and $d = (d_1, \cdots, d_N)^T$ such that

$$U_s^{1,0} - U_s^{2,0} - k + \delta_s = 0 \qquad \text{for} \qquad s \in I, \tag{3.11}$$

$$d_s = k \qquad \text{for} \qquad s \in N_1\backslash I, \tag{3.12}$$

$$d_s = k - \delta_s \qquad \text{for} \qquad s \in I. \tag{3.13}$$

Then it is easy to see from (3.8)–(3.13) that

$$d \leqslant ke, \tag{3.14}$$

$$\max((L^1U^{1,0} - F^1)_s, U_s^{1,0} - U_s^{2,0} - d_s) = 0, \quad \text{for} \quad s \in N_1. \tag{3.15}$$

It follows from (3.7), (3.15), (3.14) and the monotonicity of the solution of variational inequality respect to the obstacle vector (see Lemma 4.2 of [3]) that (3.5) holds. Similarly, we derive

$$U^{1,1,j} \geqslant U^{1,0}, \quad j = 1, \cdots, m. \tag{3.16}$$

Therefore,

$$U^{1,1} = \max_j\{U^{1,1,j}\} \geqslant U^{1,0}. \tag{3.17}$$

Now we prove

$$\max(L^1U^{1,1} - F^1, U^{1,1} - U^{2,0} - ke) \leqslant 0. \tag{3.18}$$

(3.6) and (3.4) implies

$$U_s^{1,1,1} - U_s^{2,0} - k = U_s^{1,0} - U_s^{2,0} - k \leqslant 0 \quad \text{for} \quad s \in \tilde{N}\backslash N_1. \tag{3.19}$$

For $s \in \tilde{N}\backslash N_1$ we have

$$(L^1U^{1,1,1})_{\tilde{N}\backslash N_1} = L^1_{\tilde{N}\backslash N_1, \tilde{N}\backslash N_1} U^{1,1,1}_{\tilde{N}\backslash N_1} + L^1_{\tilde{N}\backslash N_1, N_1} U^{1,1,1}_{N_1}, \tag{3.20}$$

where F_P is the subvector of vector F corresponding to the index set P, and $L_{P,Q}$ is the submatrix of matrix L corresponding to the row index set P and column index set Q. By (3.6) we know

$$L^1_{\tilde{N}\backslash N_1,\tilde{N}\backslash N_1} U^{1,1,1}_{\tilde{N}\backslash N_1} = L^1_{\tilde{N}\backslash N_1,\tilde{N}\backslash N_1} U^{1,0}_{\tilde{N}\backslash N_1}. \tag{3.21}$$

It follows from the properties of M-matrix and (3.16) that

$$L^1_{\tilde{N}\backslash N_1,N_1} U^{1,1,1}_{N_1} \leqslant L^1_{\tilde{N}\backslash N_1,N_1} U^{1,0}_{N_1},$$

which combining with (3.20) and (3.21) yields

$$(L^1 U^{1,1,1} - F^1)_{\tilde{N}\backslash N_1} \leqslant (L^1 U^{1,0} - F^1)_{\tilde{N}\backslash N_1} \leqslant 0. \tag{3.22}$$

(3.19) and (3.22) means

$$\max((L^1 U^{1,1,1} - F^1)_s, U^{1,1,1}_s - U^{2,0}_s - k) \leqslant 0 \quad \text{for} \quad s \in \tilde{N}\backslash N_1. \tag{3.23}$$

It follows from (3.7) and (3.23) that

$$\max(L^1 U^{1,1,1} - F^1, U^{1,1,1} - U^{2,0} - k) \leqslant 0.$$

Similar argument leads to

$$\max(L^1 U^{1,1,j} - F^1, U^{1,1,j} - U^{2,0} - k) \leqslant 0, \quad j = 2, \cdots, m.$$

Therefore, it is not difficult to show (ref. [7]) (3.18) holds.

By this way we obtain

$$U^{i,1} \geqslant U^{i,0}, \quad i = 1, \cdots, M, \tag{3.24}$$

$$\max(L^i U^{i,1} - F^i, U^{i,1} - U^{i+1,0} - ke) \leqslant 0, \quad i = 1, \cdots, M, \tag{3.25}$$

where $U^{M+1,0} = U^{1,0}$. (3.24) means

$$\tilde{U}^1 \geqslant \tilde{U}^0. \tag{3.26}$$

It follows from (3.24) and (3.25) that

$$\max(L^i U^{i,1} - F^i, U^{i,1} - U^{i+1,1} - ke) \leqslant 0$$

where $U^{M+1,1} = U^{1,1}$. This is

$$\tilde{U}^1 \in S_0. \tag{3.27}$$

Based on (3.26) and (3.27), we obtain by similar argument

$$\tilde{U}^{n+1} \geqslant \tilde{U}^n \geqslant \cdots \geqslant \tilde{U}^1 \geqslant \tilde{U}^0, \tag{3.28}$$

$$\{\tilde{U}^n\} \subset S_0. \tag{3.29}$$

By Lemma 2.1 and (3.29) we know

$$\tilde{U}^n \leqslant \tilde{U}_a, \quad n = 0, 1, \cdots,$$

which and (3.28) means there exists $\tilde{U}^* \in (R^N)^M$ such that

$$\lim_{n\to\infty} \tilde{U}^n = \tilde{U}^* = (U^{1,*}, \cdots, U^{M,*}).$$

Checking the above argument process we know

$$U^{i,n} \leqslant U^{i,n+1,j} \leqslant U^{i,n+1}, \quad j = 1, \cdots, m.$$

Hence we have also

$$\lim_{n\to\infty} U^{i,n+1,j} = U^{i,*}, \quad j = 1, \cdots, m,$$

which and (3.3) leads to

$$\max(L^i U^{i,*} - F^i, U^{i,*} - U^{i+1,*-ke}) = 0, \quad i = 1, \cdots, M,$$

where $U^{M+1,*} = U^{1,*}$. Hence $\tilde{U}^*$ is the unique solution of (1.2). (i) is proved.

(ii) Since $\tilde{U}^0 \in S_1$ we have

$$\max(L^i U^{i,0} - F^i, U^{i,0} - U^{i+1,0} - ke) \geqslant 0, \quad i = 1, \cdots, M, \tag{3.30}$$

where $U^{M+1,0} = U^{1,0}$. Now we prove

$$U^{1,1,1} \leqslant U^{1,0}. \tag{3.31}$$

Let $I = \{s \in \tilde{N} : U_s^{1,0} - U_s^{2,0} - k \geqslant 0\}$. Then

$$(L^1 U^{1,0} - F^1)_S \geqslant 0 \quad \text{for} \quad s \in \tilde{N} \backslash I. \tag{3.32}$$

By the definition of set I, (3.6) and (3.7) we know

$$U_s^{1,1,1} \leqslant U_s^{1,0} \quad \text{for} \quad s \in I \cup (\tilde{N} \backslash N_1). \tag{3.33}$$

Since (3.7) we have

$$(L^1 U^{1,1,1} - F^1)_s \leqslant 0 \quad \text{for} \quad s \in N_1. \tag{3.34}$$

It follows from (3.32) and (3.34) that

$$(L^1(U^{1,0} - U^{1,1,1}))_s \geqslant 0 \quad \text{for} \quad s \in (\tilde{N} \backslash I) \cap N_1,$$

which and (3.33) implies (noting L^1 is an M-matrix)

$$U_s^{1,1,1} \leqslant U_s^{1,0} \quad \text{for} \quad s \in (\tilde{N} \backslash I) \cap N_1.$$

The above inequality and (3.33) means (3.31) holds.

Similarly, we obtain

$$U^{1,1,j} \leqslant U^{1,0}, \quad j = 1, \cdots, M. \tag{3.35}$$

Therefore, we have

$$U^{1,1} = \min_j (U^{1,1,j}) \leqslant U^{1,0}. \tag{3.36}$$

Now we prove

$$\max(L^1 U^{1,1} - F^1,\ U^{1,1} - U^{2,0} - ke) \geqslant 0. \tag{3.37}$$

By (3.6) we have

$$U_s^{1,1,1} - U_s^{2,0} - k = U_s^{1,0} - U_s^{2,0} - k \quad \text{for} \quad s \in \tilde{N} \backslash N_1. \tag{3.38}$$

It follows from the properties of M-matrix and (3.31) that

$$L^1_{\tilde{N}\backslash N_1, N_1} U^{1,1,1}_{N_1} \geqslant L^1_{\tilde{N}\backslash N_1, N_1} U^{1,0}_{N_1}.$$

Together with (3.6), we then have

$$\begin{aligned}(L^1 U^{1,1,1})_{\tilde{N}\backslash N_1} &= L^1_{\tilde{N}\backslash N_1, \tilde{N}\backslash N_1} U^{1,1,1}_{\tilde{N}\backslash N_1} + L^1_{\tilde{N}\backslash N_1, N_1} U^{1,1,1}_{N_1} \\ &\geqslant L^1_{\tilde{N}\backslash N_1, \tilde{N}\backslash N_1} U^{1,0}_{\tilde{N}\backslash N_1} + L^1_{\tilde{N}\backslash N_1, N_1} U^{1,0}_{N_1} \\ &= (L^1 U^{1,0})_{\tilde{N}\backslash N_1}.\end{aligned}$$

Therefore, we obtain

$$(L^1 U^{1,1,1} - F^1)_s \geqslant (L^1 U^{1,0} - F^1)_s \quad \text{for} \quad s \in \tilde{N} \backslash N_1. \tag{3.39}$$

(3.38),(3.39) and (3.30) together imply that

$$\max((L^1 U^{1,1,1} - F^1)_s,\ U_s^{1,1,1} - U_s^{2,0} - k) \geqslant 0 \quad \text{for} \quad s \in \tilde{N} \backslash N_1. \tag{3.40}$$

(3.40) and (3.7) result in

$$\max(L^1 U^{1,1,1} - F^1, U^{1,1,1} - U^{2,0} - ke) \geqslant 0.$$

Similarly, we also have

$$\max(L^1 U^{1,1,j} - F^1,\ U^{1,1,j} - U^{2,0} - ke) \geqslant 0, \quad j = 1, \cdots, m.$$

Hence we can derive (ref. [7]) that (3.37) holds. Now we use an argument similar to that in the proof of (i) to obtain

$$\tilde{U}^1 \in S_1,$$

and

$$\tilde{U}^{n+1} \leqslant \tilde{U}^n \leqslant \cdots \leqslant \tilde{U}^1 \leqslant \tilde{U}^0, \tag{3.41}$$

$$\{\tilde{U}^n\} \subset S_1. \tag{3.42}$$

The conclusion of (ii) is an obvious result of (3.41) and (3.42). The proof is complete. □

Remark 3.3. If S_1 has a bound below, then it follows from the conclusion in (ii) that $\{\tilde{U}^n\}$ is convergent. For example, if $F^i \geqslant 0$, $i = 1, \cdots, M$ then $0 \in (R^N)^M$ is a lower bound of the set S_1. It is an open problem for general case.

References

[1] M. Boulbrachene and M. Haiour, The finite element approximation of Hamilton-Jacobi-Bellman equations. *Computer and Mathematics with Applications*, 41 (2001), 993-1007.

[2] P.L. Lions and J.L. Mendali, Optimal control of stochastic integrals and HJB equations, *SIAM Journal on Control and Optimization*, 20 (1982), 58-81.

[3] S.Z. Zhou and W.P. Zhan, A new domain decomposition method for an HJB equatoin, *Journal of Computational and Applied Mathematics*, 159 (2003), 195-204.

[4] P.G. Ciarlet, *The finite element method for elliptic problems*, North-Holland, Amsterdam, 1978.

[5] J.P. Zeng and S.Z. Zhou, On monotone and geometric convergence of Schwarz methods for two-sided obstacle problems, *SIAM Journal on Numerical Analysis*, 35 (1998), 600-616.

[6] S.Z. Zhou and G.H. Chen, A monotone iterative algorithm for a discrete HJB equation, *Mathematica Applicata*, 18 (2005), 639-645. (in Chinese)

[7] S.Z. Zhou and J.P. Zeng, Schwarz algorithm for obstacle problems with a type of nonlinear operators, *Acta Mathematicae Applicatae Sinica*, 20 (1997), 521-530 (in Chinese).

Relaxation Lax-Friedrichs Sweeping Scheme for Static Hamilton-Jacobi Equations*

P. Zhu (祝鹏)† S.Z. Zhou (周叔子)†

Abstract This paper presents a relaxation Lax-Friedrichs sweeping scheme to approximate viscosity solutions of static Hamilton Jacobi equations in any number of spatial dimensions. It is a generalization of the scheme proposed in [C.-Y. Kao, S. Osher, J. Qian, Journal of Computational Physics 196(2004) 367-391]. Numerical examples suggest that the relaxation Lax-Friedrichs sweeping scheme has smaller number of iterations than the original Lax-Friedrichs sweeping scheme when the relaxation factor ω is slightly larger than one. And first order convergence is also demonstrated by numerical results. A theoretical analysis for our scheme in a special case is given.

Keywords Static Hamilton-Jacobi equations, fast sweeping method, Lax-Friedrichs flux, SOR iteration.

1 Introduction

The Hamilton-Jacobi (H-J) equations arise in many applications such as geometrical optics, crystal growth, etching, computer vision, obstacle navigation, path planning, photolithography, and seismology. In general, these nonlinear PDEs cannot be solved analytically. The solutions usually develop singularities in their derivatives even with smooth initial conditions. In these cases, the solutions do not satisfy the equations in the classical sense. The weak solution that is usually sought is called the viscosity solution [1]. Numerically in general, one looks for a consistent and monotone scheme to construct an approximate viscosity solution [2].

In this paper, we focus on static H-J equations of the following form:

$$\begin{cases} H(x, \nabla\phi(x)) = R(x), & x \in \Omega, \\ \phi(x) = g(x), & x \in \Gamma, \end{cases} \tag{1.1}$$

where H, g, and $R > 0$ are Lipschitz continuous, and Γ is a subset of Ω. A special case of this

* 发表于：Numerical Algorithms，Vol. 54, 2010, pp. 325-342.

† College of Mathematics and Econometrics, Hunan University, Changsha, Hunan 410082, China.

type of equation is the Eikonal equation,

$$|\nabla\phi| = r(x) \tag{1.2}$$

with the same type of Dirichlet boundary conditions in (1.1).

There are mainly two classes of numerical methods for solving static H-J equations. The first class of numerical methods is based on reformulating the equations into suitable time-dependent problems. Osher [3] provides a nature link between static and time-dependent H-J equations by using the level-set idea. The zero level set of the viscosity solution ψ of the time-dependent H-J equation

$$\psi_t(x,t) + H(x, \nabla\psi(x,t)) = 0 \tag{1.3}$$

with suitable initial condition at various time t is the set of x such that $\phi(x) = t$ of (1.1), where the Hamiltonian H is homogeneous of degree one. Another approach to obtaining a time dependent H-J equation from the static H-J equation is using the so called paraxial formulation in which a preferred spatial direction is assumed in the characteristic propagation $[4-6]$. High order numerical schemes are well developed for the time dependent H-J equations (1.3) on structured and unstructured meshes $[7-13]$; see a recent review on high order numerical methods for time-dependent H-J equations by Shu [14].

The other class of numerical methods for static H-J equation is to treat the problem as a stationary boundary value problem: discretize the problem into a system of nonlinear equations and design an efficient numerical algorithm to solve the system. Among such methods are the fast marching method and the fast sweeping method. In the fast marching method [15, 16], the solution is updated pointwise in the order that the solution is strictly increasing (decreasing); hence two essential ingredients are needed in the algorithm: an upwind difference scheme and a heap-sort algorithm. In the fast sweeping method $[17-20]$, Gauss-Seidel iterations with alternating direction sweepings are incorporated into upwind finite differences. Both the fast marching method and the fast sweeping method are extremely efficient algorithms for solving static H-J equations, but they are first order schemes due to upwind discretization. Based on first order fast sweeping methods, Zhang et al.[21] incorporates high order approximations to derivatives into numerical Hamiltonian constructing high order fast sweeping methods for static H-J equations.

However, all of the above cited methods, except the methods in [21] which can deal with non-convex Hamiltonian, are designed for static H-J equations by assuming convexity and/or homogeneity of Hamiltonian. Kao et al. [17] proposes a new Gauss-Seidel (G-S) sweeping type algorithm which is based on the Lax-Friedrichs Hamiltonian. It can handle both convex and non-convex Hamiltonian, no matter how complicated they are. In this paper, we use SOR relaxation iterations with alternating direction sweeping instead of G-S iterations in [17]; i.e., we introduce a relaxation factor ω and generalize the Lax-Friedrichs sweeping scheme in [17], when $\omega = 1$ is the original method. Numerical results suggest that the number of iterations is reduced evidently when ω is slightly larger than one.

The rest of the paper is organized as follows. In section 2, we give implementation details of our algorithm, which includes SOR type update formula and computational boundary conditions. In section 3, we prove a few basic properties of the new algorithm. In section 4, we present various numerical examples to illustrate the efficiency and accuracy of the new method. In section 5, we analyze our scheme in a special case. We conclude the paper in section 6.

2 Relaxation Lax-Friedrichs Sweeping Scheme

For illustration purpose, we start with the one dimensional case. The one dimensional Lax-Friedrichs Hamiltonian is (dropping the obvious x dependence on H)

$$\hat{H}(p^-, p^+) = H\left(\frac{p^+ + p^-}{2}\right) - \frac{1}{2}\sigma_x(p^+ - p^-),$$

where $p = \phi_x, p^{\pm}$ are corresponding forward and backward difference approximations of $p = \phi_x$, and [A,B] is the range of $p^{\pm}$; σ_x is the artificial viscosity satisfying

$$\sigma_x \geqslant \max_{p \in [A,B]} |H'(p)|,$$

where H' is the derivatives of H. Consider an uniform discretization $\{x_i\}_{i=1}^m$ of Ω with mesh size Δx. At mesh point x_i we have following numerical approximation:

$$\hat{H} = R,$$

which implies

$$H(x_i, \frac{\phi_{i+1} - \phi_{i-1}}{2\Delta x}) - \sigma_x \frac{\phi_{i+1} - 2\phi_i + \phi_{i-1}}{2\Delta x} = R(x_i).$$

Solving for ϕ_i in the above, we get a simple Lax-Friedrichs (LxF) update formula,

$$\phi_i^{\text{new}} = \frac{\Delta x}{\sigma_x}\left(R(x_i) - H\left(x_i, \frac{\phi_{i+1} - \phi_{i-1}}{2\Delta x}\right)\right) + \frac{1}{2}(\phi_{i+1} + \phi_{i-1}). \tag{2.1}$$

We on purpose did not put superscripts on ϕ_{i+1} and ϕ_{i-1} since those superscripts depend on the directions of sweeping. Kao et al. [17] utilize G-S iterations with alternating sweeping, if we sweep from left to right, i.e. $i = 1 : m$,

$$\phi_i^{\text{new}} = \frac{\Delta x}{\sigma_x}\left(R(x_i) - H\left(x_i, \frac{\phi_{i+1}^{\text{old}} - \phi_{i-1}^{\text{new}}}{2\Delta x}\right)\right) + \frac{1}{2}(\phi_{i+1}^{\text{old}} + \phi_{i-1}^{\text{new}}), \tag{2.2}$$

and then from right to left, i.e. $i = m : 1$,

$$\phi_i^{\text{new}} = \frac{\Delta x}{\sigma_x}\left(R(x_i) - H\left(x_i, \frac{\phi_{i+1}^{\text{new}} - \phi_{i-1}^{\text{old}}}{2\Delta x}\right)\right) + \frac{1}{2}(\phi_{i+1}^{\text{new}} + \phi_{i-1}^{\text{old}}). \tag{2.3}$$

(2.2)–(2.3) are called G-S type LxF updated formula in our paper.

In this paper, we use the SOR iterations with alternating direction sweepings instead of G-S iterations in [17]. And then the corresponding formula are the following:
if we sweep from left to right, i.e. $i = 1 : m$,

$$\begin{cases} \phi_i^{\text{tmp}} = \dfrac{\Delta x}{\sigma_x}\left(R(x_i) - H\left(x_i, \dfrac{\phi_{i+1}^{\text{old}} - \phi_{i-1}^{\text{new}}}{2\Delta x}\right)\right) + \dfrac{1}{2}(\phi_{i+1}^{\text{old}} + \phi_{i-1}^{\text{new}}), \\ \phi_i^{\text{new}} = \omega\phi_i^{\text{tmp}} + (1-\omega)\phi_i^{\text{old}}, \end{cases} \tag{2.4}$$

and then from right to left, i.e. $i = m : 1$,

$$\begin{cases} \phi_i^{\text{tmp}} = \dfrac{\Delta x}{\sigma_x}\left(R(x_i) - H\left(x_i, \dfrac{\phi_{i+1}^{\text{new}} - \phi_{i-1}^{\text{old}}}{2\Delta x}\right)\right) + \dfrac{1}{2}(\phi_{i+1}^{\text{new}} + \phi_{i-1}^{\text{old}}), \\ \phi_i^{\text{new}} = \omega\phi_i^{\text{tmp}} + (1-\omega)\phi_i^{\text{old}}. \end{cases} \tag{2.5}$$

(2.4)–(2.5) are called SOR type LxF update formula. Obviously when $\omega = 1$ our scheme is the original method.

Remark 2.1. Dropping the superscripts on ϕ_{i+1} and ϕ_{i-1}, (2.4) and (2.5) can be rewritten as

$$\phi_i^{\text{new}} = \phi_i + \omega\frac{\Delta x}{\sigma_x}\left\{R(x_i) - \hat{H}\left(\frac{\phi_i - \phi_{i-1}}{\Delta x}, \frac{\phi_{i+1} - \phi_i}{\Delta x}\right)\right\}, \tag{2.6}$$

which is coincide with the first order version of fixed-point iterative sweeping scheme [22].

For the two dimensional case, we have similar update formula and iteration tactics. Consider a rectangular domain $[x_{min}, x_{max}] \times [y_{min}, y_{max}]$ with a uniform discretization $(x_i, y_j), i = 0, 1, \cdots, m_1, m_1 + 1$, and $j = 0, 1, \cdots, m_2, m_2 + 1$, where $x_i = (i-1)\Delta x + x_{min}, y_j = (j-1)\Delta y + y_{min}, \Delta x = (x_{max} - x_{min})/(m_1 - 1)$ and $\Delta y = (y_{max} - y_{min})/(m_2 - 1)$. In two dimensional case we have four sweeping directions: (i) from lower left to upper right; (ii) from lower right to upper left; (iii) from upper left to lower right; (iv) from upper right to lower left. For saving space, we only list update formulation in one direction. The update formulations in other sweeping directions are similar.

If we sweep from lower left to upper right, i.e. $i = 1 : m_1, j = 1 : m_2$,

$$\begin{cases} \phi_{ij}^{\text{tmp}} = \gamma\left(R_{ij} - H\left(\dfrac{\phi_{i+1,j}^{\text{old}} - \phi_{i-1,j}^{\text{new}}}{2\Delta x}, \dfrac{\phi_{i,j+1}^{\text{old}} - \phi_{i,j-1}^{\text{new}}}{2\Delta y}\right)\right) \\ \qquad\qquad + \gamma\left(\sigma_x\dfrac{\phi_{i+1,j}^{\text{old}} + \phi_{i-1,j}^{\text{new}}}{2\Delta x} + \sigma_y\dfrac{\phi_{i,j+1}^{\text{old}} + \phi_{i,j-1}^{\text{new}}}{2\Delta y}\right), \\ \phi_{ij}^{\text{new}} = \omega\phi_{ij}^{\text{tmp}} + (1-\omega)\phi_{ij}^{\text{old}} \end{cases} \tag{2.7}$$

where $\gamma=\dfrac{1}{\dfrac{\sigma_x}{\Delta x}+\dfrac{\sigma_y}{\Delta y}}$, H_p and H_q are partial derivatives of H respectively, σ_x and σ_y are artificial viscosities satisfying $\sigma_x \geqslant \max|H_p|, \sigma_y \geqslant \max|H_q|$.

We remark that no nonlinear inversion is required in the above formula, therefore the algorithm is simple to implement, no matter how complicated the Hamiltonian might be.

Because the Lax-Friedrichs Hamiltonian gives a solution depending on all of its neighbors in all Cartesian dimensions, we have to specify carefully the values of points outside the computational domain. We follow the method of Kao et al. [17] to handle the computational boundary. For simplicity, we detail boundary conditions in the two dimensional case only. Consider a rectangular domain $[x_{\min}, x_{\max}]\times[y_{\min}, y_{\max}]$ with a uniform discretization $(x_i, y_j), i=0,1,\cdots,m_1,m_1+1$, and $j=0,1,\cdots,m_2,m_2+1$, where $x_i=(i-1)\Delta x+x_{\min}, y_j=(j-1)\Delta y+y_{\min}, \Delta x=(x_{\max}-x_{\min})/(m_1-1)$ and $\Delta y=(y_{\max}-y_{\min})/(m_2-1)$. Combining extrapolation, maximization and minimization to calculate the values for points outside the computational domain, [17] imposes following conditions:

$$\begin{cases}\phi_{0,j}^{\text{new}}=\min\{\max\{2\phi_{1,j}^{\text{new}}-\phi_{2,j}^{\text{new}},\phi_{2,j}^{\text{new}}\},\phi_{0,j}^{\text{old}}\},\\ \phi_{i,0}^{\text{new}}=\min\{\max\{2\phi_{i,1}^{\text{new}}-\phi_{i,2}^{\text{new}},\phi_{i,2}^{\text{new}}\},\phi_{i,0}^{\text{old}}\},\\ \phi_{m_1+1,j}^{\text{new}}=\min\{\max\{2\phi_{m_1,j}^{\text{new}}-\phi_{m_1-1,j}^{\text{new}},\phi_{m_1-1,j}^{\text{new}}\},\phi_{m_1+1,j}^{\text{old}}\},\\ \phi_{i,m_2+1}^{\text{new}}=\min\{\max\{2\phi_{i,m_2}^{\text{new}}-\phi_{i,m_2-1}^{\text{new}},\phi_{i,m_2-1}^{\text{new}}\},\phi_{i,m_2+1}^{\text{old}}\}.\end{cases}\tag{2.8}$$

We summarize the relaxation Lax-Friedrichs sweeping scheme as following. There are essentially three steps: initialization, alternating sweepings, and enforcing computational boundary conditions. We take the two dimensional case for the sake of exposition.

Algorithm RLxFSM:

1. *Initialization.* We assign exact values or interpolated values $\phi_{i,j}^0$ at grid points or near the given boundary values Γ, and these values are fixed during the iterations. At all other grid points, we assign large positive values M to $\phi_{i,j}^0$, where M should be larger than the maximum of the true solutions, and these values will be updated in the process of iterations.

2. *Alternating sweepings.* At iteration $n+1$, we calculate ϕ^{n+1} according to (2.7) at all grid points $(x_i, y_j), 1\leqslant i\leqslant m_1, 1\leqslant j\leqslant m_2$ except for those which have assigned values and update ϕ^{n+1} only when it is less than its previous value ϕ^n. Recall that this process needs to be done in alternating sweeping directions, which means that it needs four different sweeps in the two dimensional case: (i) From lower left to upper right, $i=1:m_1, j=1:m_2$; (ii) from lower right to upper left, $i=m_1:1, j=1:m_2$; (iii) from upper left to lower right, $i=1:m_1, j=m_2:1$; and (iv) from upper right to lower left, $i=m_1:1, j=m_2:1$. In general, in the d-dimensional case, we need 2^d alternating sweeps.

3. *Enforcing computational boundary conditions.* After each sweep, we enforce computational boundary conditions according to formula (2.8), trivially modified depending on which bound-

ary we are at.

4. *Convergence criterion.* If

$$||\phi^{n+1} - \phi^n||_{L^1} \leqslant \varepsilon,$$

where ε is a given convergence threshold value, the algorithm converges and stops.

3 Basic Properties of the RLxFSM Algorithm

In this section we prove a few basic properties for the relaxation Lax-Friedrichs sweeping algorithm. We follow the framework of [18 – 20].

Proposition 3.1. (No-increasing) *The solution of the RLxFSM algorithm is non-increasing with each iteration.*

Proof. This is evident from the updating formula which only updates the current value if it is larger than newly computed value during the iterations. □

The update formula (2.4)–(2.5) or (2.6) and (2.7) are derived from the Lax-Friedrichs numerical flux, its consistency is obviously true. In the following theorem, we will adopt the framework given in [23] to show the monotonicity. We present it in one dimension but it is true in general.

Proposition 3.2. (Monotonicity) *When $0 < \omega \leqslant 1$, the RLxFSM algorithm is monotone and Lipschitz continuous, i.e.,*

$$1 > \frac{\partial \phi_i^{\text{new}}}{\partial \phi_i} \geqslant 0, \quad 1 \geqslant \frac{\partial \phi_i^{\text{new}}}{\partial \phi_{i-1}} \geqslant 0, \quad 1 \geqslant \frac{\partial \phi_i^{\text{new}}}{\partial \phi_{i+1}} \geqslant 0 \tag{3.1}$$

and

$$\frac{\partial \phi_i^{\text{new}}}{\partial \phi_i} + \frac{\partial \phi_i^{\text{new}}}{\partial \phi_{i-1}} + \frac{\partial \phi_i^{\text{new}}}{\partial \phi_{i+1}} = 1. \tag{3.2}$$

Proof. From (2.6), we have

$$\begin{aligned}
\frac{\partial \phi_i^{\text{new}}}{\partial \phi_i} &= 1 - \omega, \\
\frac{\partial \phi_i^{\text{new}}}{\partial \phi_{i-1}} &= \frac{1}{2}\omega\Big(1 + \frac{1}{\sigma_x} H'\Big(\frac{\phi_{i+1} - \phi_{i-1}}{2\Delta x}\Big)\Big), \\
\frac{\partial \phi_i^{\text{new}}}{\partial \phi_{i+1}} &= \frac{1}{2}\omega\Big(1 - \frac{1}{\sigma_x} H'\Big(\frac{\phi_{i+1} - \phi_{i-1}}{2\Delta x}\Big)\Big).
\end{aligned}$$

Obviously, (3.2) holds. Since $\sigma_x \geqslant \max|H'|$ and $0 < \omega \leqslant 1$, we easily get (3.1). □

Proposition 3.3. (Order preserving) *The RLxFSM algorithm is monotone in the initial data.*

Proof. From the monotonicity property of the solution, if $\phi_{ij} \leqslant \psi_{ij}$ at all grid points initially, then $\phi_{ij} \leqslant \psi_{ij}$ at all grid points after any number of iterations. □

4 Numerical Results and Some Observations

In this section, just as in [17], we apply the relaxation Lax-Friedrichs sweeping scheme to Wulff crystal shape problems and traveltime calculations of elastic waves. In all the examples, the threshold value at which the iteration stops is taken to be $\varepsilon = 10^{-10}$.

4.1 The Wulff crystal shape problem

The level set formulation of the Wulff crystal shape problem [17] is

$$\begin{cases} \psi_t + \gamma\Big(\dfrac{\nabla\psi}{|\nabla\psi|}\Big)|\nabla\psi| = 0, & x \in \mathbb{R}^2, t > 0, \\ \psi = 0, & x \in \Gamma, \end{cases} \tag{4.1}$$

where γ is the normal speed. The zero level set of the viscosity solution ψ of (3.1) at time t is the viscosity solution $\phi(x, y) = t$ of the following static Hamilton-Jacobi equation

$$\begin{cases} \gamma\left(\dfrac{\nabla\phi}{|\nabla\phi|}\right)|\nabla\phi| = 1, & x \in \mathbb{R}^2, \\ \phi = 0, & x \in \Gamma. \end{cases} \tag{4.2}$$

In [17], for two dimensional case γ is given as $\gamma(v)$ where v is the angle between the outward normal direction $\dfrac{\nabla\phi}{|\nabla\phi|}$ and x-axis with $-\pi \leqslant v \leqslant \pi$. Thus we have $\cos(v) = \dfrac{p}{\sqrt{p^2+q^2}}$ and $\sin(v) = \dfrac{q}{\sqrt{p^2+q^2}}$, where $p = \dfrac{\partial\phi}{\partial x}$ and $q = \dfrac{\partial\phi}{\partial y}$. Applying these trigonometric equalities to a given norm speed $\gamma(v)$, we obtain a corresponding H-J equation. For example, if $\gamma(v) = 1 + |\sin(v + \frac{\pi}{2})|$, then we have the corresponding H-J equation

$$\sqrt{p^2+q^2} + |p| = 1.$$

In each Wulff crystal shape problem, we specify the normal speed, obtain the corresponding H-J equation. In some Wulff crystal shape problems, we have term such as $\sqrt{p^2+q^2}$ in the denominator; we regularize it by adding a small quantity ϵ, e.g., $\epsilon = 10^{-6}$ in order to avoid "dividing by zero".

We apply the scheme to two dimension problems with two different types of boundary conditions:
(BC1) a single source point at the center of the domain: $\Gamma = \{(0, 0)\}$ and $\phi(\Gamma) = 0$;
(BC2) Γ being a square with length 1 in the center of the domain and $\phi(\Gamma) = 0$;
In all the cases, the computational domain is $[-1, 1] \times [-1, 1]$.

In the following tables, we list the number of iterations for some Hamiltonians.

Table 1 illustrates the number of iterations for $|p| + |q| = 1$ with different mesh sizes and different relaxation factors. In this case the Hamiltonian is very simple, we can take the artificial viscosity exactly, $\sigma_x = \sigma_y = 1$. The scheme has the least number of iterations when $\omega = 1$. And for $\omega > 1$ the scheme is not convergent. It is maybe suggested that the Lax-Friedrichs sweeping scheme in [17] is the best if we can take the artificial viscosity exactly.

Table 1 The number of iterations for $|p| + |q| = 1, \sigma_x = \sigma_y = 1$

ω	Iteration–number					
	mesh :400 × 400		mesh :200 × 200		mesh :100 × 100	
	BC1	BC2	BC1	BC2	BC1	BC2
0.7	116	67	70	42	46	28
0.8	86	50	52	31	35	21
0.9	58	35	37	22	27	15
1.0	20	2	19	3	18	3

Table 2 and Table 3 presents the number of iterations for the same H-J equation

$$\sqrt{p^2+q^2}+\sqrt{\frac{(p^2+q^2)^{3/2}-(3p^2q-q^3)}{2\sqrt{p^2+q^2}}}=1 \tag{4.3}$$

Table 2 The number of iterations for (4.3), $\sigma_x = \sigma_y = 2$

ω	Iteration–number					
	mesh :400 × 400		mesh :200 × 200		mesh :100 × 100	
	BC1	BC2	BC1	BC2	BC1	BC2
0.70	332	200	207	136	146	106
0.80	252	152	160	107	118	85
0.90	189	115	124	85	96	69
1.00	139	87	98	68	79	57
1.05	119	75	88	61	72	52
1.10	103	—	80	—	66	—
1.15	90	—	72	—	61	—
1.18	84	—	68	—	58	—
r	39.6%	13.8%	30.6%	10.3%	26.6%	8.8%

Table 3 The number of iterations for (4.3), $\sigma_x = \sigma_y = 2.5$

ω	Iteration–number					
	mesh :400 × 400		mesh :200 × 200		mesh :100 × 100	
	BC1	BC2	BC1	BC2	BC1	BC2
0.70	461	282	294	197	212	154
0.80	357	220	230	156	171	124
0.90	275	170	181	125	139	101
1.00	210	131	144	101	114	83
1.05	182	115	128	90	103	75
1.10	157	101	115	81	94	68
1.15	136	—	103	—	85	—
1.20	118	—	93	—	78	—
1.25	103	—	84	—	71	—
1.29	94	—	78	—	66	—
r	55.2%	22.9%	45.8%	19.8%	42.1%	18.1%

but with different artificial viscosities. In the table and below, r is a ratio: $r = (n_1 - n_\omega)/n_1$, where n_1 is the number of iterations for $\omega = 1$ and n_ω is the least number of iterations. We can see that the scheme has smaller number of iterations than the Lax-Friedrichs sweeping scheme when the relaxation factor ω is slightly larger than one. And they also show that the larger the artificial is, the larger r is. For complicated Hamiltonian, it is difficult to compute $\max\{|H_p|\}$ and $\max\{|H_q|\}$ exactly and we always take some upper bounds as artificial viscosities. For the same artificial viscosity, we can reduce the number of iterations by adjusting the relaxation factor ω. The meaning of "—" in the tables is that the scheme is not convergent to viscosity solution.

Table 4 shows the numbers of iterations for the H-J equation

$$\sqrt{p^2+q^2} + \left|\frac{p^3 - 3pq^2}{p^2+q^2}\right| = 1 \tag{4.4}$$

with different mesh sizes and relaxation factors. Once again, the number of iterations is reduced evidently when relaxation factor ω is slightly larger than one.

There are nine cases for the Wulff crystal shape problem in [17], but we only report three cases in this paper for saving space. For the other six cases, we have the same results.

Table 4 The number of iterations for (4.4), $\sigma_x = \sigma_y = 3$

ω	Iteration–number					
	mesh :400 × 400		mesh :200 × 200		mesh :100 × 100	
	BC1	BC2	BC1	BC2	BC1	BC2
0.70	516	321	339	228	256	187
0.80	382	239	257	174	202	148
0.90	277	174	196	134	163	119
1.00	195	125	151	106	133	97
1.05	165	107	134	94	121	88
1.10	142	92	118	85	110	80
r	27.2%	26.4%	21.9%	19.8%	17.3%	17.5%

Figure 1 and 2 illustrate the Wulff crystal shapes for (4.3) with different artificial viscosities, and the corresponding crystal shape is a triangle. In the figures and below, dotted line is where the boundary condition is prescribed. As we can see from the figures, they have good resolution as the figures in [17].

Figure 3 presents the Wulff crystal shapes for (4.4), and the corresponding crystal shape is a hexagon. The result is computed when $\omega = 1.1$. They also have good resolution.

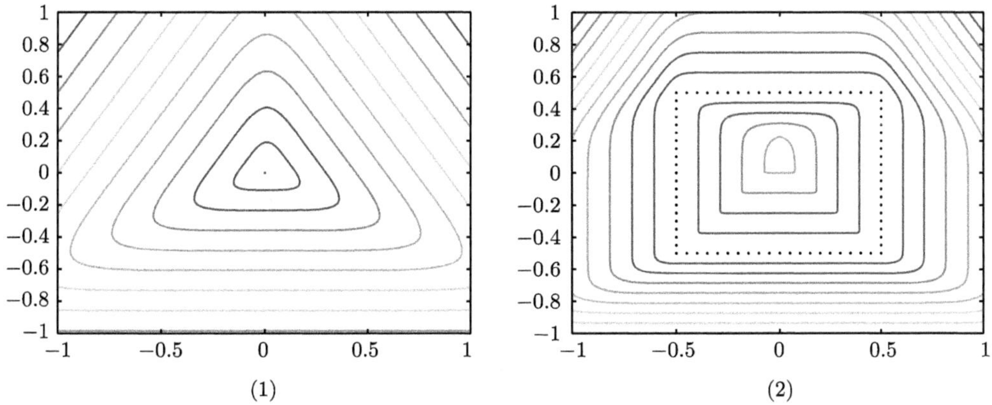

Figure 1　Contours of (4.3), $\sigma_x = \sigma_y = 2, 200 \times 200$ mesh

(1) A single source point: 68 iterations, $\omega = 1.18$; (2) Source on a square: 61 iterations, $\omega = 1.05$

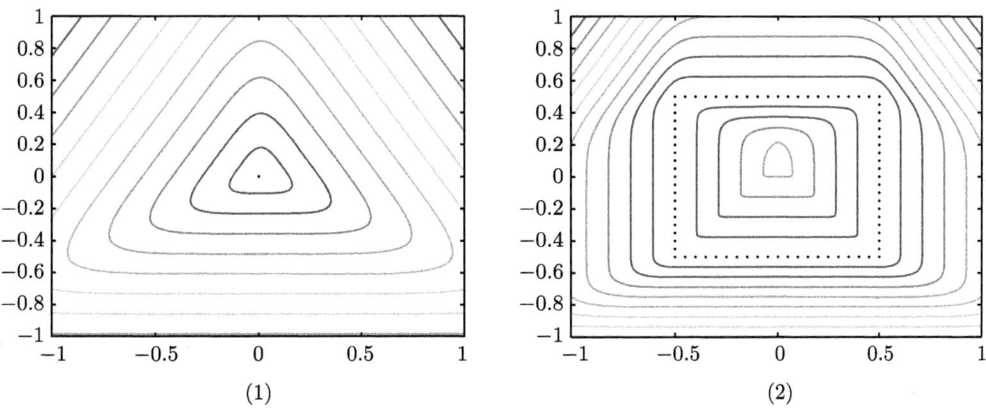

Figure 2　Contours of (4.3), $\sigma_x = \sigma_y = 2.5$, 200×200 mesh

(1) A single source point: 78 iterations, $\omega = 1.29$; (2)Source on a square: 81 iterations, $\omega = 1.10$

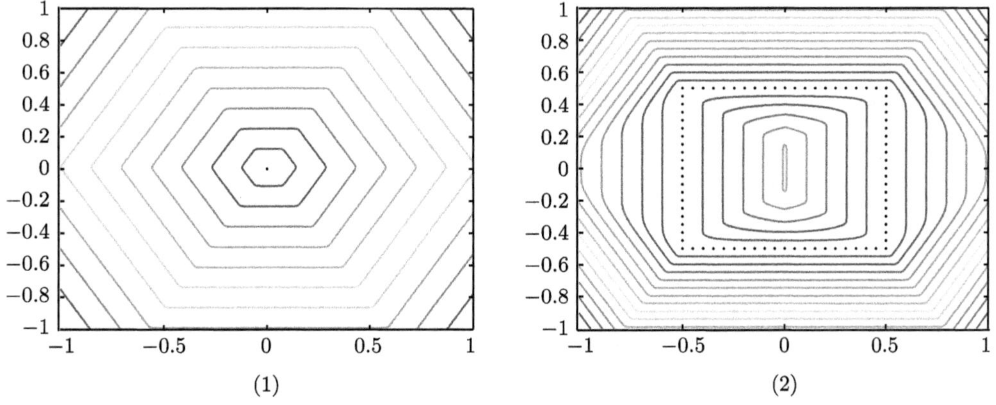

Figure 3　Contours of (4.4), $\sigma_x = \sigma_y = 3$, 200×200 mesh

(1)A single source point: 118 iterations; (2)Source on a square: 85 iterations

4.2　Traveltime computation for elastic wave propagation

In the high frequency asymptotic for linear elastic wave propagation, we need to compute traveltime functions for the quasi-P and quasi-SV wave modes. The quasi-P and the quasi-SV slowness surfaces are defined by the quadratic equation [21]:

$$c_1\phi_x^4 + c_2\phi_x^2\phi_y^2 + c_3\phi_y^4 + c_4\phi_x^2 + c_5\phi_y^2 + 1 = 0, \tag{4.5}$$

where

$$c_1 = a_{11}a_{44}, c_2 = a_{11}a_{33} + a_{44}^2 - (a_{13} + a_{44})^2,$$
$$c_3 = a_{33}a_{44}, c_4 = -(a_{11} + a_{44}), c_5 = -(a_{33} + a_{44}).$$

Here $a_{ij}s$ are given elastic parameters. The corresponding quasi-P wave Eikonal equation is

$$\sqrt{-\frac{1}{2}(c_4\phi_x^2 + c_5\phi_y^2) + \sqrt{\frac{1}{4}(c_4\phi_x^2 + c_5\phi_y^2)^2 - (c_1\phi_x^4 + c_2\phi_x^2\phi_y^2 + c_3\phi_y^4)}} = 1 \tag{4.6}$$

which is a convex Hamilton-Jacobi equation.

The corresponding quasi-SV wave Eikonal equation is

$$\sqrt{-\frac{1}{2}(c_4\phi_x^2 + c_5\phi_y^2) - \sqrt{\frac{1}{4}(c_4\phi_x^2 + c_5\phi_y^2)^2 - (c_1\phi_x^4 + c_2\phi_x^2\phi_y^2 + c_3\phi_y^4)}} = 1 \tag{4.7}$$

which is a non-convex Hamilton-Jacobi equation.

The elastic parameters are taken as the following two cases:

(1) $a_{11} = 15.0638$, $a_{33} = 10.8373$, $a_{13} = 1.6381$, $a_{44} = 3.1258$ for the whole domain;

(2) $a_{11} = 15.3871$, $a_{33} = 14.5161$, $a_{13} = 3.9321$, $a_{44} = 5.6074$ for upper half domain; $a_{11} = 15.0638$, $a_{33} = 10.8373$, $a_{13} = 1.6381$, $a_{44} = 3.1258$ for lower half domain. In this case the corresponding H-J equations have discontinuous coefficients; as in [17], this model is used to test the stability and robustness of our scheme.

The computational domain is $[-1, 1] \times [-1, 1]$, and $\Gamma = \{(0, 0)\}$.

Table 5 and 6 present the number of iterations for quasi-P and quasi-SV wave equations respectively. The elastic parameters are taken as case (1). As we can see from the table, the number of iterations can be reduced evidently if proper relaxation factor is taken. The contour plots of the solutions are shown in Figure 4.

Table 7 and 8 also show the number of iterations for quasi-P and quasi-SV wave equations respectively. The elastic parameters are taken as case (2). In this case, the corresponding H-J equations have discontinuous coefficients. However, our relaxation Lax-Friedrichs sweeping schemes also have smaller number of iterations than the Lax-Friedrichs sweeping scheme in [17] when relaxation factor is slightly larger than one. The contour plots of solutions are demonstrated in Figure 5. They have good resolution as figures in [17].

Table 5 The number of iterations for (4.6), $\sigma_x = \sigma_y = 4$, the elastic parameters are taken as case (1)

ω	Iteration–number		
	mesh :400 × 400	mesh :200 × 200	mesh :100 × 100
0.70	235	142	94
0.80	187	114	76
0.90	149	92	62
1.00	118	73	50
1.10	90	58	40
1.20	62	42	31
1.25	46	34	26
r	61.0%	53.4%	48.0%

Table 6 The number of iterations for (4.7), $\sigma_x = \sigma_y = 2.1$, the elastic parameters are taken as case (1)

ω	Iteration–number		
	mesh :400 × 400	mesh :200 × 200	mesh :100 × 100
0.70	182	111	74
0.80	143	89	60
0.90	112	71	48
1.00	87	56	39
1.05	76	49	35
1.10	66	43	31
r	23.3%	23.2%	20.5%

Table 7 The number of iterations for (4.6), $\sigma_x = \sigma_y = 4$, the elastic parameters are taken as case (2)

ω	Iteration–number		
	mesh :400 × 400	mesh :200 × 200	mesh :100 × 100
0.70	233	140	91
0.80	185	112	73
0.90	147	90	59
1.00	116	72	48
1.10	88	56	38
1.15	75	49	34
1.18	67	44	31
r	42.2%	38.9%	35.4%

Table 8 The number of iterations for(4.7), $\sigma_x = \sigma_y = 2.1$, the elastic parameters are taken as case (2)

ω	Iteration–number		
	mesh :400 × 400	mesh :200 × 200	mesh :100 × 100
0.70	174	106	70
0.80	137	84	56
0.90	107	67	45
1.00	83	53	36
1.05	73	46	32
1.09	63	41	28
r	24.1%	22.6%	22.2%

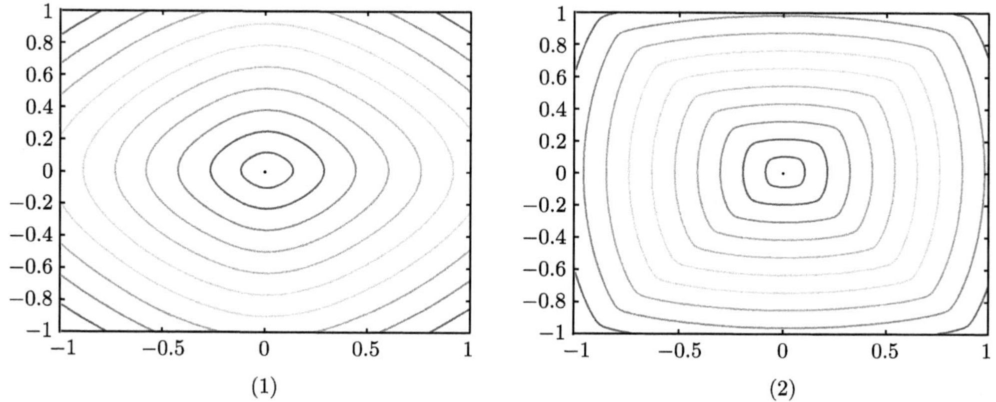

Figure 4 The elastic parameters are given as case (1), 200 × 200 mesh

(1) quasi-P: $\sigma_x = \sigma_y = 4$, $\omega = 1.25$, 34 iterations; (2) quasi-SV: $\sigma_x = \sigma_y = 2.1$, $\omega = 1.1$, 43 iterations

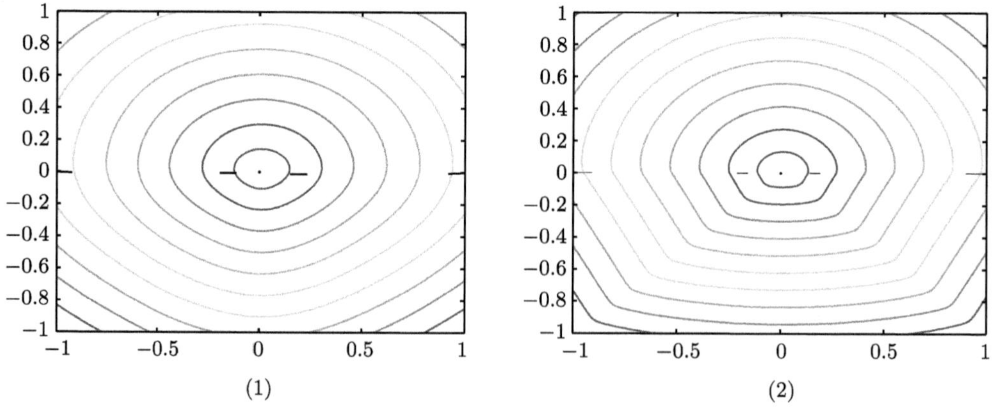

Figure 5 The elastic parameters are given as case (2), 200 × 200 mesh

(1) quasi-P: $\sigma_x = \sigma_y = 4$, $\omega = 1.18$, 44 iterations; (2) quasi-SV: $\sigma_x = \sigma_y = 2.1$, $\omega = 1.1$, 41 iterations

4.3 Convergence test

To validate the relaxation Lax-Friedrichs sweeping scheme, we apply the method to the Eikonal equation

$$\begin{cases} |\nabla\phi(x,y)| = 1, \quad (x,y) \in [-1,1]\times[-1,1], \\ \phi(0,0) = 0. \end{cases}$$

The exact solution is $\phi(x,y) = \sqrt{x^2+y^2}$. Table 9 presents the L^1 errors between numerical and exact solutions for different mesh sizes and different relaxation factors. It is suggested that on the same mesh size the error is independent on the relaxation factor and the relaxation Lax-Friedrichs sweeping scheme is a first order scheme.

Table 9 Errors and convergence order of Eikonal equation, $\sigma_x = \sigma_y = 1.5$

ω	mesh :320 × 320		mesh :160 × 160		mesh :80 × 80		mesh :40 × 40	
	Iter.	L^1 error	Iter.	L^1 error	Iter.	L^1 error	Iter.	L^1 error
0.70	255	0.122 668	163	0.220 882	117	0.395 348	96	0.706 859
0.80	205	0.122 668	132	0.220 882	95	0.395 348	78	0.706 859
0.90	166	0.122 668	107	0.220 882	78	0.395 348	64	0.706 859
1.00	134	0.122 668	87	0.220 882	64	0.395 348	53	0.706 859
1.10	106	0.122 668	71	0.220 882	52	0.395 348	44	0.706 859
1.20	82	0.122 668	56	0.220 882	42	0.395 348	36	0.706 859
1.30	58	0.122 668	42	0.220 882	33	0.395 348	29	0.706 859
1.35	44	0.122 668	34	0.220 882	29	0.395 348	26	0.706 859
Order	0.848 5		0.839 8		0.838 3		—	

5 The Analysis of Model Problem

In this section we give a theoretical analysis of our method with $0 < \omega \leqslant 1$ for one dimensional Eikonal equation.

We consider the following model problem:

$$\begin{cases} |\phi'(x)| = R(x), \quad x \in [-1,1], \\ \phi(0) = 0. \end{cases} \tag{5.1}$$

Suppose that we have a discretization $\{x_i = \dfrac{i}{m}, -m-1 \leqslant i \leqslant m+1\}$, x_{m+1} and x_{-m-1} are points outside the domain. We sweep from left to right and then from right to left. The approximation after two sweeps on the left of the center is symmetric to the approximation after one sweep on the right. Without loss of generality, we can just discuss the approximations on the right of the center. Let $(\phi^n)^+$ be the solution by the sweep from left to right at the n-th iteration and let $(\phi^n)^-$ be the sweep from right to left at the n-th iteration.

For the Eikonal equation (5.1), the SOR type LxF update formula read as for $i=1,2,\cdots,m$,

$$\begin{cases}\phi_i^{\text{tmp},+} &= \frac{\Delta x}{\sigma_x}R(x_i)-\frac{1}{2\sigma_x}|\phi_{i+1}^{n,-}-\phi_{i-1}^{n+1,+}|+\frac{1}{2\Delta x}(\phi_{i+1}^{n,-}+\phi_{i-1}^{n+1,+}),\\ \phi_i^{n+1,+} &= \omega\phi_i^{\text{tmp},+}+(1-\omega)\phi_i^{n,-},\\ \phi_0^{n+1,+} &= 0,\\ \phi_{m+1}^{n+1,+} &= \min\{\max\{2\phi_m^{n+1,+}-\phi_{m-1}^{n+1,+},\phi_{m-1}^{n+1,+}\},\phi_{m+1}^{n,+}\}\end{cases} \tag{5.2}$$

and for $i=m,m-1,\cdots,1$,

$$\begin{cases}\phi_i^{\text{tmp},-} &= \frac{\Delta x}{\sigma_x}R(x_i)-\frac{1}{2\sigma_x}|\phi_{i+1}^{n+1,-}-\phi_{i-1}^{n+1,+}|+\frac{1}{2}(\phi_{i+1}^{n+1,-}+\phi_{i-1}^{n+1,+}),\\ \phi_i^{n+1,-} &= \omega\phi_i^{\text{tmp},-}+(1-\omega)\phi_i^{n+1,+},\\ \phi_0^{n+1,-} &= 0,\\ \phi_{m+1}^{n+1,-} &= \min\{\max\{2\phi_m^{n+1,-}-\phi_{m-1}^{n+1,-},\phi_{m-1}^{n+1,-}\},\phi_{m+1}^{n+1,+}\}\end{cases} \tag{5.3}$$

Proposition 5.1. *If $\omega\in(0,1]$, the formula (5.2)-(5.3) can be reduced to*

$$\begin{cases}\phi_i^{n+1,+} &= \omega\,c_i+\omega\,b\phi_{i-1}^{n+1,+}+(1-\omega)\phi_i^{n,-}+\omega\,a\phi_{i+1}^{n,-},\\ \phi_i^{n+1,-} &= \omega\,c_i+\omega\,b\phi_{i-1}^{n+1,+}+(1-\omega)\phi_i^{n+1,+}+\omega\,a\phi_{i+1}^{n+1,-},\end{cases} \tag{5.4}$$

$i=1,2,\cdots,m$, *where*

$$a=\frac{1}{2}-\frac{1}{2\sigma_x},\quad b=\frac{1}{2}+\frac{1}{2\sigma_x},\quad c_i=\frac{\Delta x}{\sigma_x}R(x_i),\quad \sigma_x\geqslant 1.$$

Proof. By Proposition 3.1, Proposition 3.3 and the initialization step of RLxFSM algorithm, we easily obtain $\phi_{i-}^{n+1,+}\leqslant\phi_{i+1}^{n,-}$ and $\phi_{i-1}^{n+1,-}\leqslant\phi_{i+1}^{n+1,-}$ for any i and n. As a consequent, we conclude (5.4).

□

Now we can write down the update formula as the following linear system:

$$\begin{aligned}A_+(\phi^{n+1})^+ &= B_+(\phi^n)^-+C,\\ A_-(\phi^n)^- &= B_-(\phi^n)^++C.\end{aligned}$$

where $\phi^n=(\phi_1^n,\phi_2^n,\cdots,\phi_m^n,\phi_{m+1}^n)^T$, $C=(\omega\,c_1,\omega\,c_2,\cdots,\omega\,c_m,0)^T$,

$$A_+=\begin{pmatrix}1&&&&\\-\omega b&1&&&\\&\ddots&\ddots&&\\&&-\omega b&1&\\&&1&-2&1\end{pmatrix},\quad B_+=\begin{pmatrix}1-\omega&\omega a&&&\\&1-\omega&\omega a&&\\&&\ddots&\ddots&\\&&&1-\omega&\omega a\\0&\cdots&\cdots&0&0\end{pmatrix},$$

$$A_-=\begin{pmatrix}1&-\omega a&&&&\\&1&-\omega a&&&\\&&\ddots&\ddots&&\\&&&1&-\omega a&0\\0&\cdots&0&0&1&0\\0&\cdots&0&1&-2&1\end{pmatrix},B_-=\begin{pmatrix}1-\omega&&&&&\\\omega b&1-\omega&&&&\\&\ddots&\ddots&&&\\&&\omega b&1-\omega&&\\&&&\omega b&1-\omega&\omega a\\0&\cdots&\cdots&0&0&0\end{pmatrix}.$$

Thus the update formula can be written as

$$(\phi^{n+1})^+ = \widehat{B_+}(\phi^n)^- + \widehat{C_+}, \quad (\phi^n)^- = \widehat{B_-}(\phi^n)^+ + \widehat{C_-},$$

where $\widehat{B_+} = (A_+)^{-1}B_+$, $\widehat{C_+} = (A_+)^{-1}C$, $\widehat{B_-} = (A_-)^{-1}B_-$, $\widehat{C_-} = (A_-)^{-1}C$. And then

$$(\phi^{n+1})^+ = \widehat{B_+}\widehat{B_-}(\phi^n)^+ + \widehat{B_+}\widehat{C_-} + \widehat{C_+}.$$

For fixed m, $v = -\ln\rho(\widehat{B_+}\widehat{B_-})$, where $\rho(\widehat{B_+}\widehat{B_-})$ is the spectral radius of matrix $\widehat{B_+}\widehat{B_-}$, is proportional to the number of iterations of the algorithm. However, it is not obvious to estimate v theoretically. Thus, in Figure 6, we show the numerical estimation of the spectral radius and v for $\omega = 0.95$ and $\omega = 1$ when $m = 100$. As we can see from the figure, the spectral radius is larger and then v is smaller when ω is smaller than 1. This is maybe the reason that our relaxation Lax-Friedrichs sweeping scheme has more number of iterations than the Lax-Friedrichs sweeping scheme in [17] if ω is smaller than one.

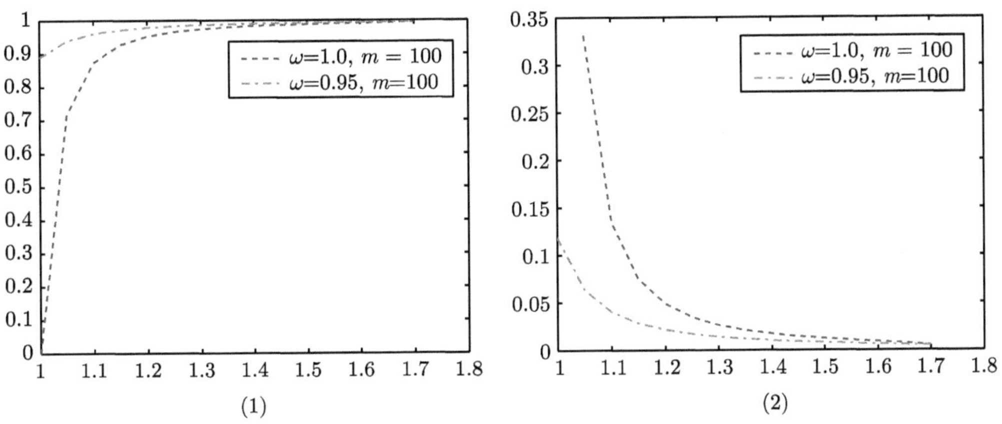

Figure 6

(1)Spectral radius of $\hat{B_+}\hat{B_-}$; (2) v of $\hat{B_+}\hat{B_-}$

Remark 5.1. Numerical examples indicate that our relaxation LxF sweeping scheme has smaller number of iterations than the Lax-Friedrichs sweeping scheme in [17] if the relaxation factor ω is slightly larger than one. But the above analysis is invalid when ω is larger than one. The theoretical analysis in this case is an open problem.

6 Conclusion

In this article, we proposed a relaxation Lax-Friedrichs sweeping scheme for static Hamilton-Jacobi equation which is a generalization of Lax-Friedrichs sweeping scheme in [17]. Numerical results indicate that the number of iterations is reduced evidently when relaxation factor ω is slightly larger than one.

Acknowledgments The authors would like to thank the reviewers for several remarks leading to a better presentation of the material in this paper. And the authors are very grateful to the help of professor Chiu Yen Kao.

References

[1] M. Crandall and P. L. Lions, Viscosity solutions of Hamilton-Jacobi equations, *Transactions of the American Mathematical Society*, 277 (1983), 1-42.

[2] M. Crandall and P. L. Lions, Two approximations of solutions of Hamilton-Jacobi equations, *Mathematics of Computation*, 43 (1984), 1-19.

[3] S. Osher, A level set formulation for the solution of the Dirichlet problems for Hamilton-Jacobi equations, *SIAM Journal on Mathematical Analysis*, 24 (1993), 1145-1152.

[4] S. Gray, W. May, Kirchhoff migration using eikonal equation travel-times, *Geophysics.* 59 (1994), 810-817.

[5] S. Kim, R. Cook, 3D traveltime computation using second-order ENO scheme, *Geophysics.* 64 (1999), 1867-1876.

[6] J. Qian, W. W. Symes, Paraxial eikonal solvers for anisotropic quasi-P travel times, *Journal of Computational Physics*, 174 (2001), 256-278.

[7] R. Abgrall, Numerical discretization of the first-order Hamilton-Jacobi equation on triangular meshes, *Communications on Pure and Applied Mathematics*, 49 (1996), 1339-1373.

[8] S. Augoula and R. Abgrall, High order numerical discretization for Hamilton-Jacobi equations on triangular meshes, *Journal of Scientific Computing*, 15 (2000), 197-229.

[9] C. Hu and C.-W. Shu, A discontinuous Galerkin finite element method for Hamilton-Jacobi equations, *SIAM Journal on Scientific Computation*, 20 (1999), 666-690.

[10] G.-S. Jiang and D. Peng, Weighted ENO schemes for Hamilton-Jacobi equations, *SIAM Journal on Scientific Computation*, 21 (2000), 2126-2143.

[11] D. Levy, S. Nayak, C.-W. Shu and Y. T. Zhang, Central WENO schemes for Hamilton-Jacobi equations on triangular meshes, *SIAM Journal on Scientific Computation*, 28 (2006), 2229-2247.

[12] S. Osher and C.-W. Shu, High-order essentially nonoscillatory schemes for Hamilton-Jacobi equations, *SIAM Journal on Numerical Analysis*, 28 (1991), 907-922.

[13] Y.-T. Zhang and C.-W. Shu, High order WENO schemes for Hamilton-Jacobi equations on triangular meshes, *SIAM Journal on Scientific Computing*, 24 (2003), 1005-1030.

[14] C.-W. Shu, High order numerical methods for time dependent Hamilton-Jacobi equations, WSPC/Lecture Notes Series, 2007.

[15] J. A. Sethian, A fast marching level set method for monotonically advancing fronts, *Proceedings of the National Academy of Sciences of USA*, 93 (1996), 1591-1595.

[16] J. N. Tsitsiklis, Efficient algorithms for globally optimal trajectories, *IEEE Transactions on Automatic Control*, 40 (1995), 1528-1538.

[17] C. Y. Kao, S. Osher, J. Qian , Lax-Friedrichs sweeping scheme for static Hamilton-Jacobi equations, *Journal of Computational Physics*, 196 (2004), 367-391.

[18] J. Qian, Y.-T. Zhang and H.-K. Zhao, Fast sweeping methods for Eikonal equations on triangular meshes, *SIAM Journal on Numerical Analysis*, 45 (2007), 83-107.

[19] J. Qian, Y.-T. Zhang and H.-K. Zhao, A fast sweeping method for static convex Hamilton-Jacobi equations, *Journal of Scientific Computing*, 31 (2007), 237-271.

[20] H.-K. Zhao, A fast sweeping method for eikonal equations, *Mathematics of Computation*, 74 (2005), 603-627.

[21] Y.-T. Zhang, H.-K. Zhao and J. Qian, High order fast sweeping methods for static Hamilton-Jacobi equations, *Journal of Scientific Computing*, 29 (2006), 25-26.

[22] Y.-T. Zhang, H.-K. Zhao and S. Chen, Fixed-point iterative sweeping methods for static Hamilton-Jacobi equations, *Methods and Applications of Analysis*, 13 (2006), 299-320.

[23] B. Cockburn and J. Qian, Continuuos dependence results for Hamilton-Jacobi equations. *Collected lectures on the preservation of stability under discretization.* Edited by D. Estep and S. Tavener. SIAM, Philadelphia, PA, 2002, pp.67-90.

A Uniformly Convergent Continuous-Discontinuous Galerkin Method for Singularly Perturbed Problems of Convection-Diffusion Type*

P. Zhu (祝鹏)† Z.Q. Xie (谢资清)‡ S.Z. Zhou (周叔子)§

Abstract In this paper, we introduce a coupled approach of local discontinuous Galerkin and standard finite element method for solving singularly perturbed convection-diffusion problems. On Shishkin mesh with linear elements, a rate $\mathcal{O}(N^{-1}\ln N)$ in an associated norm is established, where N is the number of elements. Numerical experiments complement the theoretical results. Moreover, a rate $\mathcal{O}(N^{-2}\ln^2 N)$ in a discrete L^∞ norm, and $\mathcal{O}(N^{-2})$ in L^2 norm, are observed numerically on the Shishkin mesh.

Keywords Convection diffusion equation, local discontinuous Galerkin method, finite element method, Shishkin mesh, uniform convergence.

1 Introduction

In recent years the numerical solutions of singularly perturbed boundary value problems have been received much attention and already studied in many papers and books, see for instance [1 − 4]. One of the difficulties in numerically computing the solution of singularly perturbed problems lays in the so-called boundary layer behavior, i.e., the solution varies very rapidly in a very thin layer near the boundary. Traditional methods such as finite element and finite difference methods, do not work well for these problems as they often produce oscillatory solutions which are inaccurate if the perturbed parameter ϵ is small. When ϵ approaches zero, the problem changes from an elliptic equation to a hyperbolic one. Inspired by the great success of the discontinuous Galerkin(DG) method in solving hyperbolic equations, Cockburn

* 发表于：Applied Mathematics and Computation, Vol. 217, 2011, pp. 4781-4790.

† Institute of Mathematics, Jiaxing University, Jiaxing, Zhejiang 314001, China.

‡ College of Mathematics and Computer Science, Hunan Normal University, Changsha, Hunan 410081, China.

§ College of Mathematics and Econometrics, Hunan University, Changsha, Hunan 410082, China.

and Shu [5], Celiker and Cockburn [6], Xie et al. [7, 8] and Zhang et al. [9] adopted the local discontinuous Galerkin(LDG) method to solve convection-diffusion equations and analyzed the corresponding superconvergence properties. On the other hand, nonsymmetric discontinuous Galerkin method with interior penalty (the NIPG method), originally designed for elliptic equations, is analyzed by Zarin and Roos [10] for convection-diffusion problems with parabolic layers.

A disadvantage of DG method is that the method produces more degrees of freedom than the continuous finite element method (CFEM). With this motivation, this work derives and analyzes a coupled approach of LDG and CFEM on Shishkin mesh for singularly perturbed convection diffusion problems. By splitting the domain into the coarse and the fine part, we adopt the CFEM with linear elements in the fine part where the mesh size is comparable with ϵ, and use LDG method in the coarse part for its stabilization.

The idea of combining DG and CFEM to obtain the advantages of both methods is not new. A coupled LDG-CFEM approach has also been studied by Perugia and Schötzau [11] for the modeling of elliptic problems arising in electromagnetics. Roos and Zarin [12], Zarin [13] analyzed the NIPG-CFEM coupled method on Shishkin mesh for two dimensional convection-diffusion problems with exponentially layers or characteristic layers. In the present paper, we do not use the solution decomposition, which is generally necessary to prove uniform convergence on layer-adapted meshes. Although in the present work, we only analyze the LDG case in the DG-CFEM coupled approach, our method can be generalized to all DG methods belong to the unify framework [14] including the NIPG method.

The paper is organized as follows. In Section 2, we introduce the coupled LDG and CFEM for the singularly perturbed problems. The stability and error analysis of the coupled method with linear elements on a Shishkin mesh [15] is given in Section 3. The implement of our coupled method on a Shishkin mesh is presented in Section 4. It aims to validate our theoretical result. Further, we numerically observe the uniform convergence rate $\mathcal{O}(N^{-2}\ln^2 N)$ in a discrete L^∞ norm, and $\mathcal{O}(N^{-2})$ in L^2 norm. This phenomenon is not found in [12] and [13]. We end in Section 5 with some concluding remarks.

In the sequel, with C we shall denote a generic positive constant independent of the perturbation parameter ϵ and mesh size.

2 Coupling the LDG and CFEM

Consider the following one-dimensional convection-diffusion problem,

$$\begin{cases} -\epsilon u'' + bu' + cu = f, & \text{in } \Omega = (0,1), \\ u(0) = u(1) = 0, \end{cases} \tag{2.1}$$

where $0 < \epsilon \ll 1$ is a small positive parameter, and b, c, f are sufficiently smooth functions with the following properties

$$b(x) \geqslant b_0 > 0,\ c(x) \geqslant 0,\ c(x) - \frac{1}{2}b'(x) \geqslant c_0 > 0, \quad \forall x \in \bar{\Omega}, \tag{2.2}$$

for some constants b_0 and c_0. This assumption guarantees that (2.1) has a unique solution in $H^2(\Omega) \cap H_0^1(\Omega)$ for all $f \in L^2(\Omega)$[16]. Typically the solution of (2.1) has an exponential boundary layer at $x = 1$.

The Shishkin Mesh. Let N be an even integer. Define the transition parameter

$$\tau = \min\left(\frac{1}{2}, \frac{\kappa}{b_0}\epsilon \ln N\right)$$

with $\kappa \geqslant 2$ and divide each of the subdomains $\Omega_1 = [0, 1-\tau]$ and $\Omega_2 = [1-\tau, 1]$ into $N/2$ equidistant subintervals. Notice that $\epsilon \ll 1$, here and below we take $\tau = \frac{\kappa}{b_0}\epsilon \ln N$. Now, we have

$$x_j = \begin{cases} 2(1-\tau)j/N, & j = 0, 1, \cdots, N/2, \\ 1 - 2\tau(N-j)/N, & j = N/2+1, \cdots, N. \end{cases}$$

We denote

$$H = 2(1-\tau)/N, \qquad h = 2\tau/N.$$

It can be easily shown that

$$H \leqslant CN^{-1}, \qquad h \leqslant C\epsilon N^{-1} \ln N.$$

Denote the mesh by $I_j = [x_{j-1}, x_j]$ for $j = 1, \cdots, N$ and set $\mathcal{T}_N^1 = \{I_j\}_{j=1}^{N/2}$ and $\mathcal{T}_N^2 = \{I_j\}_{j=N/2+1}^{N}$. We denote by u_j^+ and u_j^- the values of u at x_j, from the right cell and the left cell of x_j, respectively.

Weak formulation. We discretize problem (2.1) by using the LDG method in Ω_1, and CFEM in Ω_2. Let $u_i = u|_{\Omega_i}, i = 1, 2$. In order to use the LDG method in Ω_1, we introduce the auxiliary variable $q = u_1'$ in Ω_1 and rewrite problem (2.1) as the following equivalent transmission problem:

$$\begin{cases} q - u_1' = 0, & \text{in } \Omega_1, \\ -\epsilon q' + bu_1' + cu_1 = f, & \text{in } \Omega_1, \\ -\epsilon u_2'' + bu_2' + cu_2 = f, & \text{in } \Omega_2, \\ u_1(1-\tau) = u_2(1-\tau), \\ q(1-\tau) = u_2'(1-\tau), \end{cases} \tag{2.3}$$

with boundary conditions

$$u_1(0) = u_2(1) = 0. \tag{2.4}$$

Multiplying the first three equations of (2.3) by test functions w, v_1, v_2, respectively, and integrating by parts, it can be seen that the solution (q, u_1, u_2), of problem (2.3) and (2.4) satisfies, for all $I_j \in \mathcal{T}_N^1$,

$$\int_{I_j} qw\mathrm{d}x + \int_{I_j} u_1 w'\mathrm{d}x - u_{1,j}^- w_j^- + u_{1,j-1}^+ w_{j-1}^+ = 0, \tag{2.5}$$

$$\begin{aligned}&\int_{I_j} (\epsilon q - bu_1)v_1'\mathrm{d}x + \int_{I_j} (c-b')u_1 v_1 \mathrm{d}x - (\epsilon q - bu_1)_j^- v_{1,j}^- \\ &+ (\epsilon q - bu_1)_{j-1}^+ v_{1,j-1}^+ = \int_{I_j} f v_1 \mathrm{d}x,\end{aligned} \tag{2.6}$$

for piecewise smooth functions w and v_1 and,

$$\int_{\Omega_2} (\epsilon u_2' - bu_2)v_2'\mathrm{d}x + \int_{\Omega_2} (c-b')u_2 v_2\mathrm{d}x + (\epsilon u_2' - bu_2)(1-\tau)v_2(1-\tau) = \int_{\Omega_2} f v_2 \mathrm{d}x, \tag{2.7}$$

for smooth function v_2 with $v_2(1) = 0$. The above equations have to be coupled through the transmission conditions, i.e. the last two equations of (2.3). The second of these conditions is imposed by replacing $u_2'(1-\tau)$ by $q(1-\tau)$ in the last equation, which becomes

$$\int_{\Omega_2} (\epsilon u_2' - bu_2)v_2'\mathrm{d}x + \int_{\Omega_2} (c-b')u_2 v_2\mathrm{d}x + (\epsilon q - bu_1)(1-\tau)v_2(1-\tau) = \int_{\Omega_2} f v_2 \mathrm{d}x, \tag{2.8}$$

while the first one is imposed at the discrete level by choosing the numerical fluxes in the LDG method in a suitable way. (2.5), (2.6) and (2.8) are the weak formulations of problem (2.3) and (2.4).

Let us now denote by $\mathbb{P}^1(I_j)$ the space of polynomials of degree at most 1 on I_j, and define the finite element space V_N^1 and V_N^2 as follows,

$$\begin{aligned}V_N^1 &= \{u_1 \in L^2(\Omega_1) : u_1|_{I_j} \in \mathbb{P}^1(I_j), \forall I_j \in \mathcal{T}_N^1\}, \\ V_N^2 &= \{u_2 \in H^1(\Omega_2) : u_2(1) = 0, u_2|_{I_j} \in \mathbb{P}^1(I_j), \forall I_j \in \mathcal{T}_N^2\}.\end{aligned}$$

The space V_N^2 is a standard conforming finite element space, whereas the functions in V_N^1 are completely discontinuous across interelement boundaries.

We will search for approximate solutions $(q_N, u_{1,N}, u_{2,N})$ of (2.3) and (2.4) in the finite element space $V_N^1 \times V_N^1 \times V_N^2$ that satisfy (2.3) and (2.4) in a weak sense. Consider the following general formulation: Find $(q_N, u_{1,N}, u_{2,N}) \in V_N^1 \times V_N^1 \times V_N^2$ such that

$$\int_{I_j} q_N w\mathrm{d}x + \int_{I_j} u_{1,N} w'\mathrm{d}x - \widehat{u}_{1,j} w_j^- + \widehat{u}_{1,j-1} w_{j-1}^+ = 0, \tag{2.9}$$

$$\begin{aligned}&\int_{I_j} (\epsilon q_N - bu_{1,N})v_1'\mathrm{d}x + \int_{I_j} (c-b')u_{1,N} v_1 \mathrm{d}x - (\epsilon\widehat{q} - b\widetilde{u}_1)_j v_{1,j}^- \\ &+ (\epsilon\widehat{q} - b\widetilde{u}_1)_{j-1} v_{1,j-1}^+ = \int_{I_j} f v_1 \mathrm{d}x,\end{aligned} \tag{2.10}$$

for all test function $(w, v_1) \in V_N^1 \times V_N^1$ and for all $I_j \in \mathcal{T}_N^1$, and

$$\int_{\Omega_2} (\epsilon u_{2,N}' - bu_{2,N})v_2'\mathrm{d}x + \int_{\Omega_2} (c-b')u_{2,N} v_2\mathrm{d}x + (\epsilon\widehat{q} - b\widetilde{u}_1)(1-\tau)v_2(1-\tau) = \int_{\Omega_2} f v_2 \mathrm{d}x, \tag{2.11}$$

for all test function $v_2 \in V_N^2$, where $\widehat{u}_1, \widetilde{u}_1$ and $\widehat{q}$ are the numerical fluxes, which approximate the traces of $u_{1,N}$ and q_N on the boundary of the elements of $\mathcal{T}_N^1$. To complete the specification of the method, it only remains to define the numerical fluxes.

The numerical fluxes. We use the following notation to describe the numerical fluxes at the interior nodes. The *average* and *jump* of the trace of u at the interior node x_j are given by

$$\{u\}_j = \frac{1}{2}(u_j^+ + u_j^-) \qquad \text{and} \qquad [u]_j = u_j^+ - u_j^-,$$

respectively. We now define the numerical fluxes $\widehat{u}_1$ and $\widehat{q}$ by

$$\widehat{u}_1(x_j) = \begin{cases} 0, & \text{if } j = 0, \\ \{u_{1,N}\}_j - \beta[u_{1,N}]_j, & \text{if } j = 1, \cdots, N/2-1, \\ u_{2,N}(x_j), & \text{if } j = N/2, \end{cases} \tag{2.12}$$

and

$$\widehat{q}(x_j) = \begin{cases} q_N^+(x_j) + \alpha u_{1,N}^+(x_j), & \text{if } j = 0, \\ \{q_N\}_j + \beta[q_N]_j + \alpha[u_{1,N}]_j, & \text{if } j = 1, \cdots, N/2-1, \\ q_N^-(x_j) + \alpha(u_{2,N} - u_{1,N}^-)(x_j), & \text{if } j = N/2, \end{cases} \tag{2.13}$$

Here the scalar $\alpha = \alpha(x)$ and $\beta = \beta(x)$ are auxiliary parameters. Their purpose is to enhance the stability and accuracy properties of the LDG method (see [5] and [17]).

The numerical flux associated with the convection is the classical upwinding flux, namely,

$$\widetilde{u}_1(x_j) = \begin{cases} 0, & \text{if } j = 0, \\ u_{1,N}^-(x_j), & \text{if } j = 1, \cdots, N/2. \end{cases} \tag{2.14}$$

3 Stability and Error Analysis of the Coupled Method

This section is devoted to the existence and uniqueness of the solution of the coupled method (2.9)–(2.11) with numerical fluxes (2.12)–(2.14), and its corresponding error analysis. If the stabilization parameter α is taken of order $\mathcal{O}(1/H)$, we can rewrite our method in the primal form by eliminating q following Arnold et al. [14].

Primal formulation. A straightforward computation shows that for all $u \in V_N^1$ and $v \in V_N^1$,

$$\int_{\Omega_1} uv'\mathrm{d}x = -\int_{\Omega_1} u'v\mathrm{d}x - \sum_{j=1}^{N/2-1}([u]_j\{v\}_j + \{u\}_j[v]_j) + u^-(1-\tau)v^-(1-\tau) - u^+(0)v^+(0). \tag{3.1}$$

Summing up (2.9) for all $I_j \in \mathcal{T}_N^1$, and combining with (2.12) and (3.1), we obtain

$$\begin{aligned}\int_{\Omega_1}(q_N - u_{1,N}')w\mathrm{d}x = & \sum_{j=1}^{N/2-1}[u_{1,N}]_j(\{w\}_j + \beta[w]_j) \\ & + u_{1,N}^+(0)w^+(0) + (u_{2,N} - u_{1,N}^-)(1-\tau)w^-(1-\tau).\end{aligned} \tag{3.2}$$

Let us introduce the space $V_N := \{v = (v_1, v_2) : v_1 \in V_N^1, v_2 \in V_N^2\}$ and $V(N) = [H^2(\Omega) \cap H_0^1(\Omega)] + V_N$, where $H_0^1(\Omega) = \{v \in H^1(\Omega) : v|_{\Omega_1} = v_1, v|_{\Omega_2} = v_2, v_1(0^+) = 0, v_2(1) = 0\}$. For $v \in V(N)$, we define $\mathcal{L}_1(v)$ as the unique element in V_N^1 satisfying

$$\int_{\Omega_1} \mathcal{L}_1(v) r \mathrm{d}x = \sum_{j=1}^{N/2-1} [v_1]_j(\{r\}_j + \beta[r]_j) + v_1^+(0)r^+(0) + (v_2 - v_1^-)(1-\tau)r^-(1-\tau) \tag{3.3}$$

for all $r \in V_N^1$.

As a result, (3.2) can be rewritten as

$$q_N = u_{1,N}' + \mathcal{L}_1(u_N) \qquad \text{in } V_N^1. \tag{3.4}$$

Summing up (2.10) for all $I_j \in \mathcal{T}_N^1$ and adding together with (2.11), we get

$$\int_{\Omega_1} (\epsilon q_N - b u_{1,N}) v_1' \mathrm{d}x + \int_{\Omega_2} (\epsilon u_{2,N}' - b u_{2,N}) v_2' \mathrm{d}x + \int_{\Omega} (c - b') u_N v \mathrm{d}x + \sum_{j=1}^{N/2-1} (\epsilon \widehat{q} - b\widetilde{u}_1)_j [v_1]_j$$
$$+(\epsilon \widehat{q} - b\widetilde{u}_1)(1-\tau)(v_2 - v_1^-)(1-\tau) + (\epsilon \widehat{q} - b\widetilde{u}_1)(0) v_1^+(0) = \int_{\Omega} f v \mathrm{d}x.$$

Inserting (2.13) and (2.14) into the equation above, and recalling the definition of $\mathcal{L}_1(\cdot)$, we obtain

$$\int_{\Omega_1} \epsilon q_N (v_1' + \mathcal{L}_1(v)) \mathrm{d}x + \int_{\Omega_2} \epsilon u_{2,N}' v_2' \mathrm{d}x - \int_{\Omega} b u_N v' \mathrm{d}x + \int_{\Omega} (c - b') u_N v \mathrm{d}x$$
$$+ \sum_{j=1}^{N/2-1} \epsilon\alpha [u_{1,N}]_j [v_1]_j + \epsilon\alpha (u_{2,N} - u_{1,N}^-)(1-\tau)(v_2 - v_1^-)(1-\tau) + \epsilon\alpha\, u_{1,N}^+(0) v_1^+(0)$$
$$- \sum_{j=1}^{N/2-1} b(x_j) u_{1,N}^-(x_j)[v_1]_j - b(1-\tau) u_{1,N}^-(1-\tau)(v_2 - v_1^-)(1-\tau) = \int_{\Omega} f\, v \mathrm{d}x. \tag{3.5}$$

Similar to the definition of $\mathcal{L}_1(v)$, for $v \in V(N)$, we define $\mathcal{L}_2(v)$ as the unique element in V_N^1 satisfying

$$\int_{\Omega_1} b \mathcal{L}_2(v) r \mathrm{d}x = \sum_{j=1}^{N/2-1} b(x_j)[v_1]_j r^-(x_j) + b(1-\tau)(v_2 - v_1^-)(1-\tau) r^-(1-\tau) \tag{3.6}$$

for all $r \in V_N^1$.

As a consequence, (3.5) can be rewritten as

$$\int_{\Omega_1} \epsilon q_N (v_1' + \mathcal{L}_1(v)) \mathrm{d}x + \int_{\Omega_2} \epsilon u_{2,N}' v_2' \mathrm{d}x - \int_{\Omega} b u_N v' \mathrm{d}x + \int_{\Omega} (c - b') u_N v \mathrm{d}x$$
$$- \int_{\Omega_1} b \mathcal{L}_2(v) u_{1,N} \mathrm{d}x + \sum_{j=1}^{N/2-1} \epsilon\alpha [u_{1,N}]_j [v_1]_j + \epsilon\alpha (u_{2,N} - u_{1,N}^-)(1-\tau)(v_2 - v_1^-)(1-\tau)$$
$$+\epsilon\alpha\, u_{1,N}^+(0) v_1^+(0) = \int_{\Omega} f\, v \mathrm{d}x. \tag{3.7}$$

Inserting (3.4) into (3.7), we obtain the so-called primal form of our coupled method which reads: find $u_N \in V_N$ such that

$$\mathcal{A}_N(u_N, v) := \mathcal{B}_N(u_N, v) + \mathcal{C}_N(u_N, v) + \mathcal{S}_N(u_N, v) = \mathcal{F}_N(v) \qquad \forall v \in V_N \tag{3.8}$$

with

$$\begin{aligned}
\mathcal{B}_N(u,v) &= \int_\Omega \epsilon(u' + \mathcal{L}_1(u))(v' + \mathcal{L}_1(v))\mathrm{d}x, \qquad \mathcal{F}_N(v) = \int_\Omega f v \mathrm{d}x,\\
\mathcal{C}_N(u,v) &= -\int_\Omega bu(v' + \mathcal{L}_2(v))\mathrm{d}x + \int_\Omega (c - b')uv\mathrm{d}x,\\
\mathcal{S}_N(u,v) &= \sum_{j=1}^{N/2-1} \epsilon\alpha[u_1]_j[v_1]_j + \epsilon\alpha u_1(0^+)v_1(0^+) + \epsilon\alpha(u_2 - u_1^-)(1-\tau)(v_2 - v_1^-)(1-\tau).
\end{aligned}$$

Here, $\mathcal{L}_1(u)$ and $\mathcal{L}_2(u)$ have been defined in $L^2(\Omega)$ by a trivial extension.

From the following lemma, the primal formulation is consistent.

Lemma 3.1. *Let u be the exact solution of the problem (2.3) and (2.4). Then the primal form (3.8) has the Galerkin orthogonality property*

$$\mathcal{A}_N(u - u_N, v) = 0, \qquad \text{for all } v \in V_N. \tag{3.9}$$

Proof. Since u is the exact solution, we have $[u_1]_j = 0, j = 1, \cdots, N/2-1, u_1(0) = 0, u_1(1-\tau) = u_2(1-\tau)$, $u_1'(1-\tau) = u_2'(1-\tau)$. Consequently, $\mathcal{S}_N(u,v) = 0, \mathcal{L}_1(u) = 0$. Then for all $v \in V_N$, we have

$$\begin{aligned}
\mathcal{B}_N(u,v) &= \int_\Omega \epsilon u'(v' + \mathcal{L}_1(v))\mathrm{d}x\\
&= \int_{\Omega_1} \epsilon u_1' v_1' \mathrm{d}x + \int_{\Omega_1} \epsilon \mathcal{L}_1(v) u_1' \mathrm{d}x + \int_{\Omega_2} \epsilon u_2' v_2' \mathrm{d}x
\end{aligned} \tag{3.10}$$

Taking $u = \epsilon u_1'$ and $v = v_1$ in (3.1), we have

$$\begin{aligned}
\int_{\Omega_1} \epsilon u_1' v_1' \mathrm{d}x &= -\int_{\Omega_1} \epsilon u_1'' v_1 \mathrm{d}x - \sum_{j=1}^{N/2-1} \epsilon([u_1']_j\{v_1\}_j + \{u_1'\}_j[v_1]_j)\\
&\quad + \epsilon u_1'((1-\tau)^-)v_1^-(1-\tau) - \epsilon u_1'(0^+)v_1(0^+).
\end{aligned} \tag{3.11}$$

Inserting (3.11) into (3.10), integrating by parts the third term in the right hand side of (3.10), and combining with the definition of $\mathcal{L}_1(\cdot)$, we obtain

$$\mathcal{B}_N(u,v) = \int_\Omega -\epsilon u'' v \mathrm{d}x + \sum_{j=1}^{N/2-1} \epsilon[u_1']_j(\beta[v_1]_j - \{v_1\}_j) + \epsilon u_2'(1)v_2(1).$$

By the definition of V_N, we get $v_2(1) = 0$. This, together with $[u'] = 0$, yields

$$\mathcal{B}_N(u,v) = \int_\Omega -\epsilon u'' v \mathrm{d}x, \qquad \text{for all } v \in V_N.$$

Now we consider the term $\mathcal{C}_N(u,v)$. It can be rewritten as

$$\mathcal{C}_N(u,v)=-\int_{\Omega_1}bu_1(v_1'+\mathcal{L}_2(v))\mathrm{d}x-\int_{\Omega_2}bu_2v_2'\mathrm{d}x+\int_{\Omega}(c-b')uv\mathrm{d}x. \tag{3.12}$$

Taking $u=-bu_1$ and $v=v_1$ in (3.1), we have

$$\begin{aligned}-\int_{\Omega_1}bu_1v_1'\mathrm{d}x=&\int_{\Omega_1}(bu_1'+b'u_1)v_1\mathrm{d}x+\sum_{j=1}^{N/2-1}b(x_j)([u_1]_j\{v_1\}_j+\{u_1\}_j[v_1]_j)\\&-b(1-\tau)u_1^-(1-\tau)v_1^-(1-\tau)+b(0)u_1^+(0)v_1^+(0).\end{aligned} \tag{3.13}$$

Inserting (3.13) into (3.12), integrating by parts the second term in the right hand side of (3.12), and combining with the definition of $\mathcal{L}_2(\cdot)$, we obtain

$$\mathcal{C}_N(u,v)=\int_{\Omega}(bu'+cu)v\mathrm{d}x+\sum_{j=1}^{N/2-1}b(x_j)v_1^+(x_j)[u_1]_j$$

Recalling $[u_1]_j=0$ and $-\epsilon u''+bu'+cu=f$, we have

$$\mathcal{A}_N(u,v)=\int_{\Omega}f\,v\mathrm{d}x,\quad \forall v\in V_N$$

In view of (3.8), (3.9) obviously holds. □

Stability analysis. To consider the stability of the primal form $\mathcal{A}_N$, we define the following norms and seminorms for $v\in V(N)$:

$$\begin{aligned}|||v|||_\epsilon^2=&||v||_{0,\Omega}^2+\epsilon|v|_{1,N}^2+\epsilon|v|_*^2+|v|_c^2,\\|v|_{1,N}^2=&|v_2|_{1,\Omega_2}^2+\sum_{j=1}^{N/2}|v_1|_{1,I_j}^2,\quad |v|_*^2=\sum_{j=1}^{N/2-1}\alpha[v_1]_j^2+\alpha v_1^2(0^+)+\alpha(v_2-v_1^-)^2(1-\tau),\\|v|_c^2=&\frac{1}{2}\sum_{j=1}^{N/2-1}b_j[v_1]_j^2+\frac{1}{2}b(0)v_1^2(0^+)+\frac{1}{2}b(1-\tau)(v_2-v_1^-)^2(1-\tau),\end{aligned} \tag{3.14}$$

where $||\cdot||_{0,D}$ and $|\cdot|_{1,D}$ are the usual Sobolev norm and semi-norm defined on region D, respectively.

Lemma 3.2. *If $\alpha=\mathcal{O}(1/H)$, there exists a constant $C_1>0$, such that*

$$\mathcal{A}_N(u_N,u_N)\geqslant C_1|||u_N|||_\epsilon^2,\quad \forall u_N\in V_N.$$

Proof. By direct computation, we obtain, for all $u_N\in V_N$,

$$\begin{aligned}\mathcal{B}_N(u_N,u_N)=&\int_{\Omega}\epsilon(u_N'+\mathcal{L}_1(u_N))^2\mathrm{d}x,\quad \mathcal{S}_N(u_N,u_N)=\epsilon|u_N|_*^2,\\\mathcal{C}_N(u_N,u_N)=&\int_{\Omega}(c-\frac{1}{2}b')u_N^2\mathrm{d}x+|u_N|_c^2.\end{aligned}$$

Adding together the above three equality, and combining with (2.2), we have

$$\begin{aligned}\mathcal{A}_N(u_N,u_N)\geqslant&\int_\Omega \epsilon(u_N'+\mathcal{L}_1(u_N))^2\mathrm{d}x+c_0\int_\Omega u_N^2\mathrm{d}x+\epsilon|u_N|_*^2+|u_N|_c^2\\ =&\epsilon|u_N|_{1,N}^2+2\epsilon\int_\Omega u_N'\mathcal{L}_1(u_N)\mathrm{d}x+\epsilon||\mathcal{L}_1(u_N)||_{0,\Omega}^2+c_0||u_N||_{0,\Omega}^2+\epsilon|u_N|_*^2+|u_N|_c^2\end{aligned}$$

Applying the arithmetic-geometric mean inequality, we have, for every $\theta>0$,

$$\mathcal{A}_N(u_N,u_N)\geqslant\epsilon[(1-\theta)|u_N|_{1,N}^2+(1-1/\theta)||\mathcal{L}_1(u_N)||_{0,\Omega}^2]+c_0||u_N||_{0,\Omega}^2+\epsilon|u_N|_*^2+|u_N|_c^2.\quad(3.15)$$

According to [11](p.422), when the coefficient $\alpha=\mathcal{O}(H^{-1})$, there exists a constant $C>0$, such that

$$||\mathcal{L}_1(u)||_{0,\Omega}\leqslant C|u|_*,\quad \forall u\in V(N).\tag{3.16}$$

Inserting (3.16) into (3.15), for any θ satisfying $\frac{C^2}{C^2+1}<\theta<1$, we easily get,

$$\begin{aligned}\mathcal{A}_N(u_N,u_N)\geqslant&\epsilon(1-\theta)|u_N|_{1,N}^2+\epsilon[C^2(1-1/\theta)+1]|u_N|_*^2+c_0||u_N||_{0,\Omega}^2+|u_N|_c^2\\ \geqslant&\gamma(\epsilon|u_N|_{1,N}^2+\epsilon|u_N|_*^2)+c_0||u_N||_{0,\Omega}^2+|u_N|_c^2,\\ \geqslant&\min\{\gamma,c_0\}|||u_N|||_\epsilon^2,\end{aligned}$$

where $\gamma=\min\{1-\theta,C^2(1-1/\theta)+1\}$. Taking $C_1=\min\{\gamma,c_0\}$, the proof is complete. □

From Lemma 3.2, we easily get

$$|||u_N|||_\epsilon\leqslant C||f||_{0,\Omega},$$

which implies the uniqueness of the solution to (3.8). Further, since (3.8) is a linear problem over the finite-dimensional space V_N, the existence of the solution follows from its uniqueness. Consequently, by (3.4), we get the existence and uniqueness of the solution to the problem (2.9)–(2.11) with numerical fluxes (2.12)–(2.14).

Remark 3.1. In fact, following [7] or [11], for any $\alpha\geqslant 0$, we can prove the existence and uniqueness of the solution to the problem (2.9)–(2.11) with numerical fluxes (2.12)–(2.14). In this paper, we are only interested in the special case $\alpha=\mathcal{O}(H^{-1})$.

Error analysis. We are now going to provide a ϵ-uniform estimate for the error $u-u_N$ in the norm (3.14). First, we start with the error decomposition

$$u-u_N=(u-u_I)+(u_I-u_N)\equiv\eta+\xi,\tag{3.17}$$

where u_I be the standard linear interpolation of u.

The final estimate for $|||u-u_N|||_\epsilon$ will be derived by applying the triangle inequality to (3.17). For this purpose, we estimate $|||\eta|||_\epsilon$ and $|||\xi|||_\epsilon$, respectively. To bound η, we need the following lemma.

Lemma 3.3. *[18] Let u be a solution of the problem (2.1) and u_I the standard linear finite element interpolation of u on the Shishkin mesh. Then the interpolation error $\eta = u - u_I$ satisfies*

$$\begin{aligned}||\eta||_{L^\infty(\Omega_1)} &\leqslant CN^{-2},\\ ||\eta||_{L^\infty(\Omega_2)} &\leqslant CN^{-2}\ln^2 N,\\ ||\eta'||_{L^2(\Omega)} &\leqslant C\epsilon^{-1/2}||\eta||^{1/2}_{L^\infty(\Omega)}.\end{aligned}$$

The following statement is the direct consequence of Lemma 3.3.

Theorem 3.4. *Under the conditions of Lemma 3.3, we have*

$$|||\eta|||_\epsilon \leqslant CN^{-1}\ln N. \tag{3.18}$$

Proof. Since $u - u_I$ is continuous in Ω, we have $|\eta|_* = 0, |\eta|_c = 0$. Then, $|||\eta|||_\epsilon^2 = ||\eta||^2_{0,\Omega} + \epsilon|\eta|^2_{1,N}$. By Lemma 3.3, we easily conclude (3.18). □

Now we turn to estimate $|||\xi|||_\epsilon$.

Theorem 3.5. *Under the conditions of Lemma 3.3 and assume $\alpha = \mathcal{O}(1/H)$. Then $\xi = u_I - u_N$ satisfies*

$$|||\xi|||_\epsilon \leqslant CN^{-1}\ln N. \tag{3.19}$$

Proof. By Lemma 3.1 and Lemma 3.2, we first obtain

$$C_1|||\xi|||^2_\epsilon \leqslant \mathcal{A}_N(\xi,\xi) = -\mathcal{A}_N(\eta,\xi) = -\mathcal{B}_N(\eta,\xi) - \mathcal{C}_N(\eta,\xi) - \mathcal{S}_N(\eta,\xi). \tag{3.20}$$

By the definition of u_I, we have $\eta(x_j) = 0, j = 0, 1, \cdots, N$. Consequently,

$$\begin{aligned}\mathcal{B}_N(\eta,\xi) &= \int_\Omega \epsilon\eta'(\xi' + \mathcal{L}_1(\xi))\mathrm{d}x,\\ \mathcal{C}_N(\eta,\xi) &= -\int_\Omega b\eta\xi'\mathrm{d}x + \int_\Omega (c - b')\eta\xi\mathrm{d}x \equiv I_1 + I_2,\\ \mathcal{S}_N(\eta,\xi) &= 0.\end{aligned} \tag{3.21}$$

Then, by (3.16) and Lemma 3.3, we have

$$|\mathcal{B}_N(\eta,\xi)| \leqslant C\epsilon||\eta'||_{L^2(\Omega)}(|\xi|_{1,N} + |\xi|_*) \leqslant C\epsilon^{1/2}||\eta'||_{L^2(\Omega)}|||\xi|||_\epsilon \leqslant CN^{-1}\ln N\,|||\xi|||_\epsilon. \tag{3.22}$$

The first term in the right hand side of (3.21) can be estimated with,

$$|I_1| \leqslant C(||\eta||_{L^\infty(\Omega_1)}||\xi'||_{L^1(\Omega_1)} + ||\eta||_{L^\infty(\Omega_2)}||\xi'||_{L^1(\Omega_2)}).$$

On Ω_1 we use an inverse inequality to estimate

$$||\xi'||_{L^1(\Omega_1)} \leqslant CN||\xi||_{L^1(\Omega_1)} \leqslant CN|||\xi|||_\epsilon,$$

while on Ω_2 we use Cauchy-Schwarz inequality to estimate

$$||\xi'||_{L^1(\Omega_2)} \leqslant \sqrt{\tau}||\xi'||_{0,\Omega_2} \leqslant C\ln^{1/2} N|||\xi|||_\epsilon.$$

These two bounds and Lemma 3.3 yield

$$|I_1| \leqslant C(N^{-1} + N^{-2}\ln^{5/2} N)|||\xi|||_\epsilon \leqslant C(N^{-1} + N^{-1}\ln N)|||\xi|||_\epsilon, \tag{3.23}$$

where we have used the fact $(\ln N)^{3/2} \leqslant CN$.

The second term in the right hand side of (3.21) can be easily estimated with

$$|I_2| \leqslant C||\eta||_{0,\Omega}\,||\xi||_{0,\Omega} \leqslant CN^{-2}\ln^2 N|||\xi|||_\epsilon.$$

This, combined with (3.23), yields

$$|\mathcal{C}_N(\eta,\xi)| \leqslant CN^{-1}\ln N|||\xi|||_\epsilon. \tag{3.24}$$

Collecting (3.20), (3.22) and (3.24), we have (3.19). □

The combination of Theorem 3.4 and Theorem 3.5 leads to our main results directly, i.e.,

Theorem 3.6. *Let u and u_N be the solutions of the continuous problem (2.1) and the discrete problem (3.8), respectively. Assume $\alpha = \mathcal{O}(1/H)$, then*

$$|||u - u_N|||_\epsilon \leqslant CN^{-1}\ln N. \tag{3.25}$$

Corollary 3.7. *Let (q_N, u_N) be the solution obtained by the coupled method (2.9)–(2.11) with numerical fluxes (2.12)–(2.14). Under the assumption of Theorem 3.3, we have*

$$|(q - q_N, u - u_N)|_{\mathcal{A}_N} \leqslant CN^{-1}\ln N, \tag{3.26}$$

where $|(\cdot,\cdot)|_{\mathcal{A}_N}$ is a problem-related norm defined by

$$|(q,u)|^2_{\mathcal{A}_N} = ||u||^2_{0,\Omega} + \epsilon||q||^2_{0,\Omega_1} + \epsilon|u_2|^2_{1,\Omega_2} + \epsilon|u|^2_* + |u|^2_c. \tag{3.27}$$

Proof. From (3.4), we have $q - q_N = (u_1 - u_{1,N})' - \mathcal{L}_1(u_N)$. Since $\mathcal{L}_1(u) = 0$, we obtain $q - q_N = (u_1 - u_{1,N})' + \mathcal{L}_1(u - u_N)$. In terms of (3.16), we conclude $|(q - q_N, u - u_N)|_{\mathcal{A}_N} \leqslant C|||u - u_N|||_\epsilon$, which implies the conclusion. □

4 Numerical Experiments

In this section, we numerically verify the sharpness of our theoretical findings and explore some situations not covered by them. In our numerical experiments, we take $\alpha = 1/H$, $\beta = 1/2$ in (2.12) and (2.13), $\kappa = 2$ in the transition parameter of Shishkin mesh.

Example. We solve the model problem (2.1) with $b = c = 1$, and

$$f(x) = \cos x\,(1 + e^{(x-1)/\epsilon}) + (1+\epsilon)\sin x\,(1 - e^{(x-1)/\epsilon}).$$

The exact solution is $u(x) = \sin x\,(1 - e^{(x-1)/\epsilon})$, which exhibits a boundary layer with the width $\mathcal{O}(\epsilon \ln \frac{1}{\epsilon})$ at the outflow boundary $x = 1$.

We display the history of convergence of our coupled method in Table 1. We define a discrete L^∞-norm error as

$$||\widehat{e_u}||_\infty = \max_{0\leqslant j\leqslant N} |(u - \widehat{u})(x_j)|, \tag{4.1}$$

where $\widehat{u} = \widehat{u}_1$ on Ω_1 and $\widehat{u} = u_{2,N}$ on Ω_2. A Shishkin mesh with N elements is called mesh N. We also display numerical rates of convergence which are computed as follows. Let $err(N)$ denote the error of the approximate solution computed on the mesh N. Then the approximate order of convergence, i.e., $order(2N)$, is defined by

$$order(2N) := \begin{cases} \dfrac{\ln(err(N)/err(2N))}{\ln(2\ln(N)/\ln(2N))}, & \text{for the norm (3.41) and (4.1)}, \\[2ex] \dfrac{\ln(err(N)/err(2N))}{\ln(2)}, & \text{for the } L^2 \text{ norm.} \end{cases}$$

Table 1 History of convergence of the coupled method, under the norm (3.27), Shishkin mesh, $\kappa = 2$

N	$\epsilon = 1.0\text{e}-05$		$\epsilon = 1.0\text{e}-06$		$\epsilon = 1.0\text{e}-07$	
	error	order	error	order	error	order
16	1.174661e−01	—	1.174651e−01	—	1.174650e−01	—
32	7.401573e−02	0.98	7.401510e−02	0.98	7.401503e−02	0.98
64	4.456098e−02	0.99	4.456060e−02	0.99	4.456056e−02	0.99
128	2.602707e−02	1.00	2.602684e−02	1.00	2.602682e−02	1.00
256	1.487921e−02	1.00	1.487908e−02	1.00	1.487907e−02	1.00
512	8.370788e−03	1.00	8.370715e−03	1.00	8.370708e−03	1.00

Table 2 History of convergence of the coupled method, under the norm (4.1), Shishkin mesh, $\kappa = 2$

N	$\epsilon = 1.0\text{e}-05$		$\epsilon = 1.0\text{e}-06$		$\epsilon = 1.0\text{e}-07$	
	error	order	error	order	error	order
16	1.241546e−02	—	1.241538e−02	—	1.241537e−02	—
32	4.900315e−03	1.98	4.900289e−03	1.98	4.900287e−03	1.98
64	1.754352e−03	2.01	1.754345e−03	2.01	1.754344e−03	2.01
128	5.934619e−04	2.01	5.934598e−04	2.01	5.934598e−04	2.01
256	1.936661e−04	2.00	1.936654e−04	2.00	1.936654e−04	2.00
512	6.127390e−05	2.00	6.127367e−05	2.00	6.127338e−05	2.00

Table 3 History of convergence of the coupled method, under the L^2 norm , Shishkin mesh, $\kappa=2$

N	$\epsilon=1.0e-05$		$\epsilon=1.0e-06$		$\epsilon=1.0e-07$	
	error	order	error	order	error	order
16	4.981133e−04	—	4.947059e−04	—	4.943639e−04	—
32	1.262621e−04	1.98	1.241453e−04	1.99	1.239317e−04	2.00
64	3.222860e−05	1.97	3.113287e−05	2.00	3.102118e−05	2.00
128	8.313351e−06	1.95	7.813138e−06	1.99	7.761349e−06	2.00
256	2.171748e−06	1.94	1.963761e−06	1.99	1.941742e−06	2.00
512	5.749559e−07	1.92	4.946103e−07	1.99	4.858491e−07	2.00

From Table 1 and Figure 1 we observe that the numerical results for the problem-related norm (3.27) of the error agree with those predicted in Corollary 3.7. Further from Table 2 and Table 3, it is numerically observed that

$$||\widehat{e_u}||_\infty \leqslant CN^{-2}\ln^2 N, \qquad ||u-u_N||_{0,\Omega} \leqslant CN^{-2},$$

which is uniformly optimal.

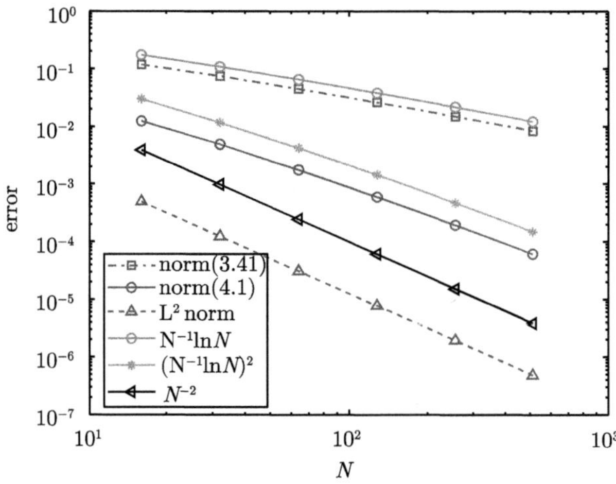

Figure 1 Convergence curve of $u-u_N$, the Shishkin mesh, $\epsilon=10^{-7}$

5 Conclusions

In this paper, we introduce a new coupled LDG-CFEM method for solving singularly perturbed convection-diffusion problems, and analyze its stability and uniform convergence property on the Shishkin mesh. For linear elements, a rate $\mathcal{O}(N^{-1}\ln N)$ in an associated norm is established. Our numerical experiments indicate the sharpness of this error estimate. Moreover, a rate $\mathcal{O}(N^{-2}\ln^2 N)$ in a discrete L^∞ norm, and $\mathcal{O}(N^{-2})$ in L^2 norm are numerically observed on the Shishkin mesh.

References

[1] T. Linß, Layer-adapted meshes for convection-diffusion problems, *Computer Methods in Applied Mechanics and Engineering*, 192 (2003), 1061-1105.

[2] H.-G. Roos, Layer-adapted grids for singular perturbation problem, *ZAMM - Journal of Applied Mathematics and Mechanics*, 78 (1998), 291-309.

[3] H.-G. Roos, M. Stynes and L. Tobiska, *Robust numerical methods for singularly perturbed differential equations*, Springer Series in Computational Mathematics, Volume 24, Springer-Verlag Berlin Heidelberg, 2008.

[4] M. Stynes and E. O'Riorddan, A uniformly convergent Galerkin method on a Shishkin mesh for a convection-diffusion problem, *Journal of Mathematical Analysis and Applications*, 214 (1997), 36-54.

[5] B. Cockburn and C.-W. Shu, The local discontinuous Galerkin finite element method for convection-diffusion systems, *SIAM Journal on Numerical Analysis*, 35 (1998), 2440-2463.

[6] F. Celiker and B. Cockburn, Superconvergence of the numerical traces of discontinuous Galerkin and Hybridized methods for convection-diffusion problems in one space dimension, *Mathematics of Computation*, 76 (2007), 67-96.

[7] Z. Q. Xie and Z. Zhang, Superconvergence of DG method for one-dimensional singularly perturbed problems, *Journal of Computational Mathematics*, 25 (2007), 185-200.

[8] Z. Q. Xie, Z. Z. Zhang, and Z. Zhang, A numerical study of uniform superconvergence for solving singularly perturbed problems, *Journal of Computational Mathematics*, 27 (2009), 280-298.

[9] Z. Z. Zhang, Z. Q. Xie, and Z. Zhang, Superconvergence of discontinuous Galerkin methods for convection diffusion problems, *Journal of Scientific Computing*, 41 (2009), 70-93.

[10] H. Zarin, and H.-G. Roos, Interior penalty discontinuous approximations of convection-diffusion problems with parabolic layers, *Numerische Mathematik*, 100 (2005), 735-759.

[11] I. Perugia and D. Schötzau, On the coupling of local discontinuous Galerkin and conforming finite element methods, *Journal of Scientific Computing*, 16 (2001), 411-433.

[12] H.-G. Roos, and H. Zarin, A supercloseness result for the discontinuous Galerkin stabilization of convection-diffusion problems on Shishkin meshes, *Numerical Methods for Partial Differential Equations*, 23 (2007), 1560-1576.

[13] H. Zarin, Continuous-discontinuous finite element method for convection-diffusion problems with characteristic layers, *Journal of Computational and Applied Mathematics*, 231 (2009), 626-636.

[14] D. N. Arnold, F. Brezzi, B. Cockburn, and L. D. Marini, Unified analysis of discontinuous Galerkin methods for elliptic problems, *SIAM Journal on Numerical Analysis*, 39 (2002), 1749-1779.

[15] G. I. Shishkin, *Grid approximation of singularly perturbed elliptic and parabolic equations*[Second doctorial thesis]. Moscow: Keldysh Institue, 1990 (in Russian).

[16] Z. Zhang, Finite element superconvergence on Shishkin mesh for 2-D convection-diffusion problems, *Mathematics of Computation*, 245 (2003), 1147-1177.

[17] P. Castillo, B. Cockburn, I. Perugia and D. Schötzau, An a priori error analysis of the local discontinuous Galerkin method for elliptic problems, *SIAM Journal on Numerical Analysis*, 38 (2000), 1676-1706.

[18] T. Linß, The necessity of Shishkin decompositions, *Applied Mathematics Letters*, 14 (2001), 891-896.

Uniform Convergence of a Coupled Method for Convection-Diffusion Problems in 2-D Shishkin Mesh*

Z.Q. Xie (谢资清)† P. Zhu (祝鹏)‡ S.Z. Zhou (周叔子)§

Abstract In this paper, we introduce a coupled approach of local discontinuous Galerkin (LDG) and continuous finite element method (CFEM) for solving singularly perturbed convection-diffusion problems. When the coupled continuous-discontinuous linear FEM is used under the Shishkin mesh, a uniform convergence rate $\mathcal{O}(N^{-1}\ln N)$ in an associated norm is established, where N is the number of elements. Numerical experiments complement the theoretical results. Moreover, a uniform convergence rate $\mathcal{O}(N^{-2})$ in L^2 norm, is observed numerically on the Shishkin mesh.

Keywords Convection diffusion equation, local discontinuous Galerkin method, finite element method, Shishkin mesh, uniform convergence.

1 Introduction

In recent years, the numerical solutions of singularly perturbed boundary value problems have been received much attention and already studied in many papers and books, see for instance [1 − 4]. One of the difficulties in numerically computing the solution of singularly perturbed problems lays in the so-called boundary layer behavior, i.e., the solution varies very rapidly in a very thin layer near the boundary. Traditional methods such as finite element and finite difference methods, do not work well for these problems as they often produce oscillatory solutions which are inaccurate if the perturbed parameter ϵ is small. When ϵ approaches zero, the problem changes from an elliptic equation to a hyperbolic one. Inspired by the great success of the discontinuous Galerkin (DG) method in solving hyperbolic equations, Cockburn

* 发表于：International Journal of Numerical Analysis & Modeling, Vol. 10, 2013, pp. 845-859.

† College of Mathematics and Computer Science, Hunan Normal University, Changsha, Hunan 410081, China.

‡ Institute of Mathematics, Jiaxing University, Jiaxing, Zhejiang 314001, China.

§ College of Mathematics and Econometrics, Hunan University, Changsha, Hunan 410082, China.

and Shu [5], Celiker and Cockburn [6], Xie et al. [7 – 9] and Zhang et al. [10] adopted the local discontinuous Galerkin (LDG) method to solve convection-diffusion equations and analyzed the corresponding convergence properties. On the other hand, nonsymmetric discontinuous Galerkin method with interior penalty (the NIPG method), originally designed for elliptic equations, is analyzed by Zarin and Roos [11] for convection-diffusion problems with parabolic layers.

A disadvantage of DG method is that it produces more degrees of freedom than the continuous finite element method (CFEM). With this motivation, our work is to derive a coupled approach of LDG and CFEM and analyze the uniform convergence in a DG-norm under Shishkin mesh for singularly perturbed convection diffusion problems. The basic idea is to decompose the domain into coarse and fine part, the latter is used to simulate the boundary layer. Then the CFEM using linear elements is adopted in the fine part where the mesh size is comparable with ϵ, and LDG method is used in the coarse part for its stabilization.

A coupled LDG-CFEM approach has also been studied by Perugia and Schötzau [12] for the modeling of elliptic problems arising in electromagnetics. Roos and Zarin [13], Zarin [14] analyzed the NIPG-CFEM coupled method on Shishkin mesh for convection-diffusion problems with exponentially layers or characteristic layers. In this paper, the coupled LDG method is used for the singularly perturbed convection-diffusion equation for the first time to our knowledge. Moreover, distinguished from the general approaches for proving uniform convergence of numerical methods for singularly perturbed problem on layer-adapted meshes, in which solution decomposition is necessary, our analysis is based on the uniform error estimates for the interpolation under the Shishkin mesh. Our method can be generalized to all DG methods belong to the unify framework in [15], including the NIPG method.

The paper is organized as follows. In Section 2, we introduce the coupled LDG and CFEM for the singularly perturbed problems. Then stability and error analysis of the coupled method on Shishkin mesh is given in Section 3. The implement of our coupled method on Shishkin mesh is presented in Section 4. It aims to validate our theoretical results. Furthermore, the uniform convergence rate $\mathcal{O}(N^{-2})$ in L^2 norm is observed numerically. This phenomenon is not found in [13] and [14].

In the sequel, with C we shall denote a generic positive constant independent of the perturbation parameter ϵ and mesh size.

2 Coupling the LDG and CFEM

Consider the following two dimensional convection-diffusion problem

$$\begin{cases} -\epsilon\Delta u + \mathbf{b}\cdot\nabla u + cu = f, & \text{in } \Omega = (0,1)^2, \\ u = 0, & \text{on } \partial\Omega, \end{cases} \tag{2.1}$$

where $0 < \epsilon \ll 1$ is a small positive parameter, $\mathbf{b}, c$, and f are sufficiently smooth functions with the following properties

$$\begin{aligned}
&\mathbf{b}(x,y) = (b_1(x,y), b_2(x,y)) \geqslant (\beta_1, \beta_2) > (0,0), \quad c(x,y) \geqslant 0, \quad \forall (x,y) \in \bar{\Omega},\\
&c_0^2(x,y) \equiv (c - \frac{1}{2}\nabla \cdot \mathbf{b})(x,y) \geqslant \gamma_0 > 0, \quad \forall (x,y) \in \bar{\Omega},\\
&f(0,0) = f(1,0) = f(0,1) = f(1,1) = 0,
\end{aligned} \tag{2.2}$$

for some constants β_1, β_2 and γ_0. With the assumptions above, it is well-known that there exists a solution u of (2.1) that in general exhibits an exponentially boundary layer near $x = 1$ and $y = 1$.

The Shishkin Mesh. Define the transition parameter

$$\tau_x = \min\left(\frac{1}{2}, \frac{\kappa}{\beta_1}\epsilon \ln N\right), \quad \tau_y = \min\left(\frac{1}{2}, \frac{\kappa}{\beta_2}\epsilon \ln N\right),$$

with $\kappa \geqslant 2$ and divide Ω into four sub-domains

$$\begin{aligned}
\Omega_0 &= [0, 1-\tau_x] \times [0, 1-\tau_y], \qquad \Omega_x = [1-\tau_x, 1] \times [0, 1-\tau_y],\\
\Omega_y &= [0, 1-\tau_x] \times [1-\tau_y, 1], \qquad \Omega_{xy} = [1-\tau_x, 1] \times [1-\tau_y, 1].
\end{aligned}$$

Each sub-domain is then decomposed into $N/2 \times N/2$ uniform rectangles (see Figure 1). Therefore, there are $(N+1)^2$ nodes (x_i, y_j), $i, j = 0, 1, \cdots, N$, and N^2 elements

$$K_{ij} = [x_{i-1}, x_i] \times [y_{j-1}, y_j], \quad i, j = 1, 2, \cdots, N.$$

Consequently,

$$x_j = \begin{cases} 2(1-\tau_x)j/N, & j = 0, 1, \cdots, N/2,\\ 1 - 2\tau_x(N-j)/N, & j = N/2+1, \cdots, N,\end{cases}$$

and

$$y_j = \begin{cases} 2(1-\tau_y)j/N, & j = 0, 1, \cdots, N/2,\\ 1 - 2\tau_y(N-j)/N, & j = N/2+1, \cdots, N.\end{cases}$$

Denote

$$\begin{aligned}
H_x &= 2(1-\tau_x)/N, \qquad h_x = 2\tau_x/N,\\
H_y &= 2(1-\tau_y)/N, \qquad h_y = 2\tau_y/N.
\end{aligned}$$

In fact we assume that $\tau_x = \dfrac{\kappa}{\beta_1}\epsilon \ln N$ and $\tau_y = \dfrac{\kappa}{\beta_1}\epsilon \ln N$, as otherwise N^{-1} is much smaller than ϵ (which is very unlikely in practice).

Obviously

$$\max\{H_x, H_y\} = \mathcal{O}(N^{-1}), \qquad \max\{h_x, h_y\} = \mathcal{O}(\epsilon N^{-1} \ln N).$$

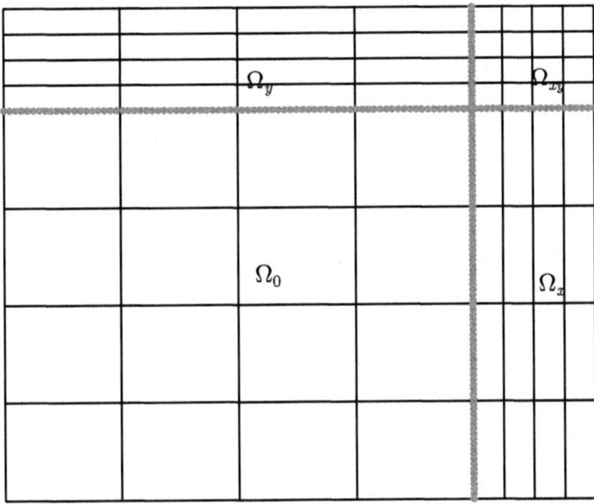

Figure 1 Shishkin Mesh with $N = 8$ and $\epsilon = 0.05$

Set $\Omega_1 = \Omega_0$ and $\Omega_2 = \Omega_x \cup \Omega_y \cup \Omega_{xy}$, and the interface $\Gamma = \partial\Omega_1 \cap \partial\Omega_2$. Let $\mathcal{T}_N^1 = \{K_{ij} : 1 \leqslant i, j \leqslant N/2\}$ and $\mathcal{T}_N^2 = \{K_{ij} : i > N/2 \,\text{or}\, j > N/2\}$. Obviously $\mathcal{T}_N^1$ and $\mathcal{T}_N^2$ are rectangular partitions of Ω_1 and Ω_2, respectively.

Weak formulation. Let $u_i = u|_{\Omega_i}, \Gamma_D^i = \partial\Omega \cap \partial\Omega_i$, $i = 1, 2$. Denote by $\mathbf{n}_i$ the unit outward normal vector to $\Gamma \cap \partial\Omega_i, i = 1, 2$, and by $\mathbf{n}$ the unit outward normal vector to $\partial\Omega$. As mentioned before, we discretize problem (2.1) by using the LDG method in Ω_1, and CFEM in Ω_2. For this purpose, we introduce an auxiliary variable $\mathbf{q} = \nabla u_1$ in Ω_1 and rewrite problem (2.1) as the following equivalent form, i.e.,

$$\begin{cases} \mathbf{q} - \nabla u_1 = 0, & \text{in } \Omega_1, \\ -\epsilon \nabla \cdot \mathbf{q} + \mathbf{b} \cdot \nabla u_1 + c u_1 = f, & \text{in } \Omega_1, \\ -\epsilon \Delta u_2 + \mathbf{b} \cdot \nabla u_2 + c u_2 = f, & \text{in } \Omega_2, \\ u_1 = u_2, & \text{on } \Gamma, \\ \mathbf{q} \cdot \mathbf{n}_1 = -\nabla u_2 \cdot \mathbf{n}_2, & \text{on } \Gamma, \end{cases} \tag{2.3}$$

with boundary conditions

$$u_i = 0 \quad \text{on} \quad \Gamma_D^i \quad i = 1, 2. \tag{2.4}$$

Multiplying the first three equations of (2.3) by test functions $\mathbf{w}, v_1, v_2$, respectively, and integrating by parts, it is noted that the solution $(\mathbf{q}, u_1, u_2)$, of problem (2.3) and (2.4) satisfies, for all $K \in \mathcal{T}_N^1$,

$$\int_K \mathbf{q} \cdot \mathbf{w} \mathrm{d}\mathbf{x} + \int_K u_1 \nabla \cdot \mathbf{w} \mathrm{d}\mathbf{x} - \int_{\partial K} u_1 \mathbf{w} \cdot \mathbf{n}_K \mathrm{d}s = 0, \tag{2.5}$$

$$\int_K (\epsilon \mathbf{q} - \mathbf{b} u_1) \cdot \nabla v_1 \mathrm{d}\mathbf{x} + \int_K (c - \nabla \cdot \mathbf{b}) u_1 v_1 \mathrm{d}\mathbf{x} - \int_{\partial K} v_1 (\epsilon \mathbf{q} - \mathbf{b} u_1) \cdot \mathbf{n}_K \mathrm{d}s = \int_K f v_1 \mathrm{d}\mathbf{x}, \tag{2.6}$$

for piecewise smooth functions $\mathbf{w}$ and v_1, with $\mathbf{n}_K$ the unit outward normal to ∂K, and

$$\int_{\Omega_2} (\epsilon \nabla u_2 - \mathbf{b} u_2) \cdot \nabla v_2 \mathrm{d}\mathbf{x} + \int_{\Omega_2} (c - \nabla \cdot \mathbf{b}) u_2 v_2 \mathrm{d}\mathbf{x} - \int_{\Gamma} v_2 (\epsilon \nabla u_2 - \mathbf{b} u_2) \cdot \mathbf{n}_2 \mathrm{d}s = \int_{\Omega_2} f v_2 \mathrm{d}\mathbf{x},$$

for smooth function v_2 with $v_2|_{\Gamma_D^2} = 0$. The above equations have to be coupled through the transmission conditions, i.e., the last two equations of (2.3). Replacing $-(\epsilon \nabla u_2 - \mathbf{b} u_2) \cdot \mathbf{n}_2$ by $(\epsilon \mathbf{q} - \mathbf{b} u_1) \cdot \mathbf{n}_1$ in the last equation, leads to

$$\int_{\Omega_2} (\epsilon \nabla u_2 - \mathbf{b} u_2) \cdot \nabla v_2 \mathrm{d}\mathbf{x} + \int_{\Omega_2} (c - \nabla \cdot \mathbf{b}) u_2 v_2 \mathrm{d}\mathbf{x} + \int_{\Gamma} v_2 (\epsilon \mathbf{q} - \mathbf{b} u_1) \cdot \mathbf{n}_1 \mathrm{d}s = \int_{\Omega_2} f v_2 \mathrm{d}\mathbf{x}, \tag{2.7}$$

while the first one is imposed at the discrete level by choosing the numerical fluxes in the LDG method in a suitable way. It will be seen that the combination of (2.5), (2.6) and (2.7) are the basis for the coupled continuous-discontinuous Galerkin approach.

Denote by $\mathbb{Q}^1(K)$ the space of bilinear functions defined on K, and define the finite element space Q_N, V_N^1 and V_N^2 as follows,

$$Q_N = \left\{ \mathbf{q} \in L^2(\Omega_1)^2 : \mathbf{q}|_K \in \mathbb{Q}^1(K)^2, \forall K \in \mathcal{T}_N^1 \right\},$$

$$V_N^1 = \left\{ u_1 \in L^2(\Omega_1) : u_1|_K \in \mathbb{Q}^1(K), \forall K \in \mathcal{T}_N^1 \right\},$$

$$V_N^2 = \left\{ u_2 \in H^1(\Omega_2) : u_2|_{\Gamma_D^2} = 0, u_2|_K \in \mathbb{Q}^1(K), \forall K \in \mathcal{T}_N^2 \right\}.$$

The space V_N^2 is a standard conforming finite element space, whereas the functions in V_N^1 are completely discontinuous across interelement boundaries.

We will search for approximate solutions $(\mathbf{q}_N, u_{1,N}, u_{2,N})$ of (2.3) and (2.4) in the finite element space $Q_N \times V_N^1 \times V_N^2$ that satisfy (2.3) and (2.4) in a weak sense, i.e., we will find $(\mathbf{q}_N, u_{1,N}, u_{2,N}) \in Q_N \times V_N^1 \times V_N^2$ such that

$$\int_K \mathbf{q}_N \cdot \mathbf{w} \mathrm{d}\mathbf{x} + \int_K u_{1,N} \nabla \cdot \mathbf{w} \mathrm{d}\mathbf{x} - \int_{\partial K} \widehat{u}_{1,N} \mathbf{w} \cdot \mathbf{n}_K \mathrm{d}s = 0, \tag{2.8}$$

$$\int_K (\epsilon \mathbf{q}_N - \mathbf{b} u_{1,N}) \cdot \nabla v_1 \mathrm{d}\mathbf{x} + \int_K (c - \nabla \cdot \mathbf{b}) u_{1,N} v_1 \mathrm{d}\mathbf{x} - \int_{\partial K} v_1 (\epsilon \widehat{\mathbf{q}}_N - \mathbf{b} \widetilde{u}_{1,N}) \cdot \mathbf{n}_K \mathrm{d}s = \int_K f v_1 \mathrm{d}\mathbf{x}, \tag{2.9}$$

for any test function $(\mathbf{w}, v_1) \in Q_N \times V_N^1$ and $K \in \mathcal{T}_N^1$, and

$$\int_{\Omega_2}(\epsilon\nabla u_{2,N}-\mathbf{b}u_{2,N})\cdot\nabla v_2\mathrm{d}\mathbf{x}+\int_{\Omega_2}(c-\nabla\cdot\mathbf{b})u_{2,N}v_2\mathrm{d}\mathbf{x}\int_{\Gamma}v_2(\epsilon\widehat{\mathbf{q}}_N-\mathbf{b}\widetilde{u}_{1,N})\cdot\mathbf{n}_1\mathrm{d}s=\int_{\Omega_2}fv_2\mathrm{d}\mathbf{x}, \tag{2.10}$$

for any test function $v_2\in V_N^2$, where $\widehat{u}_{1,N}$, $\widetilde{u}_{1,N}$ and $\widehat{\mathbf{q}}_N$ are the numerical fluxes, which approximate the traces of $u_{1,N}$ and $\mathbf{q}_N$ on the boundary of the elements of $\mathcal{T}_N^1$. To complete the specification of the method, it only remains to define the numerical fluxes.

The numerical fluxes. In order to define the numerical fluxes, we need to introduce some notations. An interior face of the triangulation $\mathcal{T}_N^1$ is defined to be the (non-empty) interior of $\partial K_+\cap\partial K_-$, where K_+ and K_- are two adjacent elements of $\mathcal{T}_N^1$. Analogously, a boundary face in $\mathcal{T}_N^1$ is the interior of $\partial K\cap\partial\Omega$ where K is a boundary element in $\mathcal{T}_N^1$. We assume the boundary faces are contained in either Γ_D^1 or Γ. Let $\mathcal{E}^o$ be the union of all interior faces, and $\mathcal{E}$ the union of all faces of $\mathcal{T}_N^1$. Let now $e\in\mathcal{E}^o$ be an interior face shared by K_+ and K_-, $\mathbf{n}_+$ and $\mathbf{n}_-$ the unit normal vectors on e pointing exterior to K_+ and K_-, respectively. The *average* and the *jump* of a scalar-valued function v on $e\in\mathcal{E}^o$ are given by

$$\{v\}:=\frac{1}{2}(v_++v_-),\qquad [v]:=v_+\mathbf{n}_++v_-\mathbf{n}_-,$$

respectively. For a vector-valued function $\mathbf{w}$, we define $\mathbf{w}_+$ and $\mathbf{w}_-$, analogously, i.e.,

$$\{\mathbf{w}\}:=\frac{1}{2}(\mathbf{w}_++\mathbf{w}_-),\qquad [\mathbf{w}]:=\mathbf{w}_+\cdot\mathbf{n}_++\mathbf{w}_-\cdot\mathbf{n}_-\qquad\text{on}\quad e\in\mathcal{E}^o.$$

For $e\subset\partial\Omega_1\setminus\Gamma$, each v and $\mathbf{w}$ has a uniquely defined restriction on e; we set

$$[v]:=v\mathbf{n}_e,\qquad \{\mathbf{w}\}:=\mathbf{w}\qquad\text{on}\quad e\subset\partial\Omega_1\setminus\Gamma,$$

where $\mathbf{n}_e$ is the unit outward normal of e. We do not require either of the quantities $\{v\}$ or $[\mathbf{w}]$ on the boundary faces, and leave them undefined. Now the numerical fluxes $\widehat{u}_{1,N}$ and $\widehat{\mathbf{q}}_{1,N}$ are defined by

$$\widehat{u}_1=\begin{cases}\{u_{1,N}\}+\beta\cdot[u_{1,N}], & \text{if}\ \ e\in\mathcal{E}^o,\\ u_{2,N}, & \text{if}\ \ e\subset\Gamma,\\ 0, & \text{if}\ \ e\subset\Gamma_D^1,\end{cases} \tag{2.11}$$

and

$$\widehat{\mathbf{q}}_{1,N}=\begin{cases}\{\mathbf{q}_N\}-\alpha[u_{1,N}]-\beta[\mathbf{q}_N], & \text{if}\ \ e\in\mathcal{E}^o,\\ \mathbf{q}_N-\alpha(u_{1,N}-u_{2,N})\mathbf{n}_1, & \text{if}\ \ e\subset\Gamma,\\ \mathbf{q}_N-\alpha\,u_{1,N}\mathbf{n}, & \text{if}\ \ e\subset\Gamma_D^1,\end{cases} \tag{2.12}$$

Here the scalar $\alpha=\alpha(\mathbf{x})$ and the vector $\beta=\beta(\mathbf{x})$ are auxiliary parameters, which play a very important role in guaranteing the stability and enhance the accuracy of the LDG scheme (see [5] and [16]).

The numerical flux associated with the convection is the classical upwinding one, which can be expressed as

$$\widetilde{u}_{1,N}=\begin{cases}\{u_{1,N}\}+\beta_0\cdot[u_{1,N}], & \text{if } e\in\mathcal{E}^o,\\ u_{1,N}, & \text{if } e\subset\Gamma,\\ 0, & \text{if } e\subset\Gamma_D^1,\end{cases}\tag{2.13}$$

where β_0 is a function on $\mathcal{E}^o$ such that, for $\mathbf{x}\in\partial K\cap\mathcal{E}^o$,

$$\beta_0\cdot\mathbf{n}_K(\mathbf{x})=\frac{1}{2}\operatorname{sign}(\mathbf{b}(\mathbf{x})\cdot\mathbf{n}_K(\mathbf{x})),\tag{2.14}$$

where $\mathbf{n}_K(\mathbf{x})$ is the unit outward normal of K at $\mathbf{x}$.

3 Stability and Error Analysis of the Coupled Method

This section is devoted to the existence and uniqueness of the solution of the coupled method (2.8)–(2.10) with numerical fluxes (2.11)–(2.13), and its corresponding error analysis. If the stabilization parameter α is taken of order $\mathcal{O}(1/H_x, 1/H_y)$, we can rewrite our method in the primal form by eliminating $\mathbf{q}$ following Arnold et al. [15].

Primal formulation. A straightforward computation shows that

$$\begin{aligned}\int_{\Omega_1}\nabla\cdot\mathbf{w}v\mathrm{d}\mathbf{x}=&-\int_{\Omega_1}\mathbf{w}\cdot\nabla v\mathrm{d}\mathbf{x}+\int_{\mathcal{E}^o}([\mathbf{w}]\{v\}+\{\mathbf{w}\}\cdot[v])\mathrm{d}s\\&+\int_{\Gamma}v\mathbf{w}\cdot\mathbf{n}_1\mathrm{d}s+\int_{\Gamma_D^1}v\mathbf{w}\cdot\mathbf{n}\mathrm{d}s,\end{aligned}\tag{3.1}$$

for $\mathbf{w}\in H^1(\mathcal{T}_N^1)\times H^1(\mathcal{T}_N^1)$ and $v\in H^1(\mathcal{T}_N^1)$, where $H^1(\mathcal{T}_N^1)$ is the piecewise Sobolev space defined by

$$H^1(\mathcal{T}_N^1):=\{\ \phi\in L^2(\Omega_1):\phi|_K\in H^1(K)\text{ for all }K\in\mathcal{T}_N^1\ \}.$$

Differential operators are understood to act on such a space piecewisely.

Summing up (2.8) for all $K\in\mathcal{T}_N^1$, combined with (2.11) and (3.1), we obtain

$$\int_{\Omega_1}(\mathbf{q}_N-\nabla u_{1,N})\cdot\mathbf{w}\mathrm{d}\mathbf{x}+\int_{\mathcal{E}^o}(\{\mathbf{w}\}-\beta[\mathbf{w}])\cdot[u_{1,N}]\mathrm{d}s+\int_{\Gamma}(u_{1,N}-u_{2,N})\mathbf{w}\cdot\mathbf{n}_1\mathrm{d}s+\int_{\Gamma_D^1}u_{1,N}\mathbf{w}\cdot\mathbf{n}\mathrm{d}s=0.\tag{3.2}$$

Denote $V(N)=\left[H^2(\Omega)\cap H^1_{\Gamma_D}(\Omega)\right]+V_N$, where $H^1_{\Gamma_D}(\Omega)=\{\ v=(v_1,v_2)\in H^1(\Omega):v_1|_{\Gamma_D^1}=0,\ v_2|_{\Gamma_D^2}=0\ \}$, $V_N:=\{\ v=(v_1,\,v_2):v_1\in V_N^1,\,v_2\in V_N^2\ \}$. For $v\in V(N)$, we define $\mathcal{L}_1(v)$ as the unique element in Q_N such that

$$\int_{\Omega_1}\mathcal{L}_1(v)\cdot\mathbf{r}\mathrm{d}\mathbf{x}=\int_{\mathcal{E}^o}(\{\mathbf{r}\}-\beta[\mathbf{r}])\cdot[v_1]\mathrm{d}s+\int_{\Gamma}(v_1-v_2)\mathbf{r}\cdot\mathbf{n}_1\mathrm{d}s+\int_{\Gamma_D^1}v_1\mathbf{r}\cdot\mathbf{n}\mathrm{d}s,\tag{3.3}$$

for all $\mathbf{r} \in Q_N$. As a result, (3.2) can be rewritten as

$$\mathbf{q}_N = \nabla u_{1,N} - \mathcal{L}_1(u_N). \tag{3.4}$$

Summing up (2.9) for all $K \in \mathcal{T}_N^1$ and then adding up (2.10), we get

$$\begin{aligned}
&\int_{\Omega_1} (\epsilon \mathbf{q}_N - \mathbf{b}u_{1,N}) \cdot \nabla v_1 \mathrm{d}\mathbf{x} + \int_{\Omega_2} (\epsilon \nabla u_{2,N} - \mathbf{b}u_{2,N}) \cdot \nabla v_2 \mathrm{d}\mathbf{x} + \int_{\Omega} (c - \nabla \cdot \mathbf{b}) u_N v \mathrm{d}\mathbf{x} \\
&\quad - \int_{\mathcal{E}\backslash\Gamma} (\epsilon \widehat{\mathbf{q}}_N - \mathbf{b}\widetilde{u}_{1,N}) \cdot [v_1] \mathrm{d}s - \int_{\Gamma} (\epsilon \widehat{\mathbf{q}}_N - \mathbf{b}\widetilde{u}_{1,N}) \cdot \mathbf{n}_1 (v_1 - v_2) \mathrm{d}s = \int_{\Omega} f v \mathrm{d}\mathbf{x}.
\end{aligned} \tag{3.5}$$

Inserting (2.12) and (2.13) into (3.5), and recalling the definition of $\mathcal{L}_1(\cdot)$, we obtain

$$\begin{aligned}
&\int_{\Omega_1} \epsilon \mathbf{q}_N \cdot (\nabla v_1 - \mathcal{L}_1(v)) \mathrm{d}\mathbf{x} + \int_{\Omega_2} \epsilon \nabla u_{2,N} \cdot \nabla v_2 \mathrm{d}\mathbf{x} - \int_{\Omega} \mathbf{b} \cdot \nabla v u_N \mathrm{d}\mathbf{x} \\
&\quad + \int_{\Omega} (c - \nabla \cdot \mathbf{b}) u_N v \mathrm{d}\mathbf{x} + \int_{\mathcal{E}\backslash\Gamma} \epsilon\alpha [u_{1,N}][v_1] \mathrm{d}s + \int_{\Gamma} \epsilon\alpha (u_{1,N} - u_{2,N})(v_1 - v_2) \mathrm{d}s \\
&\quad + \int_{\mathcal{E}^o} \mathbf{b} \cdot [v_1](\{u_{1,N}\} + \beta_0 \cdot [u_{1,N}]) \mathrm{d}s + \int_{\Gamma} \mathbf{b} \cdot \mathbf{n}_1 u_{1,N}(v_1 - v_2) \mathrm{d}s = \int_{\Omega} f v \mathrm{d}\mathbf{x}.
\end{aligned} \tag{3.6}$$

Similar to the definition of $\mathcal{L}_1(\cdot)$, for $v \in V(N)$, we define $\mathcal{L}_2(v)$ as the unique element in Q_N such that

$$\int_{\Omega_1} \mathbf{b} \cdot \mathcal{L}_2(v) u \mathrm{d}\mathbf{x} = \int_{\mathcal{E}^o} \mathbf{b} \cdot [v_1](\{u_1\} + \beta_0 \cdot [u_1]) \mathrm{d}s + \int_{\Gamma} \mathbf{b} \cdot \mathbf{n}_1 u_1 (v_1 - v_2) \mathrm{d}s, \tag{3.7}$$

for all $u \in V_N$. As a consequence, (3.6) can be rewritten as

$$\begin{aligned}
&\int_{\Omega_1} \epsilon \mathbf{q}_N \cdot (\nabla v_1 - \mathcal{L}_1(v)) \mathrm{d}\mathbf{x} + \int_{\Omega_2} \epsilon \nabla u_{2,N} \cdot \nabla v_2 \mathrm{d}\mathbf{x} - \int_{\Omega} \mathbf{b} \cdot \nabla v u_N \mathrm{d}\mathbf{x} \\
&\quad + \int_{\Omega} (c - \nabla \cdot \mathbf{b}) u_N v \mathrm{d}\mathbf{x} + \int_{\mathcal{E}\backslash\Gamma} \epsilon\alpha [u_{1,N}][v_1] \mathrm{d}s + \int_{\Gamma} \epsilon\alpha (u_{1,N} - u_{2,N})(v_1 - v_2) \mathrm{d}s \\
&\quad + \int_{\Omega_1} \mathbf{b} \cdot \mathcal{L}_2(v) u_N \mathrm{d}\mathbf{x} = \int_{\Omega} f v \mathrm{d}\mathbf{x}.
\end{aligned} \tag{3.8}$$

Inserting (3.4) into (3.8), we obtain the so-called primal form of our coupled method which reads: find $u_N \in V_N$ such that

$$\mathcal{A}_N(u_N, v) := \mathcal{B}_N(u_N, v) + \mathcal{C}_N(u_N, v) + \mathcal{S}_N(u_N, v) = \mathcal{F}_N(v) \qquad \forall v \in V_N, \tag{3.9}$$

with

$$\begin{aligned}
\mathcal{B}_N(u, v) &= \int_{\Omega} \epsilon (\nabla u - \mathcal{L}_1(u)) \cdot (\nabla v - \mathcal{L}_1(v)) \mathrm{d}\mathbf{x}, \qquad \mathcal{F}_N(v) = \int_{\Omega} f v \mathrm{d}\mathbf{x}, \\
\mathcal{C}_N(u, v) &= -\int_{\Omega} \mathbf{b}u \cdot (\nabla v - \mathcal{L}_2(v)) \mathrm{d}\mathbf{x} + \int_{\Omega} (c - \nabla \cdot \mathbf{b}) u v \mathrm{d}\mathbf{x}, \\
\mathcal{S}_N(u, v) &= \int_{\mathcal{E}\backslash\Gamma} \epsilon\alpha [u_1][v_1] \mathrm{d}s + \int_{\Gamma} \epsilon\alpha (u_1 - u_2)(v_1 - v_2) \mathrm{d}s.
\end{aligned}$$

Here, $\mathcal{L}_1(u)$ and $\mathcal{L}_2(u)$ have been defined in $L^2(\Omega)$ by a trivial extension.

From the following lemma, the primal formulation is consistent.

Lemma 3.1. *Let u be the exact solution of (2.3) and (2.4), then we have the Galerkin orthogonality property, i.e.,*

$$\mathcal{A}_N(u-u_N,v)=0,\qquad \text{for all } v\in V_N. \tag{3.10}$$

Proof. Since u is the exact solution, we have $[u_1]_e=0, [\nabla u_1]=0$ for all $e\in\mathcal{E}^o$, $u_1=0$ on Γ_D^1, and $u_1=u_2, \nabla u_1\cdot\mathbf{n}_1=-\nabla u_2\cdot\mathbf{n}_2$ on Γ. Consequently,

$$\mathcal{L}_1(u)=0,\qquad \mathcal{S}_N(u,v)=0. \tag{3.11}$$

Then, for all $v\in V_N$, we have

$$\begin{aligned}\mathcal{B}_N(u,v)=&\int_\Omega \epsilon\nabla u\cdot(\nabla v-\mathcal{L}_1(v))\mathrm{d}\mathbf{x}\\ =&\int_{\Omega_1}\epsilon\nabla u_1\cdot\nabla v_1\mathrm{d}\mathbf{x}-\int_{\Omega_1}\epsilon\mathcal{L}_1(v)\cdot\nabla u_1\mathrm{d}\mathbf{x}+\int_{\Omega_2}\epsilon\nabla u_2\cdot\nabla v_2\mathrm{d}\mathbf{x}.\end{aligned} \tag{3.12}$$

Taking $\mathbf{w}=\epsilon\nabla u_1$ and $v=v_1$ in (3.1), we have

$$\begin{aligned}\int_{\Omega_1}\epsilon\nabla u_1\cdot\nabla v_1\mathrm{d}\mathbf{x}=&-\int_{\Omega_1}\epsilon\Delta u_1 v_1\mathrm{d}\mathbf{x}+\int_{\mathcal{E}^o}\epsilon\left([\nabla u_1]\{v_1\}+\{\nabla u_1\}\cdot[v_1]\right)\mathrm{d}s\\ &+\int_\Gamma \epsilon v_1\nabla u_1\cdot\mathbf{n}_1\mathrm{d}s+\int_{\Gamma_D^1}\epsilon v_1\nabla u_1\cdot\mathbf{n}\mathrm{d}s.\end{aligned} \tag{3.13}$$

Inserting (3.13) into (3.12), and integrating by parts in the third term in the right hand side of (3.12), in terms of the definition of $\mathcal{L}_1(\cdot)$, we have

$$\mathcal{B}_N(u,v)=\int_\Omega -\epsilon\Delta u v\mathrm{d}\mathbf{x}+\int_{\mathcal{E}^o}\epsilon[\nabla u_1](\beta\cdot[v_1]+\{v_1\})\mathrm{d}s+\int_{\Gamma_D^2}\epsilon v_2\nabla u_2\cdot\mathbf{n}\mathrm{d}s.$$

By the definition of V_N, we get $v_2=0$ on Γ_D^2. This, together with $[\nabla u_1]=0$, yields

$$\mathcal{B}_N(u,v)=\int_\Omega -\epsilon\Delta u v\mathrm{d}\mathbf{x},\qquad \text{for all } v\in V_N. \tag{3.14}$$

Now we consider the term $\mathcal{C}_N(u,v)$. It can be rewritten as

$$\mathcal{C}_N(u,v)=-\int_{\Omega_1}\mathbf{b}u_1\cdot(\nabla v_1-\mathcal{L}_2(v))\mathrm{d}\mathbf{x}-\int_{\Omega_2}\mathbf{b}\cdot\nabla v_2 u_2\mathrm{d}\mathbf{x}+\int_\Omega(c-\nabla\cdot\mathbf{b})uv\mathrm{d}\mathbf{x}. \tag{3.15}$$

Taking $\mathbf{w}=-\mathbf{b}u_1$ and $v=v_1$ in (3.1), we have

$$\begin{aligned}-\int_{\Omega_1}\mathbf{b}u_1\cdot\nabla v_1\mathrm{d}\mathbf{x}=&\int_{\Omega_1}(\mathbf{b}\cdot\nabla u_1+\nabla\cdot\mathbf{b}u_1)v_1\mathrm{d}\mathbf{x}-\int_{\mathcal{E}^o}\mathbf{b}\cdot([u_1]\{v_1\}+\{u_1\}[v_1])\mathrm{d}s\\ &-\int_\Gamma \mathbf{b}\cdot\mathbf{n}_1 u_1 v_1\mathrm{d}s-\int_{\Gamma_D^1}\mathbf{b}\cdot\mathbf{n}u_1v_1\mathrm{d}s.\end{aligned} \tag{3.16}$$

Inserting (3.16) into (3.15), and integrating by parts in the second term in the right hand side of (3.15), in terms of the definition of $\mathcal{L}_2(\cdot)$, we obtain

$$
\begin{aligned}
\mathcal{C}_N(u,v) &= \int_\Omega (\mathbf{b}\cdot\nabla u + cu)v\mathrm{d}\mathbf{x} + \int_{\mathcal{E}^o} (\mathbf{b}\cdot[v_1]\beta_0\cdot[u_1] - \mathbf{b}\cdot[u_1]\{v_1\})\mathrm{d}s \\
&= \int_\Omega (\mathbf{b}\cdot\nabla u + cu)v\mathrm{d}\mathbf{x} + \int_{\mathcal{E}^o} \mathbf{b}\cdot[u_1](\beta_0\cdot[v_1] - \{v_1\})\mathrm{d}s \\
&= \int_\Omega (\mathbf{b}\cdot\nabla u + cu)v\mathrm{d}\mathbf{x},
\end{aligned}
\tag{3.17}
$$

due to the fact $[u_1]=0$ on $\mathcal{E}^o$. The combination of (3.9), (3.11), (3.14) and (3.17), leads to

$$
\begin{aligned}
\mathcal{A}_N(u,v) &= \int_\Omega (-\epsilon\Delta u + \mathbf{b}\cdot\nabla u + cu)\mathrm{d}\mathbf{x} \\
&= \int_\Omega f\,v\mathrm{d}\mathbf{x}, \quad \forall v\in V_N.
\end{aligned}
$$

In view of (3.9), (3.10) obviously holds. □

Stability analysis. To consider the stability of the primal form $\mathcal{A}_N$, define the following norms and seminorms for $v\in V(N)$:

$$
\begin{aligned}
&|||v|||_\epsilon^2 = ||v||_{0,\Omega}^2 + \epsilon|v|_{1,N}^2 + \epsilon|v|_*^2 + |v|_c^2, \qquad (3.18)\\
&|v|_{1,N}^2 = |v_2|_{1,\Omega_2}^2 + \sum_{K\in\mathcal{T}_N^1} |v_1|_{1,K}^2, \quad |v|_*^2 = \int_{\mathcal{E}\backslash\Gamma} \alpha[v_1]^2\mathrm{d}s + \int_\Gamma \alpha(v_1-v_2)^2\mathrm{d}s,\\
&|v|_c^2 = \frac{1}{2}\sum_{e\in\mathcal{E}^o}\int_e |\mathbf{b}\cdot\mathbf{n}_e|\,[v_1]^2\mathrm{d}s + \frac{1}{2}\int_\Gamma |\mathbf{b}\cdot\mathbf{n}_1|\,(v_1-v_2)^2\mathrm{d}s + \frac{1}{2}\int_{\Gamma_D}|\mathbf{b}\cdot\mathbf{n}|\,v^2\mathrm{d}s.
\end{aligned}
$$

Lemma 3.2. *There exists a constant $C_1>0$, such that*

$$
\mathcal{A}_N(u_N,u_N) \geqslant C_1|||u_N|||_\epsilon^2, \quad \forall u_N\in V_N.
$$

Proof. By direct computation, we obtain, for all $u_N\in V_N$,

$$
\begin{aligned}
\mathcal{B}_N(u_N,u_N) &= \int_\Omega \epsilon(\nabla u_N - \mathcal{L}_1(u_N))^2\mathrm{d}\mathbf{x}, \quad \mathcal{S}_N(u_N,u_N) = \epsilon|u_N|_*^2,\\
\mathcal{C}_N(u_N,u_N) &= \int_\Omega (c - \frac{1}{2}\nabla\cdot\mathbf{b})u_N^2\mathrm{d}\mathbf{x} + |u_N|_c^2.
\end{aligned}
$$

By (3.9) and (2.2), we have

$$
\begin{aligned}
\mathcal{A}_N(u_N,u_N) &= \int_\Omega \epsilon(\nabla u_N - \mathcal{L}_1(u_N))^2\mathrm{d}\mathbf{x} + \int_\Omega c_0^2u_N^2\mathrm{d}\mathbf{x} + \epsilon|u_N|_*^2 + |u_N|_c^2\\
&\geqslant \epsilon\,|u_N|_{1,N}^2 - 2\epsilon\int_\Omega \mathcal{L}_1(u_N)\cdot\nabla u_N\mathrm{d}\mathbf{x} + \epsilon||\mathcal{L}_1(u_N)||_{0,\Omega}^2 + \gamma_0||u_N||_{0,\Omega}^2 + \epsilon|u_N|_*^2 + |u_N|_c^2.
\end{aligned}
$$

Applying the arithmetic-geometric mean inequality, we have, for every $\theta>0$,

$$
\mathcal{A}_N(u_N,u_N) \geqslant \epsilon\left[(1-\theta)|u_N|_{1,N}^2 + (1-1/\theta)||\mathcal{L}_1(u_N)||_{0,\Omega}^2\right] + \gamma_0||u_N||_{0,\Omega}^2 + \epsilon|u_N|_*^2 + |u_N|_c^2. \tag{3.19}
$$

According to [12](p422), when the coefficient $\alpha = \mathcal{O}(1/H_x, 1/H_y)$, there exists a constant $C > 0$, such that

$$||\mathcal{L}_1(u)||_{0,\Omega} \leqslant C|u|_*, \quad \forall u \in V(N). \tag{3.20}$$

Inserting (3.20) into (3.19), for any θ satisfies $\frac{C^2}{C^2+1} < \theta < 1$, we easily get,

$$\begin{aligned}\mathcal{A}_N(u_N, u_N) &\geqslant \epsilon(1-\theta)|u_N|_{1,N}^2 + \epsilon[C^2(1-1/\theta)+1]|u_N|_*^2 + \gamma_0||u_N||_{0,\Omega}^2 + |u_N|_c^2 \\ &\geqslant \gamma_1(\epsilon|u_N|_{1,N}^2 + \epsilon|u_N|_*^2) + \gamma_0||u_N||_{0,\Omega}^2 + |u_N|_c^2 \\ &\geqslant \min\{\gamma_0, \gamma_1\}|||u_N|||_\epsilon^2,\end{aligned}$$

where $\gamma_1 = \min\{1-\theta, C^2(1-1/\theta)+1\}$. Taking $C_1 = \min\{\gamma_0, \gamma_1\}$, the proof is completed. □

From Lemma 3.2, we easily get

$$|||u_N|||_\epsilon \leqslant C||f||_{0,\Omega},$$

which implies the uniqueness of the solution to (3.9). Further, since (3.9) is a linear problem over the finite-dimensional space V_N, the existence of the solution follows from its uniqueness. Consequently, by (3.4), we get the existence and uniqueness of the solution to the problem (2.8)–(2.10) with numerical fluxes (2.11)–(2.13).

Remark 3.1. In fact, following [7] or [12], for any $\alpha \geqslant 0$, the existence and uniqueness of the solution to the problem (2.8)–(2.10) with numerical flux (2.11)–(2.13) can be proved. In this paper, we are only interested in the special case $\alpha = \mathcal{O}(1/H_x, 1/H_y)$.

Error analysis. We are now going to provide a ϵ-uniform estimate for the error $u - u_N$ in the norm (3.18). First, we start with the error decomposition

$$u - u_N = (u - u_I) + (u_I - u_N) \equiv \eta + \xi, \tag{3.21}$$

where u_I be the standard bilinear interpolation of u.

The final estimate for $|||u - u_N|||_\epsilon$ will be derived by applying the triangle inequality to (3.21). For this purpose, we estimate $|||\eta|||_\epsilon$ and $|||\xi|||_\epsilon$, respectively. To bound η, we need some regularity results. Denote the operator $\mathcal{L}_i$, $i = 0, 1$, by

$$\mathcal{L}_i v = \frac{\partial v}{\partial y}\frac{\partial^i}{\partial x^i}\left(\frac{b_2}{b_1}\right) + v\frac{\partial^i}{\partial x^i}\left(\frac{c}{b_1}\right).$$

Lemma 3.3. *[17]Let* **b** *and* c *be smooth, and let* $f \in C^{4,\lambda}(\bar{\Omega})$ *for some* $\lambda \in (0,1)$. *Further suppose that* f *satisfies the compatibility conditions*

$$f(0,0) = f(0,1) = f(1,0) = f(1,1) = 0,$$

and

$$\left(\frac{f}{b_1}\right)_y(0,0)=\left(\frac{f}{b_2}\right)_x(0,0),$$
$$\left(\left(\frac{f}{b_1}\right)_x-\mathcal{L}_0\left(\frac{f}{b_1}\right)\right)_y(0,0)=\left(\frac{f}{b_2}\right)_{xx}(0,0),$$
$$\left(\left(\frac{f}{b_1}\right)_{xx}-\mathcal{L}_0\left(\left(\frac{f}{b_1}\right)_x-\mathcal{L}_0\left(\frac{f}{b_1}\right)\right)-2\mathcal{L}_1\left(\frac{f}{b_1}\right)\right)_y(0,0)=\left(\frac{f}{b_2}\right)_{xxx}(0,0),$$
$$\left(b_2\left(\frac{f}{b_2}\right)_{xx}\right)(0,0)=\left(b_1\left(\frac{f}{b_1}\right)_{yy}\right)(0,0).$$

Then the boundary value problem (2.1) has a classical solution $u\in C^{3,\lambda}(\bar\Omega)$, that can be bounded by

$$\left|\frac{\partial^{i+j}u(x,y)}{\partial x^i\partial y^j}(x,y)\right|\leqslant C\left(1+\epsilon^{-i}e^{-\beta_1(1-x)/\epsilon}\right)\times\left(1+\epsilon^{-j}e^{-\beta_2(1-y)/\epsilon}\right)\tag{3.22}$$

for all $(x,y)\in\bar\Omega$ and $0\leqslant i+j\leqslant 2$.

Remark 3.2. The following convection-diffusion problem

$$-\epsilon\Delta u-\mathbf{b}\cdot\nabla u+cu=f,\tag{3.23}$$

is considered in [17], which exhibits an exponential layers near $x=0$ and $y=0$.

Let $\hat x=1-x,\hat y=1-y$ and $\hat u(\hat x,\hat y)=u(1-\hat x,1-\hat y)$, then (2.1) can be rewritten as

$$-\epsilon\Delta\hat u+\mathbf{b}\cdot\nabla\hat u+c\hat u=\hat f,$$

which is the same as (3.23). So the formula (8) in [17] is also valid for (2.1) if we replace x and y by $1-x$ and $1-y$, respectively.

Lemma 3.4. *[3] Let u be a solution of (2.1) and u_I the standard bilinear interpolation of u on the Shishkin mesh defined before. Under the conditions of Lemma 3.3 and assume $\epsilon^{1/2}\ln^2N\leqslant C$, then the interpolation error $\eta=u-u_I$ satisfies*

$$\|\eta\|_{L^\infty(\Omega_1)}\leqslant CN^{-2},\qquad \|\eta\|_{L^\infty(\Omega_2)}\leqslant CN^{-2}\ln^2N,$$
$$\|\eta\|_{L^2(\Omega_1)}\leqslant CN^{-2},\qquad \|\eta\|_{L^2(\Omega_2)}\leqslant CN^{-2}\ln^2N,$$
$$\|\nabla\eta\|_{L^2(\Omega)}\leqslant C\epsilon^{-1/2}N^{-1}\ln N.$$

Remark 3.3. According to [18], under the assumption of Lemma 3.3, the solution of (2.1) has the usual solution decomposition $u=S+E_1+E_2+E_{12}$, which satisfies

$$\left|\frac{\partial^{i+j}S(x,y)}{\partial x^i\partial y^j}\right|\leqslant C,$$

$$\left|\frac{\partial^{i+j}E_1(x,y)}{\partial x^i\partial y^j}\right|\leqslant C\varepsilon^{-i}e^{-\beta_1(1-x)/\varepsilon},$$

$$\left|\frac{\partial^{i+j}E_2(x,y)}{\partial x^i\partial y^j}\right|\leqslant C\varepsilon^{-j}e^{-\beta_2(1-y)/\varepsilon},$$

$$\left|\frac{\partial^{i+j}E_{12}(x,y)}{\partial x^i\partial y^j}\right|\leqslant C\varepsilon^{-(i+j)}e^{-(\beta_1(1-x)+\beta_2(1-y))/\varepsilon},$$

for $0 \leqslant i+j \leqslant 2$, which is stronger than (3.22) to some extent. Actually the estimates above imply (3.22). Based on the solution decomposition above, the conclusions of Lemma 3.4 can be found in [3], i.e., (3.127) of Page 384, (3.128b) and (3.128c) of Page 385 with the additional assumption $\epsilon^{1/2}\ln^2 N \leqslant C$. It is worthwhile to point out that there is no $\ln N$ factor in (3.128b) there.

The following statement represents the direct consequence of Lemma 3.4.

Proposition 3.5. *Under the conditions of Lemma 3.4, we have*

$$|||\eta|||_\epsilon \leqslant CN^{-1}\ln N. \tag{3.24}$$

Proof. Since $u-u_I$ is continuous in Ω, we have $|\eta|_* = 0, |\eta|_c = 0$. Then, $|||\eta|||_\epsilon^2 = ||\eta||_{0,\Omega}^2 + \epsilon|\eta|_{1,N}^2$. By Lemma 3.4, we easily conclude (3.24). □

Now we turn to estimate $|||\xi|||_\epsilon$.

Proposition 3.6. *Under the conditions of Lemma 3.4, assume $\alpha = \mathcal{O}(1/H_x, 1/H_y)$ and*

$$\ln^{3/2} N \leqslant CN, \tag{3.25}$$

then $\xi = u_I - u_N$ satisfies

$$|||\xi|||_\epsilon \leqslant CN^{-1}\ln N. \tag{3.26}$$

Proof. By Lemma 3.1 and Lemma 3.2, we first obtain

$$C_1|||\xi|||_\epsilon^2 \leqslant \mathcal{A}_N(\xi,\xi) = -\mathcal{A}_N(\eta,\xi) = -\mathcal{B}_N(\eta,\xi) - \mathcal{C}_N(\eta,\xi) - \mathcal{S}_N(\eta,\xi). \tag{3.27}$$

By the definition of u_I, we have $[\eta]_e = 0$ for all $e \in \mathcal{E}$. Consequently,

$$\begin{aligned}
\mathcal{B}_N(\eta,\xi) &= \int_\Omega \epsilon\nabla\eta\cdot(\nabla\xi - \mathcal{L}_1(\xi))\mathrm{d}\mathbf{x},\\
\mathcal{C}_N(\eta,\xi) &= -\int_\Omega \mathbf{b}\cdot\nabla\xi\eta\mathrm{d}\mathbf{x} + \int_{\Omega_1}\mathbf{b}\cdot\mathcal{L}_2(\xi)\eta\mathrm{d}\mathbf{x} + \int_\Omega (c-\nabla\cdot\mathbf{b})\eta\xi\mathrm{d}\mathbf{x}\\
&\equiv I_1 + I_2 + I_3, \qquad\qquad (3.28)\\
\mathcal{S}_N(\eta,\xi) &= 0.
\end{aligned}$$

Then, by (3.20) and Lemma 3.4, we have

$$|\mathcal{B}_N(\eta,\xi)| \leqslant C\epsilon||\nabla\eta||_{L^2(\Omega)}(|\xi|_{1,N}+|\xi|_*) \leqslant C\epsilon^{1/2}||\nabla\eta||_{L^2(\Omega)}|||\xi|||_\epsilon \leqslant CN^{-1}\ln N\,|||\xi|||_\epsilon. \tag{3.29}$$

The first term in the right hand side of (3.28) can be estimated by,

$$|I_1| \leqslant C\left(||\eta||_{L^2(\Omega_1)}||\nabla\xi||_{L^2(\Omega_1)} + ||\eta||_{L^\infty(\Omega_2)}||\nabla\xi||_{L^1(\Omega_2)}\right).$$

On Ω_1, the implementation of an inverse inequality leads to

$$||\nabla\xi||_{L^2(\Omega_1)} \leqslant CN||\xi||_{L^2(\Omega_1)} \leqslant CN|||\xi|||_\epsilon.$$

On the other hand, on Ω_2, the Cauchy-Schwarz inequality yields

$$||\nabla\xi||_{L^1(\Omega_2)} \leqslant |\Omega_2|^{1/2}||\nabla\xi||_{L^2(\Omega_2)} \leqslant C\tau^{1/2}||\nabla\xi||_{L^2(\Omega_2)} \leqslant C\ln^{1/2}N|||\xi|||_\epsilon.$$

Consequently,

$$|I_1| \leqslant C\left(N^{-1}+N^{-2}\ln^{5/2}N\right)|||\xi|||_\epsilon \leqslant C\left(N^{-1}+N^{-1}\ln N\right)|||\xi|||_\epsilon, \tag{3.30}$$

where we have used the assumption (3.25).

Due to $[\eta]_e = 0$ for all $e \in \mathcal{E}$, the second term in the right hand side of (3.28) can be rewritten as

$$I_2 = \int_{\mathcal{E}^o} \mathbf{b}\cdot[\xi_1]\{\eta_1\}\mathrm{d}s + \int_\Gamma \mathbf{b}\cdot\mathbf{n}_1\eta_1(\xi_1-\xi_2)\mathrm{d}s.$$

By Cauchy-Schwarz inequality and Lemma 3.4, we have

$$\begin{aligned}|I_2| \leqslant & \left(\sum_{e\in\mathcal{E}^o}\int_e |\mathbf{b}\cdot\mathbf{n}_e|\,|\{\eta\}|^2\mathrm{d}s\right)^{1/2}\left(\sum_{e\in\mathcal{E}^o}\int_e |\mathbf{b}\cdot\mathbf{n}_e||[\xi_1]|^2\mathrm{d}s\right)^{1/2}\\ &+\left(\int_\Gamma |\mathbf{b}\cdot\mathbf{n}_1||\eta_1|^2\mathrm{d}s\right)^{1/2}\left(\int_\Gamma |\mathbf{b}\cdot\mathbf{n}_1||\xi_1-\xi_2|^2\mathrm{d}s\right)^{1/2}\\ \leqslant & C||\eta||_{L^2(\mathcal{E})}|\xi|_c \leqslant C|\mathcal{E}|^{1/2}||\eta||_{L^\infty(\Omega_1)}|\xi|_c \leqslant CN^{-3/2}|||\xi|||_\epsilon,\end{aligned} \tag{3.31}$$

where $|\cdot|_{L^2(\mathcal{E})}$ means $\left(\sum_{e\in\mathcal{E}}|\cdot|^2_{L^2(e)}\right)^{1/2}$ and $|\mathcal{E}|$ means the sum of the lengths of all edges in $\mathcal{E}$, which can be estimated with

$$|\mathcal{E}| = [(1-\tau_x)+(1-\tau_y)](N/2+1) \leqslant 2N.$$

The third term in the right hand side of (3.28) can be easily estimated with

$$|I_3| \leqslant C||\eta||_{0,\Omega}\,||\xi||_{0,\Omega} \leqslant CN^{-2}\ln^2 N|||\xi|||_\epsilon.$$

This, combined with (3.30) and (3.31), yields

$$|\mathcal{C}_N(\eta,\xi)| \leqslant CN^{-1}\ln N|||\xi|||_\epsilon. \tag{3.32}$$

Collecting (3.27), (3.29) and (3.32), we have (3.26). □

The combination of Proposition 3.5 and Proposition 3.6 leads to our main result, i.e.,

Theorem 3.7. *Let u and u_N be the solutions of (2.1) and (3.9), respectively. Under the conditions of Lemma 3.4, assume $\ln^{3/2}N \leqslant CN, \epsilon^{1/2}\ln^2 N \leqslant C$ and $\alpha = \mathcal{O}(1/H_x, 1/H_y)$, then*

$$|||u-u_N|||_\epsilon \leqslant CN^{-1}\ln N. \tag{3.33}$$

Corollary 3.8. *Let* (q_N, u_N) *be the solution obtained by the coupled method (2.8)–(2.10) with numerical fluxes (2.11)–(2.13). Under the assumptions of Theorem 3.7, we have*

$$|(q - q_N, u - u_N)|_{\mathcal{A}_N} \leqslant CN^{-1} \ln N, \tag{3.34}$$

where $|(\cdot,\cdot)|_{\mathcal{A}_N}$ *is a problem-related norm defined by*

$$|(r,v)|^2_{\mathcal{A}_N} = ||v||^2_{0,\Omega} + \epsilon ||r||^2_{0,\Omega_1} + \epsilon |v_2|^2_{1,\Omega_2} + \epsilon |v|^2_* + |v|^2_c. \tag{3.35}$$

Proof. From (3.4), we have $q - q_N = \nabla(u_1 - u_{1,N}) + \mathcal{L}_1(u_N)$. Since $\mathcal{L}_1(u) = 0$, we obtain $q - q_N = \nabla(u_1 - u_{1,N}) - \mathcal{L}_1(u - u_N)$. In terms of (3.20), we conclude $|(q - q_N, u - u_N)|_{\mathcal{A}_N} \leqslant C|||u - u_N|||_\epsilon$, which implies the conclusion. □

4 Numerical Experiments

In this section, we numerically verify the sharpness of our theoretical findings and explore some situations not covered by them. Take $\alpha = 1/H_x$, $\beta = \beta_0 = \frac{1}{2}\text{sign}(\mathbf{b} \cdot \mathbf{n}_K)\mathbf{n}_K$, in (2.11) and (2.12), where $\mathbf{n}_K$ is the unit outward normal of the element K, and $\kappa = 2$ in the transition parameter for Shishkin mesh. By our numerical experiment, there is no difference between the results computed by taking $\alpha = 1/H_x$ and $\alpha = 1/H_y$.

Example. Take $\mathbf{b} = (2,3)$ and $c = 1$ in the model problem (2.1), f is chosen such that

$$u(x,y) = 2(\sin x)\,(1 - e^{-2(1-x)/\epsilon})y^2(1 - e^{-3(1-y)/\epsilon})$$

is the exact solution.

We display the history of convergence for our coupled method in Table 1. A Shishkin mesh with $N \times N$ elements is called mesh N. Let $err(N)$ denote the error of the approximation computed on the mesh N. Then the convergence order, i.e., $order(2N)$, is defined by

$$order(2N) := \begin{cases} \dfrac{\ln(err(N)/err(2N))}{\ln(2\ln(N)/\ln(2N))}, & \text{for the norm (3.35)}, \\[2ex] \dfrac{\ln(err(N)/err(2N))}{\ln(2)}, & \text{for the } L^2 \text{ norm.} \end{cases}$$

Table 1　History of convergence for the coupled method, under the norm (3.35), Shishkin mesh with $\kappa = 2$

N	$\epsilon = 1.0\text{e}-04$		$\epsilon = 1.0\text{e}-05$		$\epsilon = 1.0\text{e}-06$	
	error	order	error	order	error	order
8	4.363098e−01	—	4.363004e−01	—	4.362995e−01	—
16	2.937673e−01	0.98	2.937531e−01	0.98	2.937517e−01	0.98
32	1.847859e−01	0.99	1.847726e−01	0.99	1.847713e−01	0.99
64	1.111902e−01	0.99	1.111795e−01	0.99	1.111785e−01	0.99
128	6.493348e−02	1.00	6.492494e−02	1.00	6.492421e−02	1.00

Table 2 History of convergence for the coupled method, under the L^2 norm , Shishkin mesh with $\kappa = 2$

N	$\epsilon = 1.0\text{e}-04$		$\epsilon = 1.0\text{e}-05$		$\epsilon = 1.0\text{e}-06$	
	error	order	error	order	error	order
8	7.828466e−03	—	7.823300e−03	—	7.822785e−03	—
16	2.004742e−03	1.97	1.998638e−03	1.97	1.998030e−03	1.97
32	5.090079e−04	1.98	5.047762e−04	1.99	5.043584e−04	1.99
64	1.293231e−04	1.98	1.269031e−04	1.99	1.266759e−04	1.99
128	3.343957e−05	1.95	3.185281e−05	1.99	3.174415e−05	2.00

From Table 1 and Figure 2, it is observed that the numerical results for the problem-related norm (3.35) of the error agree with those predicted in Corollary 3.8 . Further, from Table 2 and Figure 1, it is numerically predicted that

$$||u - u_N||_{0,\Omega} \leqslant CN^{-2},$$

which is uniformly optimal. The reason behind it is worthwhile to investigate.

5 Conclusions

In this paper, we introduce a coupled LDG-CFEM method for solving two-dimensional singularly perturbed convection-diffusion problems, whose stability and uniform convergence property on the Shishkin mesh is investiaged. For bilinear element, a rate $\mathcal{O}(N^{-1}\ln N)$ in an associated norm is established. Our numerical experiments indicate the sharpness of this error estimate. Moreover, a rate $\mathcal{O}(N^{-2})$ in L^2 norm are numerically observed on the Shishkin mesh.

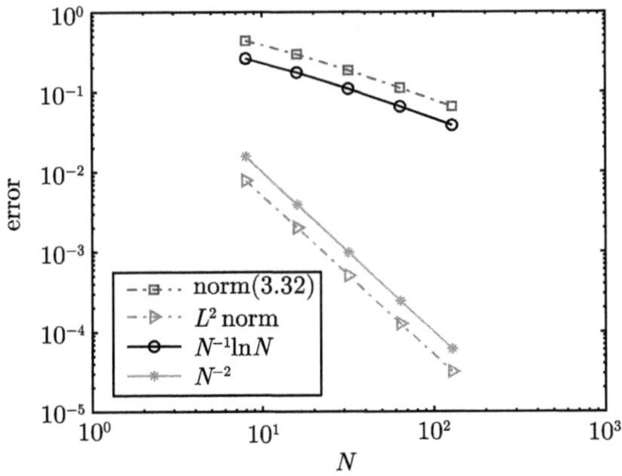

Figure 2 Convergence curve of $u - u_N$, the Shishkin mesh, $\epsilon = 10^{-6}$

References

[1] T. Linß, Layer-adapted meshes for convection-diffusion problems, *Computer Methods in Applied Mechanics & Engineering*, 192 (2003), 1061-1105.

[2] H.-G. Roos, Layer-adapted grids for singular perturbation problem, *ZAMM - Journal of Applied Mathematics and Mechanics*, 78 (1998), 291-309.

[3] H.-G. Roos, M. Stynes and L. Tobiska, *Robust numerical methods for singularly perturbed differential equations*, Springer Series in Computational Mathematics, Volume 24, Springer-Verlag Berlin Heidelberg, 2008.

[4] M. Stynes and E. O'Riorddan, A uniformly convergent Galerkin method on a Shishkin mesh for a convection-diffusion problem, *Journal of Mathematical Analysis and Applications*, 214 (1997), 36-54.

[5] B. Cockburn and C.-W. Shu, The local discontinuous Galerkin finite element method for convection-diffusion systems, *SIAM Journal on Numerical Analysis*, 35 (1998), 2440-2463.

[6] F. Celiker and B. Cockburn, Superconvergence of the numerical traces of discontinuous Galerkin and Hybridized methods for convection-diffusion problems in one space dimension, *Mathematics of Computation*, 76 (2007), 67-96.

[7] Z. Q. Xie and Z. Zhang, Superconvergence of DG method for one-dimensional singularly perturbed problems, *J. Journal of Computational Mathematics*, 25 (2007), 185-200.

[8] Z.,Q. Xie, Z. Z. Zhang, and Z. Zhang, A numerical study of uniform superconvergence for solving singularly perturbed problems, *Journal of Computational Mathematics*, 27 (2009), 280-298.

[9] Z. Q. Xie, Z. Zhang, Uniform superconvergence analysis of the discontinuous Galerkin method for a singularly perturbed problem in 1-D, *Mathematics of Computation*, 79 (2010), 35-45.

[10] Z. Z. Zhang, Z. Q. Xie, and Z. Zhang, Superconvergence of discontinuous Galerkin methods for convection diffusion problems, *Journal of Scientific Computing*, 41 (2009), 70-93.

[11] H. Zarin, and H.-G. Roos, Interior penalty discontinuous approximations of convection-diffusion problems with parabolic layers, *Numerische Mathematik*, 100 (2005), 735-759.

[12] I. Perugia and D. Schötzau, On the coupling of local discontinuous Galerkin and conforming finite element methods, *Journal of Scientific Computing*, 16 (2001), 411-433.

[13] H.-G. Roos, and H. Zarin, A supercloseness result for the discontinuous Galerkin stabilization of convection-diffusion problems on Shishkin meshes, *Numerical Methods for Partial Differential Equations*, 23 (2007), 1560-1576.

[14] H. Zarin, Continuous-discontinuous finite element method for convection-diffusion problems with characteristic layers, *Journal of Computational and Applied Mathematics*, 231 (2009), 626-636.

[15] D. N. Arnold, F. Brezzi, B. Cockburn, and L. D. Marini, Unified analysis of discontinuous Galerkin methods for elliptic problems, *SIAM Journal on Numerical Analysis*, 39 (2002), 1749-1779.

[16] P. Castillo, B. Cockburn, I. Perugia and D. Schötzau, An a priori error analysis of the local discontinuous Galerkin method for elliptic problems, *SIAM Journal on Numerical Analysis*, 38 (2000), 1676-1706.

[17] T. Linß, The necessity of Shishkin decompositions, *Applied Mathematics Letters*, 14 (2001), 891-896.

[18] T. Linß and M. Stynes, Asymptotic analysis and Shishkin-type decomposition for an elliptic convection-diffusion problem, *Journal of Mathematical Analysis and Application*, 261(2001), 604-632.

第二篇 最优化问题数值算法

A Direct Method for the Linear Complementarity Problem*

S.Z. Zhou (周叔子)†

One of the effective approaches for finding the numerical solutions of some free boundary problems is to reduce the problems into corresponding variational inequalities and then to discretize them by finite difference methods or finite element methods (see, for instance, [1-7]). Following this way we often obtain the so-called linear complementarity problem: find $w \in R^n$ such that

$$Aw \geqslant p, \quad w \geqslant q, (Aw-p)^T(w-q)=0, \tag{1}$$

where A is an $n \times n$ real matrix, and $p, q \in R^n$. Let $u = w - q$. Then (1) is reduced following form: find $u \in R^n$ such that

$$Au \geqslant b, \quad u \geqslant 0, \quad u^T(Au-b)=0. \tag{2}$$

Most of the conventional algorithms for solving Problem (2) are iterative methods, of the exisiting direct methods for (2) are polynomial time algorithms [3] . We propose paper a new direct method for (2) which is a polynomial time method provided matrix A is a Stieltjes matrix.

Denote by v_i the i-th component of vector v, and by a_{ij} the element of matrix A. The subscript set $N = \{1, \cdots, n\}$. Suppose $B \subset N$. Denote by $v(B)$ and $A(B)$ respectively subvector of v and the principal submatrix of A corresponding to the subscript set B. Let u be the solution of Problem (2). Denote by P and Z the subscript sets corresponding the positive components and zero components of u respectively, i.e.,

$$P = \{i \in N: \ u_i > 0\}, \quad Z = \{i \in N: \ u_i = 0\}.$$

It is easy to see that

$$A(P)u(P) = b(P). \tag{3}$$

If $u(P)$ is known, then so is u. But there is a difficulty for finding $u(P)$-the subscript set P is known. This difficult is similar to appearing in solving free boundary problems. In the latter case the free boundary is unknown.

Suppose A is a Stieltjes matrix (or simply S-matrix). It means that : (i) A is symmetric and positive definite; and (ii) $a_{ij} \leqslant 0$ for $i \neq j$. It is well known that (2) has a unique solution in this case. We now establish a lemma.

* 发表于：Journal of Computational Mathematics, Vol. 8, 1990, pp. 178-182.

† Department of Mathematics, Hunan University, Changsha 410082, China.

Lemma 1. *If A is a S-matrix and u is the solution of (2), then*

$$Z^* \supset P \supset P_0, \tag{4}$$

where Z^* is the subscript set corresponding to the zero components of $Au - b$, and P_0 the subscript set corresponding to the positive components of b.

Proof. It follows from (2) that if $u_i > 0$, then $(Au-b)_i = 0$. This means $Z^* \supset P$. If $b_i > 0$, then it follows from $(Au-b)_i \geqslant 0$, $u \geqslant 0$ and $a_{ij} \leqslant 0$ for $i \neq j$ that

$$u_i \geqslant a_{ii}^{-1}\Big(-\sum_{i\neq j} a_{ij}u_j + b_i\Big) \geqslant a_{ii}^{-1}b_i > 0.$$

It means $P \supset P_0$. The lemma has been proved. □

The basic idea for constructing our new direct method is as follows: starting from the given subscript set P_0, expand it step by step and obtain finally the subscript set P in finite steps. Now we give the inductive definition of the algorithm.

Algorithm DM.

(a) Suppose a subscript set P_k is known. Solve the following system by Gauss elimination:

$$A(P_k)u^{(k)}(P_k) = b(P_k). \tag{5}$$

Let

$$u_i^{(k)} = \begin{cases} u_i^{(k)}(P_k), & i \in P_k, \\ 0, & i \in N\backslash P_k. \end{cases} \tag{6}$$

(b) Calculate for $i \in N\backslash P_k$

$$C_i^{(k)} = (Au^{(k)} - b)_i. \tag{7}$$

(c) Let $M_k = \{i \in N\backslash P_k :\ C_i^{(k)} < 0\}$. If $M_k = \varnothing$, then stop computing and output $u^{(k)}$. If $M_k \neq \varnothing$, then let $P_{k+1} = P_k \cup M_k$ and repeat (a)–(c) after replacing k by $k+1$.

The following two theorems indicate that solution u of (2) is obtained in finite steps of Algorithm DM.

Theorem 2. *Suppose A is a S-matrix and M_k is defined in Algorithm DM. If $M_k = \varnothing$, then $u^{(k)} = u$.*

Theorem 3. *Suppose A is a S-matrix and m_0, m are respectively the cardinals of P_0 and P. Then there exists a nonnegative integer $k_0 \leqslant m - m_0$ such that $M_{k_0} = \varnothing$.*

These theorems also indicate that we may find solution u of (2) by solving system (5) for $k = 0, 1, \cdots, k_0$. Because $k_0 \leqslant n$, Algorithm DM is a polynomial time algorithm. In order to prove these theorems we need the following lemmas.

Lemma 4. *Suppose A is a S-matrix and $u^{(0)}(P_0)$ is defined by (5) for $k = 0$. Then $u^{(0)}(P_0) > 0$.*

Proof. Since A is a S-matrix, so is its principal submatrix $A(P_0)$. Suppose $A(P_0)$ is reducible. Because $A(P_0)$ is symmetric, we may reduce the system

$$A(P_0)u^{(0)}(P_0) = b(P_0)$$

to several irreducible subsystems

$$A(Q_l)u^{(0)}(Q_l) = b(Q_l), \quad l = 1, 2, \cdots, r.$$

where $Q_l^{'}$s are subscript sets $Q_i \cap Q_j = \emptyset$ and $Q_1 \cup Q_2 \cup \cdots \cup Q_r = P_0$. Since $A(Q_i)$ is still a principal submatrix of A, it is an irreducible S-matrix. So we have $A^{-1}(Q_l) > 0$ ([8, p.61]) and

$$u^{(0)}(Q_l) = A^{-1}(Q_l)b(Q_l) > 0.$$

The proof is complete. □

Lemma 5. *Suppose A is a S-matrix and $u^{(k)}$ is defined by Algorithm DM. Then we have*

$$u^{(k)} \geqslant u^{(k-1)} \geqslant 0.$$

Proof. According to Lemma 4, we have $u^{(0)}(P_0) > 0$. By (6)

$$u_i^{(0)} = \begin{cases} u_i^{(0)}(P_0), & i \in P_0, \\ 0, & i \in N \backslash P_0. \end{cases}$$

So $u^{(0)} \geqslant 0$. Letting $k = 0$ in (5), we obtain

$$\sum_{i \in P_0} a_{ij} u_j^{(0)} = b_i, \qquad i \in P_0.$$

It may be written as

$$\sum_{i \in P_1} a_{ij} u_j^{(0)} = b_i, \qquad i \in P_0 \tag{8}$$

because $P_1 = P_0 \cup M_0 \supset P_0$ and $u_j^{(0)} = 0$ for $j \in P_1 \backslash P_0$. It follows from the definition of M_0 that we have for $i \in P_1 \backslash P_0 = M_0$

$$\sum_{i=1}^{n} a_{ij} u_j^{(0)} = c_i^{(0)} + b_i < b_i.$$

Then

$$\sum_{i \in P_1} a_{ij} u_j^{(0)} < b_i, \qquad i \in P_1 \backslash P_0. \tag{9}$$

Combining (8) and (9) we obtain

$$A(P_1)u^{(0)}(P_1) < b(P_1).$$

On the other hand, we know that

$$A(P_1)u^{(1)}(P_1) = b(P_1).$$

So it is clear that

$$A(P_1)[u^{(1)}(P_1) - u^{(0))}(P_1)] > 0.$$

Since $A(P_1)$ is a S-matrix, it must be a monotone matrix [8]. Then we have $u^{(1)}(P_1) \geqslant u^{(0)}(P_1)$ and

$$u^{(1)} \geqslant u^{(0)} \geqslant 0.$$

The proof may be completed by induction. □

Lemma 6. *Suppose the conditions of Lemma 5 hold. Then $P_k \subset P$ and $u^{(k)} \leqslant u$.*

Proof. We have $P_0 \subset P$ by Lemma 1. Noting that $u_j \geqslant 0$ and $a_{ij} \leqslant 0$ for $j \neq i$, we obtain for $i \in P_0$ that

$$(A(P_0)u(P_0))_i = \sum_{j\in P_0} a_{ij}u_j = \sum_{j\in P} a_{ij}u_j - \sum_{j\in P\backslash P_0} a_{ij}u_j \geqslant \sum_{j\in P} a_{ij}u_j = b_i.$$

It means

$$A(P_0)u(P_0) \geqslant b(P_0).$$

It follows from this fact and (5) that

$$A(P_0)[u(P_0) - u^{(0)}(P_0)] \geqslant 0.$$

Since $A(P_0)$ is a monotone matrix, we obtain

$$u(P_0) \geqslant u^{(0)}(P_0).$$

On the other hand $u_i \geqslant 0 = u_i^{(0}$ for $i \in N\backslash P_0$. So $u \geqslant u^{(0)}$

Now we show that $P_1 \subset P$. Suppose it is not true. Then there exists a subscript i such that $i \in P_1$ and $i \notin P$. Since $P_1 = P_0 \cup M_0$ and $P_0 \subset P$, we have $i \in M_0$. It follows from $i \notin P$ that $u_i = 0$. By the definition of M_0 we know $u_i^{(0)} = 0$. We have proved $u \geqslant u^{(0)}$. So

$$a_{ij}u_j^{(0)} \geqslant a_{ij}u_j \quad \text{for } j \neq i.$$

Then we have

$$C_i^{(0)} = \sum_{j=1} a_{ij}u_j^{(0)} - b_i = \sum_{j\neq i} a_{ij}u_j^{(0)} - b_i \geqslant \sum_{j\neq i} a_{ij}u_j - b_i = (Au-b)_i \geqslant 0.$$

It contradicts $i \in M_0$. This contradiction proves $P_1 \subset P$.

We conclude by similar argument and induction that $P_k \subset P$ and $u^{(k)} \leqslant u$. □

Proof of Theorem 2. Since $M_k = \emptyset$, we hav

$$C_i^{(k)} = (Au^{(k)} - b)_i \geqslant 0, \quad i \in N\backslash P_k.$$

It follows from (5) and (6) that

$$(Au^{(k)} - b)_i = 0, \quad i \in P_k.$$

So we obtain $Au^{(k)} \geqslant b$. On the other hand, we know that $u^{(k)} \geqslant 0$ by Lemma 5. Then $u^{(k)}$ satisfies the first and second inequalities of (2). It follows from (5) and (6) that $(Au^{(k)} - b)_i = 0$ for $i \in P_k$ and $u_i^{(k)} = 0$ for $i \in N \backslash P_k$. So $u^{(k)}$ satisfies the third equation of (2). According to the uniqueness of the solution of (2) we conclude that $u^{(k)} = u$.

Proof of Theorem 3. If it is not the case, then we have

$$M_k = \emptyset, \quad k = 0, 1, \cdots, m - m_0.$$

Denote by m_k the cardinal of P_k. Since the intersection of P_k and M_k is empty, and since

$$P_{k+1} = P_k \cup M_k,$$

we obtain

$$m_{k+1} \geqslant m_k + 1, \quad k = 0, 1, \cdots, m - m_0.$$

We derive from it that

$$m_{m-m_0+1} \geqslant m_{m-m_0} + 1 \geqslant \cdots \geqslant m_0 + m - m_0 + 1 = m + 1. \tag{10}$$

On the other hand, by Lemma 6 we have

$$P_{m-m_0+1} \subset P.$$

So

$$m_{m-m_0+1} \leqslant m.$$

It contradicts (10) and the theorem has been proved.

Remark 1. We may construct a similar algorithm for Problem (1) and it is not necessary to reduce this problem into Problem (2).

Remark 2. Numerical experiments indicate that we may replace P_0 by any given subset of P in Algorithm DM. But it is an open problem to prove the reasonableness of this replacement.

References

[1] Feng Kang, Unilateral variational principle, *Computer Applications and Applied Math*, 12, 1975, 1-8.

[2] R. Glowinski et al., *Numerical Analysis of Variational Inequalities*, North-Holland, 1981.

[3] R.W. Cottle et al. (eds.), *Variational Inequalities and Complementarity Problem*, Wiley, 1980.

[4] R.W. Cottle et al., On the solution of large, structured linear complementarity problems, the partitioned case, *Appl. Math. Optim.*, 4 (1978), 347-363.

[5] J.X. Wang et al., Multigrid method for elasticity problems, *JCM*, 4 (1986), 154-163.

[6] C.W. Cryer, The efficient solution of linear complementarity problems for tridiagonal Minkowski matrices, *ACM Trans. Math. Software*, 9 (1983), 199-214.

[7] S.Z. Zhou, On the convergence of an algorithm of Uzawa's type for Saddle-point problems, *Kexue Tongbao*, 30 (1985), 1531-1534.

[8] D. M. Young, *Iterative Solution of Large Linear Systems*, AP, 1971.

The Kantorovich Theorem for Nonlinear Complementarity Problems*

S.Z. Zhou (周叔子) Q.R. Yan (严钦容)†

Keywords Nonlinear complementarity problems, Newton's method, quasi-Newton method, semilocal convergence.

Nonlinear complementarity problems (NCP) are a kind of important problem presenting in mathematical physics and economic management, whose numerical solution has recently been paid more attention to (see Refs.[1–5] and their references). Newton's method and quasi-Newton methods are considerable approaches for solving NCP[1]. There is a perfect semilocal convergence theory of the Newton method and quasi-Newton methods for solving the system of nonlinear equations [6, 7].

In this note, we establish some basic results of the semilocal convergence theory of Newton's method and quasi-Newton methods for solving NCP. We also improve the form of a condition in the Kantorovich theorem and further give an expression for the domain in which the derivative satisfies the Lipschitz condition.

Let $K = R^n_+$, F be a $K \to R^n$ mapping. The so-called nonlinear complementarity problem is as follows: find $x \in K$ such that

$$F(x) \geqslant 0,\ (F(x),\, x) = 0, \tag{1}$$

where for $x, y \in R^n$, $(x,\, y) = x^T y$ denotes the inner product of x and y. Its equivalent problem is the following variational inequality: find $x \in K$ such that

$$(F(x),\, y - x) \geqslant 0,\ \forall\, y \in K. \tag{2}$$

The quasi-Newton methods for solving (1) or (2) are as follows: given an initial vector $x^0 \in K$ and a matrix $B_0 \in R^{n\times n}$, find successively the solution x^{k+1} of the following linear complementarity problem

$$(F(x^k) + B_k(x^{k+1} - x^k),\ y - x^{k+1}) \geqslant 0,\ \forall\, y \in K, \tag{3}$$

* 发表于: Chinese Science Bulletin, Vol. 37, 1992, pp. 529-533.

† Department of Applied Mathematics, Hunan University, Changsha 410082, China.

where B_k is determined by B_{k-1}, q_k and s_k, $s_k = x^{k-1} - x^k$, $q_k = F(x^k + p_k) - F(x^k)$, p_k is defined by x_k as well as x^{k-1} and satisfies the quasi-Newton equation $B_k p_k = q_k$.

Replacing B_k by $F'(x^k)$ in the above algorithm, we obtain the Newton method for solving (1) or (2).

Throughout the note we denote by $\|\cdot\|$ the Euclid norm of a vector or the corresponding matrix norm, and $S(x^0, t^*)$ an n-dimensional open sphere with center x^0 and radius t^*.

Theorem 1. *Suppose that the following conditions hold:*

(i) $F \in C^1(K)$;

(ii) $x^0 \in K$, $B_0 \in R^{n\times n}$ *and* $F(x^0) \geqslant 0$, $\|B_0 - F'(x^0)\| \leqslant \delta$, $0 \leqslant (F(x^0), x^0) \leqslant \eta^2$, $\beta^2 = \min(B_0 y, y)\big/\|y\|^2$, *where* δ, η *and* β *are positive constants with* $\beta^2 - 3\delta > 0$;

(iii) *there exists a constant* $\gamma \geqslant 1$ *such that*

$$\|F'(x) - F'(y)\| \leqslant \gamma \|x - y\|,\ \forall\, x,\ y \in K_0,$$

where $K_0 = \{x \in K : \|x\| \leqslant \|x^0\| + (\beta^2 - 3\delta)/3\}$;

(iv) *there exists a constant* $\lambda \geqslant 1$ *such that* $\{B_k\}$ *and the sequence* $\{x^k\}$ *obtained by (3) satisfy*

$$\|B_k - F'(x^k)\| \leqslant \|B_{k-1} - F'(x^{k-1})\| + \lambda\gamma \|x^k - x^{k-1}\|,\ k = 0,\ 1, \cdots;$$

(v) $h = \gamma\eta(\lambda(\beta^2 - \delta) + \delta)\big/\beta(\beta^2 - 3\delta)^2 \leqslant \lambda/2(2\lambda + 1)$.

Then:

(a) $\{x^k\} \subset S(x^0, t^*) \cap K$ *and* $\{x^k\}$ *converges to the solution* x^* *of problem (2), where*

$$t^* = (\beta^2 - 3\delta)(1 - \sqrt{1 - 2(2\lambda + 1)h/\lambda}\,)\big/(2\lambda + 1)\gamma;$$

(b) *problem (2) has a unique solution in* $\overline{S}(x^0, t^*) \cap K$.

Proof. We construct the following sequence $\{t_k\}$: $t_0 = 0$,

$$t_{k+1} = t_k + (2\lambda + 1)f(t_k)/(\lambda(\beta^2 - \delta) + \delta),\ k = 0,\ 1,\ \cdots, \tag{4}$$

where

$$f(t) = \frac{\lambda\gamma}{2}t^2 + \frac{(3\delta - \beta^2)\lambda}{2\lambda + 1}t + \frac{(\lambda(\beta^2 - \delta) + \delta)\eta}{(2\lambda + 1)\beta}. \tag{5}$$

It is easy to verify

$$f(t^*) = 0. \tag{6}$$

We shall prove that $\{t_k\}$ is a majorizing sequence of $\{x^k\}$. It follows from a well-known lemma[7, pp. 416-417] that $\{t_k\}$ is a solution of the following difference equation problem

$$\begin{cases} t_{k+1} - t_k = \dfrac{2\lambda + 1}{\lambda(\beta^2 - \delta) + \delta}\left(\dfrac{\lambda\gamma}{2}(t_k - t_{k-1}) + \lambda\gamma t_{k-1} + \delta\right)(t_k - t_{k-1}), \\ t_0 = 0,\ t_1 = \eta/\beta. \end{cases} \tag{7}$$

It is easy to see by induction that

$$t_{k+1} > t_k,\ t_k < t^*,\ k = 0,\ 1,\ \cdots, \tag{8}$$

and

$$t^* = \lim_{k\to\infty} t_k. \tag{9}$$

Now we prove for $k = 0,\ 1,\ \cdots$ that

$$\text{problem (3) has a unique solution } x^{k+1}, \tag{10}$$

$$\left\|x^{k+1} - x^k\right\| \leqslant t_{k+1} - t_k, \tag{11}$$

$$x^{k+1} \in S(x^0, t^*) \cap K. \tag{12}$$

By induction again, it is not difficult to show that (10)–(12) hold for $k = 0$. Suppose (10)–(12) are true for $k \leqslant m$. Then we know by (iv) and (ii) that

$$\left\|B_{m+1} - F'(x^{m+1})\right\| \leqslant \delta + \lambda\gamma \sum_{j=1}^{m+1} (t_j - t_{j-1}) \leqslant \delta + \lambda\gamma t^*, \tag{13}$$

which and (ii), (iii) as well as (11) for $k \leqslant m$ yield

$$\|B_{m+1} - B_0\| \leqslant 2\delta + (\lambda+1)(\beta^2 - \lambda\delta)\big/(2\lambda+1).$$

combining this inequality and (ii), we have for $y \in R^n$

$$(B_{m+1}y, y) \geqslant (\delta + \lambda(\beta^2 - 3\delta))\|y\|^2\big/(2\lambda+1). \tag{14}$$

Hence B_{m+1} is positive definite and (10) holds for $k = m+1$, i.e.

$$(F(x^{m+1}) + B_{m+1}(x^{m+2} - x^{m+1}), y - x^{m+2}) \geqslant 0,\ \forall\, y \in K.$$

But

$$(F(x^m) + B_m(x^{m+1} - x^m), y - x^{m+1}) \geqslant 0,\ \forall\, y \in K.$$

Taking $y = x^{m+1}$, x^{m+2} respectively in the last two inequalities and performing addition of the resultant inequalities, we get

$$\begin{aligned}(B_{m+1}(x^{m+2} - x^{m+1}), x^{m+2} - x^{m+1}) &\leqslant (F(x^m) \\ -F(x^{m+1}) - B_m(x^m - x^{m+1}), x^{m+2} - x^{m+1}).\end{aligned} \tag{15}$$

It follows from (iii), (13) and (14) that

$$\left\|F(x^m) - F(x^{m+1}) - B_m(x^m - x^{m+1})\right\| \leqslant \frac{\gamma}{2}\left\|x^{m+1} - x^m\right\|^2 + (\delta + \lambda\gamma t_m)\left\|x^{m+1} - x^m\right\|,$$

which together with (7), (14) and (15) yields

$$\begin{aligned}\left\|x^{m+2}-x^{m+1}\right\| &\leqslant \frac{2\lambda+1}{\lambda(\beta^2-\delta)+\delta}\left(\frac{\lambda\gamma}{2}(t_{m+1}+t_m)+\lambda\gamma t_m+\delta\right)(t_{m+1}-t_m)\\ &= t_{m+2}-t_{m+1}.\end{aligned}$$

So, (11) holds for $k=m+1$. It is obvious that

$$\left\|x^{m+2}\right\| \leqslant \left\|x^0\right\|+\left\|x^1-x^0\right\|+\cdots+\left\|x^{m+2}-x^{m+1}\right\| \leqslant \left\|x^0\right\|+t^*,$$

and (12) holds for $k=m+1$.

We known from (11) that $\{t_k\}$ is a majorizing sequence of $\{x^k\}$. Hence $\{x^k\}$ converges to a limit x^*. It's easy to prove that $\{\|B_k\|\}$ is bounded. Letting $k\to\infty$ in (3) we conclude that x^* is a solution of (2).

It remains to prove (b). Obviously, $x^*\in\overline{S}(x^0,t^*)\cap K$. If (b) is not true then there exists $y^*\in\overline{S}(x^0,t^*)\cap K$ such that $y^*\neq x^*$ and

$$(F(y^*),y-y^*)\geqslant 0,\ \forall\, y\in K.$$

So,

$$(F(y^*),x^k-y^*)\geqslant 0.$$

On the other hand, by (3) we have

$$(F(x^{k-1})+B_{k-1}(x^k-x^{k-1}),y^*-x^k)\geqslant 0.$$

Adding the last two inequalities we obtain

$$(B_{k-1}(y^*-x^k),y^*-x^k)\leqslant (F(x^{k-1})-F(y^*)-B_{k-1}(x^{k-1}-y^*),y^*-x^k).$$

Similar to (11), we may derive

$$\left\|y^*-x^k\right\|\leqslant t^*-t_k,\ k=0,\ 1,\ 2,\ \cdots.$$

Hence

$$\|y^*-x^*\|\leqslant \left\|x^*-x^k\right\|+t^*-t_k.$$

Letting $k\to\infty$ in the above inequality, we get $y^*=x^*$, which contradicts the assumption $y^*\neq x^*$. The proof is complete. □

Remark 1. It is easy to prove that condition (iv) holds by Broyden's rank-one method, the row-column-update quasi-Newton method[8] and the column-update quasi-Newton method[9].

Remark 2. The function set will be very small if the set K_0 in condition (iii) is replaced by $K=R^n_+$. For example, cubic polynomials do not satisfy condition (iii) with $K_0=R^n_+$. The set K_0 in Theorem 1 makes the set of functions satisfying (iii) fairly large.

Theorem 2. *Suppose that*

(i) $F \in C^1(K)$;

(ii) $x^0 \in K$, $F(x^0) \geqslant 0$, $0 \leqslant (F(x^0), x^0) \leqslant \eta^2$, $\beta^2 = \min (F'(x^0)y, y) \Big/ \|y\|^2$, *where the constants* $\eta > 0$, $\beta > 0$;

(iii) *there exists a constant* $\gamma \geqslant 1$ *such that*

$$\|F'(x) - F'(y)\| \leqslant \gamma \|x - y\|, \ \forall \, x, \ y \in K_0,$$

where $K_0 = \{x \in K : \|x\| \leqslant \|x^0\| + \beta^2/2\}$;

(iv) $h = \gamma\eta/\beta^3 \leqslant 1/4$.

Then we conclude that

(a) $\{x^k\} \subset S(x^0, t^*) \cap K$ *and* $\{x^k\}$ *converges to a solution* x^* *of* (2), *where* $t^* = \beta^2(1 - \sqrt{1-4h}\,)/2\gamma$;

(b) x^* *is the unique solution of (2) in* $\overline{S}(x^0, t^*) \cap K$.

The proof of Theorem 2 is similar to that of Theorem 1 and more easier. Take the majorizing sequence $t_{k+1} = t_k + f(t_k)/\beta^2$, $f(t) = \gamma t^2 - \beta^2 t + \beta\eta$. The corresponding difference equation problem is: $t_{k+1} - t_k = \gamma(t_k^2 - t_{k-1}^2)/\beta^2$, $t_0 = 0$, $t_1 = \eta/\beta$.

References

[1] J.S. Pang and D. Chan, Iterative methods for variational and complementarity problems, *Mathematical Programming*, 24 (1982), 284-313.

[2] R.H.W. Hoppe, Multigrid algorithms for variational inequalities, *SIAM Journal on Numerical Analysis*, 24 (1987), 1046-1065.

[3] L.H. Wang, Error estimates for the finite element solutions of some variational inequalities with nonlinear monotone operator, *Journal of Computational Mathematics*, 1 (1983), 99-105.

[4] Z.Y. You and S.Q. Zhu, On global convergence and approximate iteration of the linear approximation method for solving variational inequalities, *Journal of Computational Mathematics*, 4 (1986), 289-297.

[5] 周叔子, 椭圆变分不等式离散问题的数值解法, 高校应用数学学报, 4 (1989), 576-589.

[6] 李庆扬, 莫孜中, 祁力群, 非线性方程组的数值解法, 科学出版社, 北京, 1987.

[7] J. Ortega and W.C. Rheinboldt, *Iterative Solutions of Nonlinear Equations in Several Variables*, Academic Press, New York, 1970.

[8] 冯果忱, 李光烨, 一种保持对称性、稀疏性的拟 Newton 法, 高等学校计算数学学报, 5 (1983), 36-41.

[9] 冯果忱, 李光烨, 关于换列拟 Newton 法, 高等学校计算数学学报, 5 (1983), 139-147.

A Trust Region Method for a Semismooth Reformulation to Variational Inequality Problems*

Y.F. Yang (杨余飞)† D.H. Li (李董辉)† S.Z. Zhou (周叔子)†

Abstract We present a well defined trust region method for solving the KKT system of a variational inequality problem based on its semismooth reformulation. The proposed method solves subproblems inexactly. We show that the proposed method converges globally without monotone assumption on the problems, and the convergent rate is superlinear/quadratic even if strict complementarity does not hold at the solution. Moreover, some numerical results are presented.

Keywords Semismoothness, trust region method, variational inequality problem, global convergence, superlinear convergence.

1 Introduction

We consider the variational inequality problem (denoted by VI(X, F)): find an $x^* \in X$ such that

$$F(x^*)^T(x - x^*) \geqslant 0, \quad \forall x \in X,$$

where $X \subseteq R^n$ is a closed convex set and $F: R^n \to R^n$ is a continuously differentiable function.

Variational inequality problems have many applications in economics, engineering and various equilibrium models. Iterative methods are a very important class of numerical methods for solving variational inequality problems. There have been developed many good iterative methods. We refer to [1] for a comprehensive review about the early studies for the iterative methods. In recent years, much attention have been paid to develop globally convergent method. One way to develop globalization methods is to introduce a line search process acting on certain merit functions derived from different reformulations of VI(X, F). Indeed there have been established many kinds of damped Newton methods with global convergence (see, e.g., [2–13]).

* 发表于: Optimization Methods and Software, Vol.14, 2000, pp. 139-157.

† Department of Applied Mathematics, Hunan University, Changsha 410082, P.R. China.

Another way is to use a trust region strategy. Trust region methods are an efficient class of iterative methods for solving optimization problems and have been received much attention (see, e.g., [14 – 17]). However, for solving VI(X, F), the study is relatively less. Recently, Jiang et al. [18] established a trust region method for solving generalized complementarity problems. The method converges globally under suitable conditions. An attract advantage of the method is that it converges superlinearly without assuming that strict complementarity holds at the solution. In this paper, we devote to develop a trust region method for solving VI(X, F). We establish the method based on an inexact Newton method proposed by Facchinei et al.[3]. Unlike the method in [18], the proposed trust region method solves subproblems inexactly. Under mild conditions, the proposed method converges globally and superlinearly/ quadratically.

The organization of the paper is as follows. In the next section, we introduce the concept of semismooth functions and related properties. In Section 3, we propose a trust region method and prove its well definiteness. In Section 4, we prove global and superlinear convergence for the proposed method. Some numerical results are given in Section 5.

Notation: Throughout the paper, $\|\cdot\|$ denotes the Euclidean norm or the subordinate matrix norm. For a differentiable function $f: R^n \to R^m$, f' denotes its Jacobian matrix. If $f: R \to R^n$ is twice differentiable, then $\nabla^2 f(x)$ denotes the Hessian matrix of f at x.

2 Preliminaries

In this section, we review the concept of a semismooth operator and related properties. More details can be found in [19 – 21].

Let $G: R^n \to R^n$ be locally Lipschitzian continuous. According to Rademacher's theorem, G is differentiable almost everywhere. Denote the set of points at which G is differentiable by D_G. Then the B-differential of G at a point $x \in R^n$ is defined by

$$\partial_B G(x) = \{V \mid V = \lim_{\substack{x^k \to x \\ x^k \in D_G}} G'(x^k)\},$$

where $G'(x)$ denotes the Jacobian of G at x. The generalized Jacobian of G at x in the sense of Clarke is defined by

$$\partial G(x) = \text{conv}\partial_B G(x).$$

Definition 2.1. *$G: R^n \to R^n$ is called semismooth at $x \in R^n$ if G is locally Lipschitzian at x and*

$$\lim_{\substack{V \in \partial G(x+th') \\ h' \to h, t \downarrow 0}} \{Vh'\}$$

exists for any $h \in R^n$.

It is known that a semismooth function is directional differentiable. The following lemma is due to Pang and Qi [19] (see also Proposition 2.2 in [3]).

Lemma 2.2. *Assume that $G : R^n \to R^n$ is semismooth at x. Then*

$$\lim_{\substack{V \in \partial G(x+h) \\ h \to 0}} \frac{\|G(x+h) - G(x) - Vh\|}{\|h\|} = 0. \tag{2.1}$$

A slightly stronger notion than semismoothness is strong semismoothness, we give its definition and a property as follows (see [20, 21]).

Definition 2.3. *If $G : R^n \to R^n$ is locally Lipschitzian at $x \in R^n$ and for any $h \to 0$ and any $V \in \partial G(x+h)$, we have*

$$Vh - G'(x; h) = O(\|h\|^2),$$

then we say that G is strongly semismooth at x, where $G'(x;h)$ denotes the directional derivative of G at x along direction h.

Lemma 2.4. *If G is strongly semismooth at x, then for any $h \to 0$,*

$$G(x+h) - G(x) - G'(x; h) = O(\|h\|^2).$$

The following definition and property of BD-regular is due to Qi [20] and is very important in the analysis of convergent rate for the proposed method.

Definition 2.5. *We say that a Lipschitzian function $G : R^n \to R^n$ is BD-regular at a point $x \in R^n$ if all elements in the B-differential $\partial_B G(x)$ are nonsingular.*

Lemma 2.6. *Assume that $G : R^n \to R^n$ is BD-regular at $x^* \in R^n$. Then there is a neighborhood Ω of x^* and a constant $c > 0$ such that, for all $x \in \Omega$ and all $H \in \partial_B G(x)$, H is nonsingular and $\|H^{-1}\| \leqslant c$.*

We introduce another useful lemma. The proof can be found in [3].

Lemma 2.7. *Assume that $G : R^n \to R^n$ is semismooth and that $x^* \in R^n$ is a BD-regular solution of the system $G(x) = 0$. Suppose that there are two sequences $\{x^k\}$ and $\{d^k\}$ such that*

$$x^k \to x^* \quad \text{and} \quad \lim_{k\to\infty} \frac{\|x^k + d^k - x^*\|}{\|x^k - x^*\|} = 0.$$

Then

$$\lim_{k\to\infty} \frac{\|G(x^k + d^k)\|}{\|G(x^k)\|} = 0.$$

3 Reformulation and Algorithm

In this section we consider the variational inequality problem VI(X, F) with X being specified by

$$X := \{x \in R^n \mid g(x) \geqslant 0, h(x) = 0\},$$

where $g : R^n \to R^m$ and $h : R^n \to R^l$ are assumed to be continuously differentiable. Accordingly, the KKT system of the above VI(X, F) is

$$\begin{cases} F(x) - g'(x)^T y + h'(x)^T z = 0, \\ h(x) = 0, \\ g(x) \geqslant 0, \ \ y \geqslant 0, \ \ y^T g(x) = 0. \end{cases} \tag{3.1}$$

The vectors y and z in (3.1) are called multipliers and the vector $w = (x, y, z) \in R^n \times R^m \times R^l$ is called a KKT point of VI(X, F). It is well known (see [1]) that under suitable constraint qualifications, if x^* is a solution of VI(X, F), then there exist multipliers $y^* \in R^m$ and $z^* \in R^l$, such that the vector $\omega^* = (x^*, y^*, z^*) \in R^n \times R^m \times R^l$ is a KKT point of VI(X, F). Conversely, if $\omega^* = (x^*, y^*, z^*)$ is a KKT point of VI(X, F), g_i, $i = 1, 2, \cdots, m$ are concave function, and h_j, $j = 1, 2, \cdots, l$ are affine, then x^* is a solution of VI(X, F).

We establish a trust region method based on an equivalent semismooth equation for (3.1). We first recall the Fischer-Bumeister function $\varphi : R^2 \to R$ defined by

$$\varphi(a, b) := \sqrt{a^2 + b^2} - a - b.$$

A well known property for function φ is

$$\varphi(a, b) = 0 \iff a \geqslant 0, b \geqslant 0, ab = 0.$$

By means of Fischer-Bumeister function, it is easy to see that the last relationship in (3.1) can be written in a compact form: $\varphi(g_i(x), y_i) = 0$, $i = 1, 2, \cdots, m$. Define a vector value function $\psi(x, y) = (\psi_1(x, y), \psi_2(x, y), \cdots, \psi_m(x, y))^T : R^{n+m} \to R^m$ with elements $\psi_i(x, y) = \varphi(g_i(x), y_i)$, $i = 1, 2, \cdots, m$. Then the KKT system (3.1) can be recast as the nonlinear system of equations

$$\Phi(\omega) := \Phi(x, y, z) := \begin{pmatrix} L(x, y, z) \\ h(x) \\ \psi(x, y) \end{pmatrix} = 0, \tag{3.2}$$

where $\omega = (x, y, z)$ and

$$L(x, y, z) = F(x) - g'(x)^T y + h'(x)^T z.$$

The reformulation (3.2) was used by many authors. We summarize some useful properties as follows. For the proof, see [3, 4, 10].

Lemma 3.1. *Let F be continuously differentiable and g_i, $i = 1, 2, \cdots, m$, and h_j, $j = 1, 2, \cdots, l$ be twice continuously differentiable. Then Φ is semismooth. If in addition, F', $\nabla^2 g_i$, $i = 1, 2, \cdots, m$, and $\nabla^2 h_j$, $j = 1, 2, \cdots, l$ are Lipschitzian continuous, then Φ is strongly semismooth.*

Define function $\theta : R^{n+m+l} \to R$ by

$$\theta(\omega) := \frac{1}{2}\Phi(\omega)^T\Phi(\omega) = \frac{1}{2}\|\Phi(\omega)\|^2. \tag{3.3}$$

It is obvious that ω^* is a KKT point of VI(X, F) if and only if ω^* is a global minima of the minimization problem $\min \theta(\omega)$ and $\theta(\omega^*) = 0$. Note that although function ψ is not differentiable, θ is continuously differentiable. Specifically, we have (see [3])

Lemma 3.2. *Assume that F is continuously differentiable and g_i, $i = 1, 2, \cdots, m$, and h_j, $j = 1, 2, \cdots, l$ are twice continuously differentiable. Then θ is continuously differentiable. Moreover, $\nabla\theta(\omega) = H^T\Phi(\omega)$ holds for any $H \in \partial\Phi(\omega)$.*

Now we present a trust region algorithm for solving the KKT system (3.1).

Algorithm 1.

Step 0. Given positive constants $\alpha < 1$, $\beta < 1$, $\gamma < 1$, $\sigma < 1$, $p > 2$ and a positive sequence $\{\eta_k\}$. Choose initial point $\omega^0 = (x^0, y^0, z^0) \in R^{n+m+l}$. Let $k := 0$.

Step 1. If $\nabla\theta(\omega^k) = 0$, stop.

Step 2. Select an element $H_k \in \partial_B\Phi(\omega^k)$. Find a solution d^k of the system

$$H_k d = -\Phi(\omega^k) + r^k \tag{3.4}$$

such that $\|r^k\| \leqslant \eta_k\|\Phi(\omega^k)\|$.

Step 3. If Step 2 is not possible, set $\delta^k := \|\nabla\theta(\omega^k)\|$ and go to Step 5.

Step 4. If

$$\theta(\omega^k + d^k) \leqslant \theta(\omega^k) - \alpha\|d^k\|^p, \tag{3.5}$$

let $\omega^{k+1} := \omega^k + d^k$, $k := k+1$ and go to Step 1. Otherwise, set $\delta^k := \|d^k\|$.

Step 5. Let $\Delta := \delta^k$.

Step 6. Let $s(\Delta)$ be an approximation solution of the following problem

$$\min_{s.t.\ \|s\|\leqslant\Delta} Q_k(s) = \frac{1}{2}\|\Phi(\omega^k) + H_k s\|^2 - \frac{1}{2}\|\Phi(\omega^k)\|^2 \tag{3.6}$$

satisfying

$$Q_k(s(\Delta)) \leqslant \beta Q_k(s^c(\Delta)) \quad \text{and} \quad \|s(\Delta)\| \leqslant \Delta, \tag{3.7}$$

where $s^c(\Delta)$ is the Cauchy step of (3.6), i.e., the minimal point of $Q_k(s)$ along its steepest descent direction within the trust region.

Step 7. If

$$\theta(\omega^k + s(\Delta)) \leqslant \theta(\omega^k) + \gamma\nabla\theta(\omega^k)^T s(\Delta), \tag{3.8}$$

set $\Delta_k := \Delta, s^k = s(\Delta_k), \omega^{k+1} := \omega^k + s^k, k = k+1$, go to Step 1. Otherwise, set $\Delta := \sigma\Delta$, go to Step 6.

Remark 3.1. (i) From the definition of the Cauchy step, we can deduce that

$$s^c(\Delta)=\begin{cases} -\dfrac{\|\nabla\theta(\omega^k)\|^2}{\|H_k\nabla\theta(\omega^k)\|^2}\nabla\theta(\omega^k), & \text{if}\dfrac{\|\nabla\theta(\omega^k)\|^3}{\|H_k\nabla\theta(\omega^k)\|^2}<\Delta, \\ -\dfrac{\Delta}{\|\nabla\theta(\omega^k)\|}\nabla\theta(\omega^k), & \text{otherwise.} \end{cases} \tag{3.9}$$

It implies

$$\frac{s^c(\Delta)}{\|s^c(\Delta)\|}=-\frac{\nabla\theta(\omega^k)}{\|\nabla\theta(\omega^k)\|} \quad \text{and} \quad \frac{\nabla\theta(\omega^k)^T s^c(\Delta)}{\|s^c(\Delta)\|}=-\|\nabla\theta(\omega^k)\|. \tag{3.10}$$

(ii) The Step 6 in Algorithm 1 is well defined. In other words, there exists at least one vector $s(\Delta)$ satisfying (3.7). Indeed, by the definition of the Cauchy step, we have $Q_k(s^c(\Delta))\leqslant \beta Q_k(s^c(\Delta))\leqslant 0$. Since $s=0$ is an admissible point, let $s^*(\Delta)$ be a solution of (3.6). Then it is obvious that $Q_k(s^*(\Delta))\leqslant Q_k(s^c(\Delta))\leqslant \beta Q_k(s^c(\Delta))$. This means that $s^c(\Delta)$ and $s^*(\Delta)$ are two solutions of (3.7).

(iii) From Lemma 3.2, it is easy to deduce that

$$\begin{aligned} Q_k(s)=&\frac{1}{2}\|\Phi(\omega^k)+H_k s\|^2-\frac{1}{2}\|\Phi(\omega^k)\|^2 \\ =&\nabla\theta(\omega^k)^T s+\frac{1}{2}s^T H_k^T H_k s. \end{aligned} \tag{3.11}$$

The following lemma shows that Algorithm 1 is well defined.

Lemma 3.3. *Let F be continuously differentiable and g_i, $i=1,2,\cdots,m$ and h_j, $j=1,2,\cdots,l$ be twice continuously differentiable. Then Algorithm 1 is well defined. That is, if $\nabla\theta(\omega^k)\neq 0$, then ω^{k+1} can be obtained either in Step 4 or by repeating Step 6 and Step 7 finitely.*

Proof. We only need to show that if ω^{k+1} is not obtained at Step 4, then ω^{k+1} can be obtained by repeating Step 6 and Step 7 finitely.

Set

$$\rho_k(\Delta)=\frac{\theta(\omega^k+s(\Delta))-\theta(\omega^k)}{\nabla\theta(\omega^k)^T s(\Delta)}. \tag{3.12}$$

By (3.10) and (3.11), we have

$$\begin{aligned} \nabla\theta(\omega^k)^T s(\Delta)\leqslant& Q_k(s(\Delta))\leqslant \beta Q_k(s^c(\Delta))\leqslant \beta Q_k\Big(\frac{\Delta}{\|s^c(\Delta)\|}s^c(\Delta)\Big) \\ =&\beta\Delta\frac{\nabla\theta(\omega^k)^T s^c(\Delta)}{\|s^c(\Delta)\|}+\frac{1}{2}\beta\frac{\Delta^2}{\|s^c(\Delta)\|^2}(s^c(\Delta))^T H_k^T H_k s^c(\Delta) \\ =&-\beta\Delta\|\nabla\theta(\omega^k)\|+\frac{1}{2}\beta\frac{\Delta^2}{\|s^c(\Delta)\|^2}(s^c(\Delta))^T H_k^T H_k s^c(\Delta), \end{aligned}$$

where the last equality follows from (3.10). It then follows that

$$\limsup_{\Delta\to 0}\frac{\nabla\theta(\omega^k)^T s(\Delta)}{\Delta}\leqslant -\beta\|\nabla\theta(\omega^k)\|<0.$$

Therefore, there exists $\triangle_k^* > 0$ such that when $\triangle \leqslant \triangle_k^*$,

$$\nabla\theta(\omega^k)^T s(\triangle) \leqslant -\frac{1}{2}\beta\|\nabla\theta(\omega^k)\|\triangle.$$

By the mean value theorem, we have

$$\theta(\omega^k + s(\triangle)) - \theta(\omega^k) = \nabla\theta(\omega^k)^T s(\triangle) + o(\|s(\triangle)\|).$$

Thus we get

$$|\rho_k(\triangle) - 1| = \frac{o(\|s(\triangle)\|)}{|\nabla\theta(\omega^k)^T s(\triangle)|} \leqslant \frac{2o(\triangle)}{\beta\|\nabla\theta(\omega^k)\|\triangle}.$$

Taking limits in both sides yields that $\lim\limits_{\triangle\to 0+} \rho_k(\triangle) = 1$. This means that when $\triangle > 0$ is sufficiently small $\rho_k(\triangle) \geqslant \gamma$. Consequently, (3.8) is satisfied for all $\triangle > 0$ sufficiently small. This shows that the cycling between Step 6 and Step 7 is finite. The proof is completed. □

4 Global and Superlinear Convergence

This section devotes to prove global and superlinear/quadratic convergence of Algorithm 1. We show global convergence of Algorithm 1 without monotone assumption on F. Moreover, we show that Algorithm 1 converges superlinearly/quadratically even if the strict complementarity does not hold at the solution.

Throughout this section, we assume that F is continuously differentiable and g_i, $i = 1, 2, \cdots, m$ and h_j, $j = 1, 2, \cdots, l$ are twice continuously differentiable.

Theorem 4.1. *Let $\{\omega^k\}$ be generated by Algorithm 1. If in Algorithm 1, we choose the sequence $\{\eta_k\}$ satisfying that $\eta_k \leqslant \bar{\eta}$ for all k sufficiently large with some constant $\bar{\eta} \in (0, 1)$, then every accumulation point of $\{\omega^k\}$ is a stationary point of θ.*

Proof. Notice that for each k, we have by Algorithm 1 that either (3.5) or (3.8) holds. Both cases imply that the function sequence $\{\theta(\omega^k)\}$ is nonincreasing, which in turn implies that

$$\lim_{k\to\infty} \{\theta(\omega^{k+1}) - \theta(\omega^k)\} = 0. \tag{4.1}$$

Let ω^* be an arbitrary accumulation point of $\{\omega^k\}$ and $\{\omega^k\}_{k\in K} \to \omega^*$. We will show that $\nabla\theta(\omega^*) = 0$.

First we consider the case that (3.5) is satisfied for infinite many $k \in K$. Without loss of generality, we assume that (3.5) holds for all $k \in K$, i.e.,

$$\theta(\omega^{k+1}) \leqslant \theta(\omega^k) - \alpha\|d^k\|^p, \quad \forall k \in K.$$

It then follows from (4.1) that

$$\lim_{k\in K} \|d^k\| = 0. \tag{4.2}$$

On the other hand, by (3.4), we have

$$\begin{aligned}(1-\bar{\eta})\|\Phi(\omega^k)\| &\leqslant \|\Phi(\omega^k)\| - \|r^k\| \\ &\leqslant \|\Phi(\omega^k) - r^k\| = \|H_k d^k\| \\ &\leqslant \|H_k\|\|d^k\|. \end{aligned} \tag{4.3}$$

Since $\{H_k\}_{k\in K}$ is bounded on the bounded sequence $\{\omega^k\}_{k\in K}$ by a property of the generalized Jacobian. By means of (4.2), taking limits on both sides of (4.3), we get $\Phi(\omega^*) = 0$, which shows that ω^* is a global minima of θ and hence a stationary point of θ.

Now we consider the case that (3.5) holds for only finite number of $k \in K$. In other words, for all $k \in K$ sufficiently large, ω^{k+1} is determined by Step 7. In this case, we get by Step 7 that for all $k \in K$ sufficiently large, $\omega^{k+1} = \omega^k + s(\Delta_k)$ and

$$\theta(\omega^{k+1}) \leqslant \theta(\omega^k) + \gamma\nabla\theta(\omega^k)^T s(\triangle_k). \tag{4.4}$$

Without loss of generality, we assume that (4.4) holds for all $k \in K$. We show $\nabla\theta(\omega^*) = 0$ by two cases.

Case I. $\inf\limits_{k\in K} \delta^k = 0$. If there are infinite many indices $k \in K$ such that the associated δ^k is obtained by Step 3, then it is obvious that $\nabla\theta(\omega^*) = 0$, i.e., ω^* is a stationary point of θ. Otherwise, $\delta^k = \|d^k\|$ for all $k \in K$ sufficiently large, which implies that $\inf\limits_{k\in K} \|d^k\| = 0$. By (4.3) again, we get $\Phi(\omega^*) = 0$, which also shows that ω^* is a stationary point of θ.

Case II. $\inf\limits_{k\in K} \delta^k = c_1 > 0$. If $\inf\limits_{k\in K} \triangle_k = 0$, without loss of generality, we assume that $\lim\limits_{k\in K} \triangle_k = 0$, which implies that $\triangle_k < \delta_k$ for all $k \in K$ sufficiently large. By Step 7 in Algorithm 1, for every $k \in K$ large enough, there exists an $s(\bar{\triangle}_k)$ such that (3.7) holds but (3.8) does not hold, where $\bar{\triangle}_k = \triangle_k/\sigma$. That is, $Q_k(s(\bar{\triangle}_k)) \leqslant \beta Q_k(s^c(\bar{\triangle}_k))$, but

$$\theta(\omega^k + s(\bar{\triangle}_k)) > \theta(\omega^k) + \gamma\nabla\theta(\omega^k)^T s(\bar{\triangle}_k). \tag{4.5}$$

On the other hand, it is not difficult to see from (3.11) that

$$\nabla\theta(\omega^k)^T s(\bar{\triangle}_k) \leqslant Q_k(s(\bar{\triangle}_k)) \leqslant \beta Q_k(s^c(\bar{\triangle}_k)) \leqslant \beta Q_k\Big(\frac{\bar{\triangle}_k}{\|s^c(\bar{\triangle}_k)\|} s^c(\bar{\triangle}_k)\Big).$$

Noting that $\|s^c(\bar{\triangle}_k)\| \leqslant \bar{\triangle}_k = \dfrac{\triangle_k}{\sigma} \to 0$ as $k \to \infty$ with $k \in K$. In a similar way to the proof of Lemma 3.3 we can show that if $\nabla\theta(\omega^*) \neq 0$, then $\lim\limits_{k\in K} \rho_k = 1$, where $\rho_k = \rho_k(\bar{\Delta}_k)$ is defined by (3.12). This implies that when $k \in K$ is sufficiently large, $\rho_k \geqslant \gamma$, which contradicts (4.5). The contradiction shows that ω^* is a stationary point of θ.

If $\inf\limits_{k\in K} \triangle_k = \tilde{\triangle} > 0$, without loss of generality, we assume that $\lim\limits_{k\in K} \triangle_k = \tilde{\triangle} > 0$. Since the sequences $\{H_k\}_{k\in K}$, $\{s(\triangle_k)\}_{k\in K}$ and $\{s^c(\triangle_k)\}_{k\in K}$ are bounded, there exist an index set $K_1 \subseteq K$, matrix $H^* \in R^{(n+m+l)\times(n+m+l)}$ and vectors $\tilde{s}, s^c \in R^{n+m+l}$, such that

$$\lim_{k\in K_1} H_k = H^*, \quad \lim_{k\in K_1} s(\triangle_k) = \tilde{s} \quad \text{and} \quad \lim_{k\in K_1} s^c(\triangle_k) = s^c.$$

It is not difficult to show that $H^* \in \partial_B \Phi(\omega^*)$ and s^c is the Cauchy step of the problem

$$\min_{\text{s.t. } \|s\| \leqslant \tilde{\Delta}} Q(s) = \frac{1}{2}\|\Phi(\omega^*) + H^* s\|^2 - \frac{1}{2}\|\Phi(\omega^*)\|^2.$$

Recall that $\{\theta(\omega^k)\}$ is nonincreasing, which implies that $\lim_{k\to\infty}\Big(\theta(\omega^{k+1}) - \theta(\omega^k)\Big) = 0$. Taking limits in (4.4) yields $\nabla\theta(\omega^*)^T\tilde{s} \geqslant 0$.

By (3.11) and the definition of $s^c(\Delta_k)$, we have

$$\nabla\theta(\omega^k)^T s(\triangle_k) \leqslant Q_k(s(\triangle_k)) \leqslant \beta Q_k(s^c(\triangle_k)) \leqslant 0, \quad \forall k \in K.$$

This together with the fact that $\nabla\theta(\omega^*)^T\tilde{s} \geqslant 0$ implies that $Q(s^c) = 0$. By the definition of the Cauchy step, we get $\nabla Q(s) = 0$, that is, $\nabla\theta(\omega^*) = 0$. □

Theorem 4.1 shows that under reasonable conditions, Algorithm 1 converges to a stationary point of the merit function θ defined by (3.3). Some conditions ensuring that the stationary point of θ is a KKT point of VI(X, F) were given in Section 5 of [22]. In particular, from Proposition 5.1 in [22], we have the following result.

Corollary 4.2. *Let $\{\omega^k\}$ be generated by Algorithm 1 and $\{\eta_k\}$ satisfy that $\eta_k \leqslant \bar{\eta}$ holds for all k sufficiently large with some constant $\bar{\eta} \in (0,1)$. If there is an accumulation point ω^* of $\{\omega^k\}$ such that $\partial\Phi(\omega^*)$ contains at least one nonsingular element, then every accumulation point of $\{\omega^k\}$ is a KKT point of VI(X, F).*

Proof. From Proposition 5.1 in [22] and Theorem 4.1, we claim that there is at least one accumulation point $\bar{\omega}$ of $\{\omega^k\}$ that is a KKT point of VI(X, F), or equivalently, $\theta(\bar{\omega}) = 0$. Since $\{\theta(\omega^k)\}$ converges, we have $\lim_{k\to\infty}\theta(\omega^k) = 0$. This means that every accumulation point of $\{\omega^k\}$ is a KKT point of VI(X, F). □

Lemma 4.3. *Let a sequence $\{z^k\} \subset R^{n+m+l}$ have an isolated accumulation point z^*. Let ϵ be a positive constant such that z^* is the only accumulation point of $\{z^k\}$ in the ball $\{z \mid \|z - z^*\| \leqslant \epsilon\}$. Denote $K = \{k \mid \|z^k - z^*\| \leqslant \epsilon\}$. If*

$$\lim_{\substack{k \in K \\ k \to \infty}} \|z^{k+1} - z^k\| = 0.$$

then the sequence $\{z^k\}$ itself converges to z^*.

Proof. By the conditions of the lemma, it is clear that

$$\lim_{\substack{k \in K \\ k \to \infty}} z^k = z^*.$$

It follows that there is an index $\bar{k}$ such that

$$\|z^k - z^*\| < \frac{\epsilon}{2} \quad \text{and} \quad \|z^{k+1} - z^k\| < \frac{\epsilon}{2}, \quad \forall k \in K, k \geqslant \bar{k}.$$

Therefore, we have for any $k \in K$ with $k \geqslant \bar{k}$

$$\|z^{k+1} - z^*\| \leqslant \|z^{k+1} - z^k\| + \|z^k - z^*\| < \epsilon.$$

This means that $k+1 \in K$. By induction, we claim that subsequence $\{z^k\}_{k\in K}$ contains $\{z^k\}_{k\geqslant\bar{k}}$. Consequently, z^* is the unique accumulation point of $\{z^k\}$, or equivalently $\lim_{k\to\infty} z^k = z^*$. □

By using Theorem 4.1 and Lemma 4.3, we strengthen convergence result of Algorithm 1.

Theorem 4.4. *If there is an accumulation point, ω^*, of $\{\omega^k\}$ which is a BD-regular solution of the system $\Phi(\omega) = 0$, then $\{\omega^k\}$ itself converges to ω^*.*

Proof. Note that the BD-regularity assumption implies, by Proposition 3 in [19], ω^* is an isolated solution of the system $\Phi(\omega) = 0$. So, there is a positive constant $\epsilon > 0$ such that ω^* is the unique accumulation point of $\{\omega^k\}$ on $\{\omega \mid \|\omega-\omega^*\| \leqslant \epsilon\}$. Denote $K = \{k \mid \|\omega^k-\omega^*\| \leqslant \epsilon\}$. Then, by Lemma 4.3, it suffices to show that

$$\lim_{\substack{k \in K \\ k \to \infty}} \|\omega^{k+1} - \omega^k\| = 0. \tag{4.6}$$

We notice that ω^{k+1} is generated by Step 4 or Step 7. In the first case, we have $\omega^{k+1}-\omega^k = d^k$. In the later case, we have $\omega^{k+1} - \omega^k = s^k$ and $\|s^k\| \leqslant \|\delta^k\| \leqslant \max\{\|\nabla\theta(\omega^k)\|, \|d^k\|\}$. Since we have shown in Theorem 4.1 that $\{\nabla\theta(\omega^k)\}_{k\in K} \to 0$, to show (4.6), it suffices to verify that $d^k \to 0$ as $k \to \infty$ with $k \in K$.

By Lemma 2.6, there exists a positive number c_2 such that $\|H_k^{-1}\| \leqslant c_2$ for all $k \in K$ sufficiently large. This in turn implies that there is a positive constant m such that $\|H_k v\| \geqslant m\|v\|$, $\forall v \in R^{n+m+l}$. It then follows from (3.4) that for all $k \in K$ sufficiently large,

$$m\|d^k\| \leqslant \|H_k d^k\| = \|\Phi(\omega^k) - r^k\| \leqslant (1 + \bar{\eta})\|\Phi(\omega^k)\|. \tag{4.7}$$

Consequently, we get $\{d^k\} \to 0$, as $k \to \infty$ with $k \in K$ and complete the proof. □

The following theorem shows superlinear/quadratic convergence of Algorithm 1.

Theorem 4.5. *Let the condition in Theorem 4.4 hold. If in Algorithm 1, $\{\eta_k\}$ is chosen to satisfy $\{\eta_k\} \to 0$, then $\{\omega^k\}$ converges to ω^* superlinearly. If in addition, $\{\eta_k\}$ is chosen to satisfy $\eta_k = O(\|\Phi(\omega^k)\|)$, and ∇F, $\nabla^2 g_i$, $i = 1,2,\cdots,m$, and $\nabla^2 h_j$, $j = 1,2,\cdots,l$ are Lipschitzian continuous, then the convergence rate is quadratic.*

Proof. We have shown in Theorem 4.4 that $lim_{k\to\infty}\omega^k = \omega^*$. Since ω^* is assumed to be a BD-regular solution, by Lemma 2.6, H_k^{-1} exists for any k sufficiently large, and there exists a positive number c_3 such that $\|H_k^{-1}\| \leqslant c_3$ for all k sufficiently large. This particularly implies that (3.4) is solvable for every k sufficiently large. By an elementary deduction, we get from

(3.4) that

$$\begin{aligned}\|\omega^k+d^k-\omega^*\| &= \|H_k^{-1}\Big(-\Phi(\omega^k)+H_k(\omega^k-\omega^*)+r^k\Big)\| \\ &\leqslant \|H_k^{-1}\|\,\|-\Phi(\omega^k)+H_k(\omega^k-\omega^*)+r^k\| \\ &= \|H_k^{-1}\|\,\|-\Phi(\omega^k)+\Phi(\omega^*)+H_k(\omega^k-\omega^*)+r^k\| \\ &\leqslant c_3\Big(\|\Phi(\omega^k)-\Phi(\omega^*)-H_k(\omega^k-\omega^*)\|+\|r^k\|\Big) \\ &\leqslant c_3\Big(\|\Phi(\omega^k)-\Phi(\omega^*)-H_k(\omega^k-\omega^*)\|+\eta_k\|\Phi(\omega^k)-\Phi(\omega^*)\|\Big) \\ &\leqslant c_3\Big(\|\Phi(\omega^k)-\Phi(\omega^*)-H_k(\omega^k-\omega^*)\|+L\eta_k\|\omega^k-\omega^*\|\Big), \end{aligned} \tag{4.8}$$

where $L>0$ stands for the Lipschitzian constant of Φ at ω^*. If $\lim_{k\to\infty}\eta_k=0$, then by means of (2.1), we get from (4.8) that

$$\lim_{k\to\infty}\frac{\|\omega^k+d^k-\omega^*\|}{\|\omega^k-\omega^*\|}\leqslant c_3\lim_{k\to\infty}\frac{\|\Phi(\omega^k)-\Phi(\omega^*)-H_k(\omega^k-\omega^*)\|}{\|\omega^k-\omega^*\|}+c_3L\lim_{k\to\infty}\eta_k=0. \tag{4.9}$$

By Lemma 2.7, we have $\|\theta(\omega^k+d^k)\|=o(\|\theta(\omega^k)\|)$. On the other hand, in a similar way to the proof of Theorem 4.4 we can show that (4.7) holds for all k sufficiently large with a suitable constant $m>0$. Thus we have

$$\theta(\omega^k+d^k)-\theta(\omega^k)\leqslant-\frac{1}{2}\theta(\omega^k)=-\frac{1}{4}\|\Phi(\omega^k)\|^2\leqslant-\frac{m^2}{16}\|d^k\|^2 \tag{4.10}$$

Noting that $\{\|d^k\|\}\to 0$ by (4.7), it is easy to see that (4.10) eventually implies (3.5) for any $p>2$ and any positive constant α. From this and the nonsingularity of H_k, we claim that when k is sufficiently large, w^{k+1} is always determined by Step 4. Consequently, (4.9) shows the superlinear convergence of $\{\omega^k\}$.

If $\eta_k=O(\|\Phi(\omega^k)\|)$, that is,

$$\eta_k\leqslant C\|\Phi(\omega^k)\|=C\|\Phi(\omega^k)-\Phi(\omega^*)\|\leqslant CL\|\omega^k-\omega^*\|$$

holds for all k with some constant $C>0$, we get from(4.8) that

$$\|\omega^k+d^k-\omega^*\|\leqslant c_3\Big(\|\Phi(\omega^k)-\Phi(\omega^*)-H_k(\omega^k-\omega^*)\|+LC\|\omega^k-\omega^*\|^2\Big).$$

Since Φ is strongly semismooth by Lemma 3.1, we get from Lemma 2.4 and the last inequality that $\|\omega^k+d^k-\omega^*\|=O(\|\omega^k-\omega^*\|^2)$, i.e., $\{\omega^k\}$ converges quadratically. □

5 Numerical Results

In this section, we report some numerical results to show viability of Algorithm 1.

Throughout the computational experiments, the parameters used in Algorithm 1 were set as $\sigma = 0.5$, $p = 2.1$, $\alpha = 10^{-4}$, $\gamma = 10^{-4}$ and $\beta = 0.95$. The residual vector r^k was the accumulation of rounding errors. The stopping criterion for Algorithm 1 was $\|\nabla\theta(\omega^k)\| \leqslant 10^{-10}$. Iter, NF and Ntr represented the number of iterations, the number of F-evaluations and the number of solving the subproblem (3.6) and (3.7), respectively. All computer programs were coded in FORTRAN 77 and the run was made in double precision on a RUN workstation in the State Key Laboratory of Scientific and Engineering Computing, Academic Sinica, Beijing, P. R. China.

We report the results for the following four test problems. The first test problem is due to [23], the remaining test problems are derived from Problem 14, 38 and 60 in [24] via the Karush-Kuhn-Tucker optimality condition.

Example 1.

$$F(x) = \begin{pmatrix} 2x_1 + 0.2x_1^3 - 0.5x_2 + 0.1x_3 - 4 \\ -0.5x_1 + x_2 + 0.1x_2^3 + 0.5 \\ 0.5x_1 - 0.2x_2 + 2x_3 - 0.5 \end{pmatrix}$$

$$X = \{x \in R^3 \mid x_1^2 + 0.4x_2^2 + 0.6x_3^2 - 1 \leqslant 0\}.$$

This problem has the unique solution $x^* = (1,0,0)^T$ and its results are shown in Table 1.

Table 1 Results for Example 1

Starting point ω^0	Iter	NF	Ntr	$\|\nabla\theta\|$	θ
$(1,1,1,0)$	8	15	0	0.134D-14	0.207D-30
$(2,2,2,0)$	9	17	0	0.214D-14	0.102D-30
$(0,0,0,0)$	8	16	1	0.598D-11	0.132D-23
$(-1,-1,-1,0)$	13	34	9	0.432D-14	0.395D-30
$(10,10,10,0)$	11	21	0	0.215D-14	0.988D-31
$(-10,-10,-10,0)$	15	34	5	0.109D-14	0.249D-31
$(100,100,100,0)$	16	31	0	0.329D-14	0.225D-30
$(10^4,10^4,10^4,0)$	26	51	0	0.108D-10	0.468D-23
$(10^5,10^5,10^5,0)$	32	63	0	0.327D-16	0.581D-34
$(-10^4,-10^4,-10^4,0)$	30	62	3	0.640D-12	0.147D-25
$(-10^5,-10^5,-10^5,0)$	38	84	9	0.108D-14	0.247D-31
$(10^8,10^8,10^8,0)$	48	95	0	0.253D-15	0.330D-32
$(-10^9,-10^9,-10^9,0)$	57	117	4	0.115D-14	0.229D-31

Example 2.

$$F(x) = \begin{pmatrix} 2x_1 - 4 \\ 2x_2 - 2 \end{pmatrix}$$

$$X = \{x \in R^2 \mid \; x_1 - 2x_2 + 1 = 0, \; -\frac{1}{4}x_1^2 - x_2^2 + 1 \geqslant 0\}.$$

This problem has the unique solution $x^* = (\frac{1}{2}(\sqrt{7}-1), \frac{1}{4}(\sqrt{7}+1))^T$ and its results are shown in Table 2.

Table 2 Results for Example 2

Starting point ω^0	Iter	NF	Ntr	$\|\nabla\theta\|$	θ
$(1,1,0,0)$	5	9	0	0.720D-11	0.791D-24
$(2,2,0,0)$	8	15	0	0.307D-14	0.105D-30
$(3,3,0,0)$	7	14	1	0.486D-13	0.361D-28
$(4,4,0,0)$	7	14	1	0.670D-15	0.315D-31
$(5,5,0,0)$	7	14	1	0.164D-11	0.409D-25
$(8,8,0,0)$	8	15	1	0.315D-14	0.161D-30
$(10,10,0,0)$	8	16	1	0.307D-14	0.105D-30
$(100,100,0,0)$	9	18	1	0.670D-15	0.315D-31
$(10^3,10^3,0,0)$	9	18	1	0.125D-12	0.238D-27
$(10^5,10^5,0,0)$	9	18	1	0.103D-11	0.162D-25
$(10^8,10^8,0,0)$	9	18	1	0.105D-11	0.169D-25

Example 3.

$$F(x) = \begin{pmatrix} -400(x_2 - x_1^2)x_1 - 2(1-x_1) \\ -200x_1^2 + 220.2x_2 + 19.8x_4 - 40 \\ -360(x_4 - x_3^2)x_3 - 2(1-x_3) \\ -180x_3^2 + 200.2x_4 + 19.8x_2 - 40 \end{pmatrix}$$

$$X = \{x \in R^4 \mid -10 \leqslant x_i \leqslant 10, \quad i = 1,2,3,4\}.$$

This problem has the unique solution $x^* = (1,1,1,1)^T$ and its results are shown in Table 3.

Table 3 Results for Example 3

Starting point ω^0	Iter	NF	Ntr	$\|\nabla\theta\|$	θ
$(0,0,0,0,0,0,0,0,0,0,0,0)$	6	12	1	0.564D-13	0.181D-30
$(2,2,2,2,0,0,0,0,0,0,0,0)$	10	19	0	0.477D-11	0.115D-27
$(3,3,3,3,0,0,0,0,0,0,0,0)$	12	24	1	0.110D-10	0.670D-27
$(4,4,4,4,0,0,0,0,0,0,0,0)$	13	26	1	0.110D-10	0.670D-27
$(10,10,10,10,0,0,0,0,0,0,0,0)$	16	31	0	0.982D-10	0.581D-26
$(-1,-1,-1,-1,0,0,0,0,0,0,0,0)$	8	15	0	0.328D-10	0.805D-27
$(100,100,100,100,0,0,0,0,0,0,0,0)$	22	43	0	0.998D-10	0.492D-26
$(10^4,10^4,10^4,10^4,0,0,0,0,0,0,0,0)$	28	62	7	0.993D-11	0.202D-27

Example 4.

$$F(x) = \begin{pmatrix} 4x_1 - 2x_2 - 2 \\ -2x_1 + 2x_2 + 4(x_2 - x_3)^3 \\ -4(x_2 - x_3)^3 \end{pmatrix}$$

$$X = \{x \in R^3 \mid \quad x_1(1 + x_2^2) + x_3^4 - 4 - 3\sqrt{2} = 0, \\ -10 \leqslant x_i \leqslant 10, \quad i = 1,2,3\}.$$

This problem has the unique solution $x^* = (1.104859024, 1.196674194, 1.535262257)^T$ and its results are shown in Table 4.

Table 4 Results for Example 4

Starting point ω^0	Iter	NF	Ntr	$\|\nabla\theta\|$	θ
$(1,1,1,0,0,0,0,0,0,0)$	6	13	2	0.626D-12	0.163D-26
$(2,2,2,0,0,0,0,0,0,0)$	8	15	0	0.936D-12	0.364D-26
$(3,3,3,0,0,0,0,0,0,0)$	10	19	0	0.130D-14	0.792D-32
$(4,4,4,0,0,0,0,0,0,0)$	11	21	0	0.101D-14	0.667D-32
$(5,5,5,0,0,0,0,0,0,0)$	11	21	0	0.435D-10	0.776D-23
$(8,8,8,0,0,0,0,0,0,0)$	13	25	0	0.101D-14	0.667D-32
$(9,9,9,0,0,0,0,0,0,0)$	13	25	0	0.129D-12	0.776D-28

Acknowledgments The authors thank Professor Ya-xiang Yuan for his helpful comments regarding the implementation of the algorithm.

References

[1] P.T. Harker and J.S. Pang, Finite-dimensional variational inequality and nonlinear complementarity problems: a survey of theory, algorithms and applications, *Mathematical Programming*, 48 (1990), 161-220.

[2] T.De. Luca, F. Facchinei and C. Kanzow, A semismooth equation approach to the solution of nonlinear complementarity problems, *Mathematical Programming*, 75 (1996), 407-439.

[3] F. Facchinei, A. Fischer and C. Kanzow, Inexact Newton methods for semismooth equations with applications to variational inequality problems, in: *Nonlinear Optimization and Application*, Di Pillo, G. and Giannesi, F. Eds., Plennm Press, New York, 1996, pp.125-139.

[4] F. Facchinei and J. Soares, A new merit function for nonlinear complementarity problems and a related algorithm, *SIAM Journal on Optimization*, 7 (1997), 225-247.

[5] A. Fischer, A special Newton-type optimization method, *Optimization*, 24 (1992), 269-284.

[6] A. Fischer, An NCP-function and its use for the solution of complementarity problems, In: *Recent Advances in Nonsmooth Optimization*, Du D. Z., Qi, L. and Womersley, R. S. Eds., World Scientific Publishers, Singapore, 1995, pp.88-105.

[7] M. Fukushima, Equivalent differentiable optimization problems and descent methods for asymmetric variational inequality problems, *Mathematical Programming*, 53 (1992), 99-110.

[8] S.A. Gabriel and J.S. Pang, An inexact NE/SQP method for solving the nonlinear complementarity problem, *Computational Optimization and Applications*, 1 (1992), 67-92.

[9] C. Geiger and C. Kanzow, On the resolution of monotone complementarity problems, *Computational Optimization and Applications*, 5 (1996), 155-173.

[10] H. Jiang and L. Qi, A new nonsmooth equations approach to nonlinear complementarity problems, *SIAM Journal on Control and Optimization*, 35 (1997), 178-193.

[11] J.M. Martínez and L. Qi, Inexact Newton methods for solving nonsmooth equations, *Journal of Computational and Applied Mathematics*, 60 (1995), 127-145.

[12] J.M. Peng, Global method for monotone variational inequality problems with inequality constraints, *Journal of Optimization Theory and Applications*, 95 (1997), 419-430.

[13] J.H. Wu, M. Florian and P. Marcotte, A general descent framework for the monotone variational inequality problem, *Mathematical Programming*, 61 (1993), 281-300.

[14] J.M. Martínez and A.C. Moretti, A trust region method for minimization of nonsmooth functions with linear constraints, *Mathematical Programming*, 76 (1997), 431-449.

[15] J.J. Moré, Recent developments in algorithms and software for trust region methods, In: *Mathematical Programming*, Bachem, A. Grötschel, M. and Korte, B. Eds., The State of the Art (Springer-Verlag, Berlin, 1983), pp. 258-287.

[16] M.J.D. Powell, On the global convergence of trust region algorithms for unconstrained minimization, *Mathematical Programming*, 29 (1984), 297-303.

[17] Y. Yuan, Conditions for convergence of trust region algorithms for nonsmooth optimization, *Mathematical Programming*, 31 (1985), 220-228.

[18] H. Jiang, M. Fukushima, L. Qi and D. Sun, A trust region method for solving generalized complementarity problems, *SIAM Journal on Optimization*, 8 (1998), 140-157.

[19] J.S. Pang and L. Qi, Nonsmooth equations: Motivation and algorithms, *SIAM Journal on Optimization*, 3 (1993), 443-465.

[20] L. Qi, Convergence analysis of some algorithms for solving nonsmooth equations, *Mathematics of Operations Research*, 18 (1993), 227-244.

[21] L. Qi and J. Sun, A nonsmooth version of Newton's method, *Mathematical Programming*, 58 (1993), 353-367.

[22] F. Facchinei, A. Fischer, and C. Kanzow, A semismooth Newton method for variational inequalities: Theoretical results and preliminary numerical experience, To appear, in: *SIAM Journal on Optimization*.

[23] M. Fukushima, A relaxed projection method for variational inequalities, *Mathematical Programming*, 35 (1986), 58-70.

[24] W. Hock and K. Schittkowski, Test examples for nonlinear programming codes. Lecture Notes, in: *Economic and Mathematical Systems*, Vol. 187, Springer-Verlag, Berlin, Heidelberg, New York, 1981.

Descent Directions of Quasi-Newton Methods for Symmetric Nonlinear Equations*

G.Z. Gu (顾广泽)† D.H. Li (李董辉)†

L.Q. Qi (祁力群)‡ S.Z. Zhou (周叔子)†

Abstract In general, when a quasi-Newton method is applied to solve a system of nonlinear equations, the quasi-Newton direction is not necessarily a descent direction for the norm function. In this paper, we show that when applied to solve symmetric nonlinear equations, a quasi-Newton method with positive definite iterative matrices may generate descent directions for the norm function. On the basis of a Gauss-Newton based BFGS method [D. H. Li and M. Fukushima, SIAM Journal on Numerical Analysis, 37 (1999), 152-172], we develop a norm descent BFGS method for solving symmetric nonlinear equations. Under mild conditions, we establish the global and superlinear convergence of the method. The proposed method shares some favorable properties of the BFGS method for solving unconstrained optimization problems: (a) the generated sequence of the quasi-Newton matrices is positive definite; (b) the generated sequence of iterates is norm descent; (c) a global convergence theorem is established without nonsingularity assumption on the Jacobian. Preliminary numerical results are reported, which positively support the method.

Keywords BFGS method, norm descent direction, global convergence, superlinear convergence.

1 Introduction

Let $F : R^n \to R^n$ be continuously differentiable. A general quasi-Newton method for solving the system of nonlinear equations

$$F(x) = 0 \tag{1.1}$$

generates a sequence of iterates $\{x_k\}$ by letting $x_{k+1} = x_k + d_k$, where d_k is a solution of the

* 发表于：SIAM Journal on Numerical Analysis, Vol. 40, 2003, pp. 1763-1774.

† Institute of Applied Mathematics, Hunan University, Changsha 410082, China.

‡ Department of Applied Mathematics, The Hong Kong Polytechnic University, Hung Hom, Kowloon, Hong Kong.

following system of linear equations:

$$B_k d + F(x_k) = 0. \tag{1.2}$$

If in (1.2), matrix B_k is replaced by $F'(x_k)$, the Jacobian of the function F at x_k, the method reduces to the well-known Newton method. An attractive feature of a quasi-Newton method is its local superlinear convergence property without computation of Jacobians. To enlarge the convergence domain of a quasi-Newton method, line search technique or trust region strategy can be exploited. In this paper, we use a backtracking line search technique to globalize a quasi-Newton method. A line search step at iteration k of an iterative method determines a scalar $\lambda_k > 0$ which satisfies

$$\|F(x_k + \lambda_k d_k)\| < \|F(x_k)\|. \tag{1.3}$$

The next iterate is then determined by letting $x_{k+1} = x_k + \lambda_k d_k$. The scalar $\lambda_k > 0$ is called the steplength. Let θ be the norm function defined by

$$\theta(x) = \frac{1}{2}\|F(x)\|^2. \tag{1.4}$$

Then the nonlinear equation problem (1.1) is equivalent to the following global optimization problem:

$$\min \theta(x), \quad x \in R^n, \tag{1.5}$$

and condition (1.3) is equivalent to

$$\theta(x_k + \lambda_k d_k) < \theta(x_k). \tag{1.6}$$

An iterative method that generates a sequence $\{x_k\}$ satisfying (1.3) or (1.6) is called a norm descent method. If d_k is a descent direction of θ at x_k, then inequality (1.6) holds for all $\lambda_k > 0$ sufficiently small. Accordingly, the related iterative method is a norm descent method. In particular, Newton's method with line search is norm descent. For a quasi-Newton method, however, d_k may not be a descent direction of θ at x_k even if B_k is symmetric and positive definite. To globalize a quasi-Newton method, Li and Fukushima [1] proposed an approximately norm descent line search technique and established global and superlinear convergence of a Gauss-Newton based BFGS method for solving symmetric nonlinear equations. The method in [1] is not norm descent. In addition, the global convergence theorem is established under the assumption that $F'(x)$ is uniformly nonsingular.

The purpose of this paper is to develop a norm descent Gauss-Newton based BFGS method. We adjust the steplength and the search direction simultaneously so that the generated iterate sequence satisfies (1.6). We update B_k by combining a modified BFGS formula [2] or the cautious BFGS update rule with the Gauss-Newton based BFGS method [1] such that B_{k+1} inherits positive definiteness of B_k no matter whatever line search is used. Under mild conditions, we establish a global convergence theorem which shows that there exists an accumulation point that is a stationary point of problem (1.5) even if $F'(x)$ is singular everywhere. We also get the superlinear convergence of the proposed method.

In the next section, we describe how to generate a quasi-Newton direction that is descent for θ. We also state the steps of the proposed method. In section 3, we establish the global and superlinear convergence of the proposed method. In section 4, we present some numerical results.

2 Descent Direction in a Quasi-Newton Method

In this section, we describe a way to generate a descent quasi-Newton direction for θ and then propose a norm descent BFGS method for solving (1.1). We assume that the function F is continuously differentiable, and its Jacobian $F'(x)$ is symmetric for every $x \in R^n$. Recall that in Newton's method, the Newton direction is a solution of the Newton equation

$$F'(x_k)d + F(x_k) = 0. \tag{2.1}$$

Equation (2.1) may have no solution if $F'(x_k)$ is singular. In the case where the solution set of (2.1) is empty, instead of solving (2.1), we may solve the least squares problem

$$\min \frac{1}{2}\|F'(x_k)d + F(x_k)\|^2$$

to get a direction d_k, which results in the so-called Gauss-Newton equation

$$F'(x_k)^2 d + F'(x_k)F(x_k) = 0. \tag{2.2}$$

Here we have used the symmetry of $F'(x_k)$. On the other hand, if $F'(x_k)$ is nonsingular, (2.2) is equivalent to (2.1). In [1], a Gauss-Newton based quasi-Newton method was proposed in which the quasi-Newton direction is the solution of the following system of linear equations:

$$B_k d + \bar{q}_k = 0, \tag{2.3}$$

where B_k is an approximation of matrix $F'(x_k)^2$, and $\bar{q}_k$ is an approximation of vector $F'(x_k) \cdot F(x_k)$. Specifically, let λ_{k-1} be the steplength used at the previous iteration. Then, vector $\bar{q}_k$ is defined by

$$\bar{q}_k = (F(x_k + \lambda_{k-1}F(x_k)) - F(x_k))/\lambda_{k-1} \approx F'(x_k)F(x_k),$$

and matrix B_k is updated by the BFGS formula

$$B_{k+1} = B_k - \frac{B_k s_k s_k^T B_k}{s_k^T B_k s_k} + \frac{y_k y_k^T}{y_k^T s_k}, \tag{2.4}$$

where $s_k = x_{k+1} - x_k$, $y_k = F(x_k + \delta_k) - F(x_k)$, and $\delta_k = F(x_{k+1}) - F(x_k)$. It is clear that if $\|s_k\|$ is small, then $B_{k+1}s_k = y_k \approx F'(x_{k+1})^2 s_k$. Since the solution d_k of (2.4) may not be a descent direction of θ at x_k when x_k is far away from a solution of (1.1), it is generally not possible to get a steplength $\lambda_k > 0$ satisfying (1.6). Taking this into account, Li and Fukushima [1] proposed a nondescent line search in which the steplength $\lambda_k > 0$ satisfies the following inequality:

$$\theta(x_k + \lambda_k d_k) - \theta(x_k) \leqslant -\sigma_1 \|\lambda_k d_k\|^2 - \sigma_2 \|\lambda_k F(x_k)\|^2 + \epsilon_k \|F(x_k)\|^2, \tag{2.5}$$

where σ_1 and σ_2 are positive constants, and $\epsilon_k > 0$ satisfies

$$\sum_{k=0}^{\infty} \epsilon_k < \infty.$$

Since ϵ_k is small, $\{x_k\}$ is approximately norm descent.

The purpose of this paper is to develop a norm descent BFGS method. In other words, we want to construct a system of linear equations like (2.4) such that its solution provides a descent direction of θ at x_k.

Observe that

$$\lim_{\lambda_{k-1} \to 0^+} \bar{q}_k = F'(x_k) F(x_k) \triangleq \tilde{q}_k.$$

Accordingly, the solution of (2.4) with $\tilde{q}_k$ instead of $\bar{q}_k$ is $\tilde{d}_k = -B_k^{-1} F'(x_k) F(x_k)$. If B_k is positive definite and $F'(x_k)$ is symmetric, then $\tilde{d}_k$ is a descent direction of θ at x_k. This observation prompts us to regard λ_{k-1} as a parameter. When this parameter is adjusted to be small enough, the solution of (2.4) is a descent direction of θ at x_k. The following process gives details to realize it.

Let

$$q_k(\lambda) = (F(x_k + \lambda F(x_k)) - F(x_k))/\lambda. \tag{2.6}$$

Consider the system of linear equations with parameter λ:

$$B_k d + q_k(\lambda) = 0. \tag{2.7}$$

Let $d(\lambda)$ be the solution of (2.7). The following lemma shows that when $\lambda > 0$ is sufficiently small, every solution of (2.7) is a descent direction of θ at x_k.

Lemma 2.1. *Let σ_1 and σ_2 be positive constants and B_k be a symmetric and positive definite matrix. If x_k is not a stationary point of (1.5), then there exists a constant $\bar{\lambda} > 0$ depending on k such that when $\lambda \in (0, \bar{\lambda})$, the unique solution $d(\lambda)$ of (2.7) satisfies*

$$\nabla \theta(x_k)^T d(\lambda) < 0. \tag{2.8}$$

Moreover, inequality

$$\theta(x_k + \lambda d(\lambda)) - \theta(x_k) \leqslant -\sigma_1 \|\lambda d(\lambda)\|^2 - \sigma_2 \|\lambda F(x_k)\|^2 \tag{2.9}$$

holds for all $\lambda > 0$ sufficiently small.

Proof. It is clear that

$$\lim_{\lambda\to 0} q_k(\lambda) = F'(x_k)F(x_k).$$

Therefore, we get from (2.7) that

$$\begin{aligned}\lim_{\lambda\to 0+} \nabla\theta(x_k)^T d(\lambda) &= -\lim_{\lambda\to 0+} F(x_k)^T F'(x_k)B_k^{-1}q_k(\lambda)\\ &= -F(x_k)^T F'(x_k)B_k^{-1}F'(x_k)F(x_k).\end{aligned}$$

Since $F'(x_k)$ is symmetric and $F'(x_k)F(x_k)\neq 0$ as x_k is not a stationary point of (1.5), the last equality and the positive definiteness of B_k imply (2.8). We turn to verifying (2.9).

Notice that

$$\lim_{\lambda\to 0+}(\theta(x_k+\lambda d(\lambda))-\theta(x_k))/\lambda = \lim_{\lambda\to 0+}\nabla\theta(x_k)^T d(\lambda) = -F(x_k)^T F'(x_k)B_k^{-1}F'(x_k)F(x_k).$$

However, the right-hand side of (2.9) is $o(\lambda)$. Therefore, inequality (2.9) holds for all $\lambda > 0$ sufficiently small. □

Lemma 2.1 motivates us to find a descent quasi-Newton direction by adjusting parameter λ.

Procedure 1. Let constant $\rho\in(0,1)$ be given. Let i_k be the smallest nonnegative integer such that inequality (2.9) holds with $\lambda=\rho^i$, $i=0,1,\cdots$. Let $d_k=d(\rho^{i_k})$, and $q_k=q_k(\rho^{i_k})$.

Procedure 1 ensures that the value of θ at $x_k+\rho^{i_k}d_k$ is less than that of θ at x_k, though d_k may not necessarily be a descent direction of θ at x_k. It is reasonable to let the scalar ρ^{i_k} be the steplength. However, this steplength may be very small if i_k is large. To enlarge steplength, we exploit the following forward procedure.

Procedure 2. Let i_k and d_k be determined by Procedure 1. If $i_k=0$, let $\lambda_k=1$. Otherwise, let j_k be the largest positive integer $j\in\{0,1,2,\cdots,i_k-1\}$ satisfying

$$\theta(x_k+\rho^{i_k-j}d_k)-\theta(x_k)\leqslant\sigma_1\|\rho^{i_k-j}d_k\|^2-\sigma_2\|\rho^{i_k-j}F(x_k)\|^2. \tag{2.10}$$

Note that (2.10) is satisfied with $j=0$. Therefore, Procedure 2 is well defined.

Procedures 1 and 2 describe a way to generate d_k and λ_k. It is easy to see from Procedures 1 and 2 that

$$\theta(x_k+\lambda_k d_k)-\theta(x_k)\leqslant\sigma_1\|\lambda_k d_k\|^2-\sigma_2\|\lambda_k F(x_k)\|^2. \tag{2.11}$$

which corresponds to (2.5) with $\epsilon_k=0$. It is also easy to see that if $\lambda_k\neq 1$, then $\lambda_k'=\lambda_k/\rho$ stisfies

$$\theta(x_k+\lambda_k' d_k)-\theta(x_k)\ >\sigma_1\|\lambda_k' d_k\|^2-\sigma_2\|\lambda_k' F(x_k)\|^2. \tag{2.12}$$

Notice that Procedure 1 generates a direction d_k which satisfies

$$B_k d_k + q_k = 0, \tag{2.13}$$

where $q_k=q_k(\rho^{i_k})$. Vector q_k differs from $q_k(\lambda_k)$ if $j_k\neq 0$.

Based on the above process, we propose a norm descent Gauss-Newton based BFGS method as follows.

Algorithm 1. (A descent BFGS method).

Initial. Let $B_0 \in R^{n\times n}$ be symmetric and positive definite. Let $x_0 \in R^n$. Set $k = 0$.

Step 1. Determine d_k and λ_k by Procedures 1 and 2. Let $x_{k+1} = x_k + \lambda_k d_k$.

Step 2. Update B_k to get B_{k+1} by the modified BFGS formula

$$B_{k+1} = B_k - \frac{B_k s_k s_k^T B_k}{s_k^T B_k s_k} + \frac{y_k y_k^T}{y_k^T s_k}, \tag{2.14}$$

where $s_k = x_{k+1} - x_k$,

$$y_k = \gamma_k + \left(\max\left\{0,\ -\frac{\gamma_k^T s_k}{\|s_k\|^2}\right\} + \phi(\|F(x_k)\|)\right) s_k.$$

$\gamma_k = F(x_k + \delta_k) - F(x_k)$, $\delta_k = F(x_{k+1}) - F(x_k)$, and function $\phi : R \to R$ satisfies (i) $\phi(t) > 0$ for all $t > 0$, (ii) $\phi(t) = 0$ if and only if $t = 0$, (iii) $\phi(t)$ is bounded if t is in a bounded set.

Step 3. Let $k := k + 1$ and go to Step 1.

In Step 2 of Algorithm 1, we use a modified BFGS update formula instead of the ordinary BFGS formula. The modified BFGS update formula was proposed by Li and Fukushima [2], where $\phi(t) = \mu t$ with some constant $\mu > 0$. A favorable property for this modification is that B_{k+1} inherits positive definiteness of B_k whatever line search is used [2]. Indeed, it is not difficult to get that

$$y_k^T s_k \geqslant \max\left\{\gamma_k^T s_k,\ \phi(\|F(x_k)\|)\|s_k\|^2\right\} > 0, \tag{2.15}$$

which is sufficient to guarantee positive definiteness of B_{k+1} as long as B_k is positive definite. Suppose that $\{x_k\}$ is contained in a bounded set at which F is continuously differentiable. It is not difficult to deduce that

$$\|y_k\| \leqslant 2\|\gamma_k\| + \phi(\|F(x_k)\|)\|s_k\| \leqslant 2L\|\delta_k\| + M\|s_k\| \leqslant (2L^2 + M)\|s_k\|, \tag{2.16}$$

where $M > 0$ is an upper bound of $\phi(\|F(x_k)\|)$ and $L > 0$ is a Lipschitz constant of F. Inequalities (2.15) and (2.16) imply that

$$\max\left\{\gamma_k^T s_k,\ \phi(\|F(x_k)\|)\|s_k\|^2\right\} \leqslant y_k^T s_k \leqslant (2L^2 + M)\|s_k\|^2. \tag{2.17}$$

Another way to develop quasi-Newton methods is to adopt the so-called cautious update rule proposed by Li and Fukushima [3]. The steps of the related BFGS algorithm is stated as follows.

Algorithm 2. (A descent cautious BFGS method).

Initial. Let $B_0 \in R^{n\times n}$ be symmetric and positive definite. Let $x_0 \in R^n$. Set $k = 0$.

Step 1. Determine d_k and λ_k by Procedures 1 and 2. Let $x_{k+1} = x_k + \lambda_k d_k$.

Step 2. Update B_k to get B_{k+1} by the modified BFGS formula

$$B_{k+1} = \begin{cases} B_k - \dfrac{B_k s_k s_k^T B_k}{s_k^T B_k s_k} + \dfrac{y_k y_k^T}{y_k^T s_k}, & \text{if } \dfrac{\gamma_k^T s_k}{\|s_k\|^2} \geqslant \phi(\|F(x_k)\|), \\ B_k, & \text{otherwise}, \end{cases} \tag{2.18}$$

where γ_k and ϕ are the same as those in Algorithm 1.

Step 3. Let $k := k+1$ and go to Step 1.

The only difference between Algorithms 1 and 2 is the update formula. The cautious BFGS method possesses similar properties of the modified BFGS method. For details, we refer to [3].

3 Global and Superlinear Convergence

In this section, we prove the global and superlinear convergence of Algorithm 1. The global convergence of Algorithm 2 can be obtained in a similar way. Without specification, we let $\{x_k\}$ and $\{B_k\}$ stand for the sequences of iterates and matrices generated by Algorithm 1, respectively. The following lemma is straightforward from Algorithm 1.

Lemma 3.1. *The sequence $\{\theta(x_k)\}$ is strictly decreasing. In addition, the following inequalities hold:*

$$\sum_{k=0}^{\infty} \|s_k\|^2 < \infty, \qquad \sum_{k=0}^{\infty} \|\lambda_k F(x_k)\|^2 < \infty. \tag{3.1}$$

We summarize the condition needed for the global convergence of Algorithm 1 as follows.

Assumption A.

(i) The level set

$$\Omega = \{x \in R^n \mid \theta(x) \leqslant \theta(x_0)\}$$

is bounded.

(ii) Function F is continuously differentiable on Ω, and $F'(x)$ is symmetric for every $x \in \Omega$.

It is clear that under condition (i) in Assumption A, sequence $\{x_k\} \subset \Omega$ is bounded.

We are going to establish a global convergence theorem of Algorithm 1 to show that under Assumption A, there exists an accumulation point of $\{x_k\}$ which is a stationary point of (1.5), namely,

$$\liminf_{k\to\infty} \|\nabla\theta(x_k)\| = 0. \tag{3.2}$$

It is easy to see from Lemma 3.1 that if $\limsup_{k\to\infty} \lambda_k > 0$, then $\liminf_{k\to\infty} \|F(x_k)\| = 0$ and, hence, (3.2) holds. So, we need only to show (3.2) for the case $\lim_{k\to\infty} \lambda_k = 0$. We do it by assuming

$$\liminf_{k\to\infty} \|\nabla\theta(x_k)\| > 0 \tag{3.3}$$

to deduce a contradiction.

Notice that (3.3) particularly implies that there is a constant $\eta > 0$ such that $\|F(x_k)\| \geqslant \eta$ for all k. It follows from (2.17) and the properties of ϕ that if (3.3) holds, then there are positive constants $c \leqslant C$ such that

$$c\|s_k\|^2 \leqslant y_k^T s_k \leqslant C\|s_k\|^2. \tag{3.4}$$

Therefore, we get the following lemma from (2.16), (3.4), and [4, Theorem 2.1].

Lemma 3.2. *If (3.3) holds, then there are positive constants β_i, $i = 1, 2, 3$, such that for any positive integer k, inequalities*

$$\|B_i s_i\| \leqslant \beta_1 \|s_i\|, \quad \beta_2 \|s_i\|^2 \leqslant s_i^T B_i s_i \leqslant \beta_3 \|s_i\|^2. \tag{3.5}$$

hold for at least $\lceil k/2 \rceil$ many $i \leqslant k$.

Inequalities (3.5) together with (2.13) imply that there are at least $\lceil k/2 \rceil$ many $i \leqslant k$ satisfying

$$\|q_i\| = \|B_i d_i\| \leqslant \beta_1 \|d_i\|, \quad \|d_i\| \leqslant \beta_2^{-1} \|q_i\|. \tag{3.6}$$

We now prove the global convergence of Algorithm 1.

Theorem 3.3. *Let Assumption A hold and $\{x_k\}$ be generated by Algorithm 1. Then (3.2) holds.*

Proof. We need only to show (3.2) for the case $\lim_{k\to\infty} \lambda_k = 0$. In this case, inequality (2.12) holds for all k sufficiently large. Suppose contrarily that (3.2) does not hold or, equivalently, (3.3) holds. Denote by K the set of indices i such that (3.5) holds. Then K is infinite. Since $\{x_k\} \subset \Omega$ is bounded, it is clear that sequences $\{q_k\}_{k\in K}$ and $\{d_k\}_{k\in K}$ are bounded. Let $K_1 \subseteq K$ and subsequences $\{x_k\}_{k\in K_1}$ and $\{d_k\}_{k\in K_1}$ converge to x^* and d^*, respectively. Then we have

$$\lim_{k\in K_1} q_k = \nabla\theta(x^*). \tag{3.7}$$

Dividing both sides of (2.12) by λ_k' and then taking limits as $k \to \infty$ with $k \in K_1$, we get

$$\nabla\theta(x^*)^T d^* \geqslant 0. \tag{3.8}$$

On the other hand, taking the inner product with d_k in (2.13), we get

$$0 = d_k^T B_k d_k + q_k^T d_k \geqslant \beta_2 \|d_k\|^2 + q_k^T d_k.$$

Taking limits in both sides as $k \to \infty$ with $k \in K_1$ yields

$$\nabla\theta(x^*)^T d^* \leqslant -\beta_2 \|d^*\|^2.$$

This together with (3.8) implies that $d^* = 0$. It then follows from (3.6) that $\lim_{k\in K_1} q_k = 0$, which together with (3.7) yields a contradiction with (3.3). The contradiction proves (3.2). □

Remark 3.1. In [5] the global convergence of Broyden's class of variable metric methods except for DFP was proved. The proof there depends on the convexity of the objective function. A similar result was obtained by Powell [6] when the BFGS method is applied to convex minimization problems. For nonconvex minimization problems, no theory exists to support the global convergence of the BFGS method. On the contrary, an example has been constructed [7] recently, which shows that the ordinary BFGS method with the Wolfe line search may fail to converge to a stationary point of a nonconvex unconstrained minimization.

On the other hand, a modified BFGS method was proposed by Li and Fukushima [2]. In the modified BFGS method, the iterative matrix B_k is always positive definite whatever line search is used as long as B_0 is positive definite. Moreover, a liminf result was obtained for nonconvex unconstrained minimization. Besides, another modified BFGS method called the cautious BFGS method was proposed by Li and Fukushima [3]. The cautious BFGS method also possesses global convergence in the sense $\liminf_{k\to\infty} \|\nabla f(x_k)\| = 0$ when it is applied to $\min f(x)$. In both papers, the results were obtained without the requirement of nonsingular Hessian. These two papers show the possibility to improve the unconstrained minimization result by Byrd, Nocedal, and Yuan [5] and Powell [6].

This paper adopts a similar updating technique as used in [8] and [9]. Consequently, we established Theorem 3.3, which shows that the iterative sequence has an accumulation point which is a stationary point of problem $\min \theta(x) = \dfrac{1}{2}\|F(x)\|^2$. It may not be a solution of the nonlinear equation (1.1) if the Jacobian is singular at that point. The next theorem shows a strong convergence property of Algorithm 1.

Theorem 3.4. *Let Assumption A hold. Suppose that the sequence $\{x_k\}$ generated by Algorithm 1 has a subsequence converging to a stationary x^* at which $F'(x^*)$ is nonsingular. Then x^* is a solution of (1.1). Moreover, the whole sequence $\{x_k\}$ converges to x^*.*

Proof. Since x^* satisfies $\nabla\theta(x^*) = F'(x^*)F(x^*) = 0$, we obviously have $F(x^*) = 0$ if $F'(x^*)$ is nonsingular. Since $\{\theta(x_k)\}$ converges, every accumulation point of $\{x_k\}$ is a solution of (1.1). By the nonsingularity of $F'(x^*)$ again, x^* is an isolated limit point of $\{x_k\}$. However, we have from (3.1) that $x_{k+1} - x_k \to 0$ as $k \to \infty$. Therefore, the whole sequence $\{x_k\}$ converges to x^*. □

In a way similar to the proof of [2, Theorem 3.8], it is not difficult to prove the superlinear convergence of Algorithm 1. We state the theorem as follows but omit the proof.

Theorem 3.5. *Let the conditions of Theorem 3.4 hold. Suppose further that F' is Lipschitz continuous. Then $\{x_k\}$ is superlinearly convergent.*

Similar to the above argument, we can establish the global and superlinear convergence of Algorithm 2. We state the results as follows but omit the proof.

Theorem 3.6. *Let Assumption A hold and $\{x_k\}$ be generated by Algorithm 2. Then (3.2) holds. If the sequence $\{x_k\}$ has a subsequence converging to a stationary x^* at which $F'(x^*)$ is* *nonsingular, then x^* is a solution of (1.1). Moreover, the whole sequence $\{x_k\}$ converges to x^*. If we further suppose that F' is Lipschitz continuous, then $\{x_k\}$ is superlinearly convergent.*

4 Numerical Results

In this section, we test the proposed descent BFGS methods on nonlinear equation problems obtained from [1, 10] and the unconstrained optimization problems obtained from the website

ftp://ftp.mathworks.com/pub/contrib/v4/optim/uncprobs/.

We call Algorithms 1 and 2 the DBFGS (descent BFGS) method and the CBFGS (cautious BFGS) method, respectively, and call the BFGS method based on the Gauss-Newton approach and the nondescent line search [1] the NBFGS (nondescent BFGS) method. Then we compare their performance.

The parameters are specified as follows. We take $\rho = 0.1$ and $\sigma_1 = \sigma_2 = 10^{-5}$. in (2.9). The initial quasi-Newton matrices are set to be $B_0 = A$ [1] for nonlinear equation problems and $B_0 = I$ for unconstrained optimization problems. The function ϕ is determined by

$$\phi(t) = \begin{cases} Ct^2, & \text{if } t \leqslant 1, \\ Ct^{0.1}, & \text{otherwise,} \end{cases}$$

where $C = 10^{-5}$. For the NBFGS method, we update B_k by the BFGS formula [1] if $y_k^T s_k \geqslant 10^{-5}$. Otherwise, we let $B_{k+1} = B_k$. We stop the iteration process if $\|F(x_k)\| \leqslant 10^{-4}$.

The tested results are listed in Tables 1 and 2. Table 3 gives the average performance of the three methods for solving nonlinear equation problems. The columns of the tables have the following meaning:

Dim: the dimension of the problem.

Method: the name of the algorithm.

Init: the initial point, namely, integer l in Table 1 meaning $x_0 = (l, l, \cdots, l)^T$.

Iter: the total number of iterations.

Inner: for the NBFGS method, the number of iterations at which $y_k^T s_k \geqslant 10^{-5}$ is satisfied;

for the DBFGS method and the CBFGS method, the maximum number of inner iterations

to generate the descent direction d_k.

Numf: the number of the function evaluations.

Fnorm: the final value of $\|F(x_k)\|$.

Table 1　Test results for nonlinear equation problems $B_0 = A$

Dim	Method	Init	Iter	Inner	Numf	Fnorm	Dim	Method	Init	Iter	Inner	Numf	Fnorm
10	BFGS	0	6	0	19	2.9e−06	50	BFGS	0	25	0	76	3.2e−05
	DBFGS		6	1	19	2.9e−06		DBFGS		25	1	76	3.9e−05
	CBFGS		6	1	19	2.9e−06		CBFGS		25	1	76	3.2e−05

Continued

Dim	Method	Init	Iter	Inner	Numf	Fnorm	Dim	Method	Init	Iter	Inner	Numf	Fnorm
10	BFGS	1	11	0	35	9.1e−05	50	BFGS	1	37	0	114	3.3e−05
	DBFGS		10	2	42	3.5e−05		DBFGS		36	2	120	1.6e−05
	CBFGS		10	2	42	3.5e−05		CBFGS		36	2	120	1.6e−05
	BFGS	−1	12	0	38	2.0e−05		BFGS	−1	37	0	114	1.8e−05
	DBFGS		10	2	36	1.1e−05		DBFGS		36	2	120	1.0e−05
	CBFGS		10	2	36	1.1e−05		CBFGS		36	2	120	1.0e−05
	BFGS	10	13	0	41	4.1e−05		BFGS	10	38	0	117	6.0e−05
	DBFGS		13	2	52	2.6e−05		DBFGS		38	2	133	4.9e−05
	CBFGS		13	2	52	2.6e−05		CBFGS		38	2	133	4.8e−05
	BFGS	−10	13	0	41	3.9e−04		BFGS	−10	38	0	117	6.1e−05
	DBFGS		13	2	52	2.5e−05		DBFGS		38	2	133	5.3e−05
	CBFGS		13	2	52	2.5e−05		CBFGS		38	2	133	4.9e−05
	BFGS	10^2	14	0	44	1.4e−05		BFGS	10^2	41	0	127	4.5e−05
	DBFGS		12	2	48	1.3e−05		BFGS		40	2	132	1.5e−05
	CBFGS		12	2	48	1.3e−05		CBFGS		40	2	132	1.5e−05
	BFGS	-10^2	14	0	44	2.0e−05		BFGS	-10^2	41	0	127	5.0e−05
	DBFGS		12	2	48	1.1e−05		DBFGS		38	2	126	8.6e−05
	CBFGS		12	2	48	1.1e−05		CBFGS		38	2	126	8.6e−05
	BFGS	10^3	16	0	50	1.8e−06		BFGS	10^3	44	0	136	4.1e−05
	DBFGS		13	2	51	3.9e−06		DBFGS		40	2	137	9.9e−05
	CBFGS		13	2	51	3.9e−06		CBFGS		40	2	137	9.9e−05
	BFGS	-10^3	16	0	50	7.5e−07		BFGS	-10^3	44	0	136	3.4e−05
	DBFGS		13	2	51	5.0e−06		DBFGS		40	2	137	7.9e−05
	CBFGS		13	2	51	5.0e−06		CBFGS		40	2	137	7.9e−05
	BFGS	10^4	16	0	50	1.1e−05		BFGS	10^4	49	0	152	5.4e−05
	DBFGS		14	2	54	5.7e−06		DBFGS		44	2	160	4.9e−05
	CBFGS		14	2	54	5.7e−06		CBFGS		44	2	160	4.9e−05
	BFGS	-10^4	16	0	50	8.8e−06		BFGS	-10^4	49	0	144	9.6e−05
	DBFGS		14	2	54	5.1e−06		DBFGS		44	2	152	6.9e−05
	CBFGS		14	2	54	5.1e−06		CBFGS		44	2	152	6.9e−05
100	BFGS	0	50	0	151	5.7e−06	200	BFGS	0	501	0	2181	8.9e−05
	DBFGS		50	1	151	5.7e−06		DBFGS		155	3	1142	9.5e−05
	CBFGS		50	1	151	5.7e−06		CBFGS		151	3	1104	9.5e−05
	BFGS	1	63	0	192	1.0e−04		BFGS	1	2191	1	8661	1.0e−04
	DBFGS		62	2	198	6.8e−05		DBFGS		116	3	383	2.2e−05
	CBFGS		62	2	198	6.8e−05		CBFGS		116	3	383	2.2e−05
	BFGS	−1	63	0	192	5.8e−05		BFGS	−1	1971	1	7835	1.0e−04
	DBFGS		62	2	198	5.4e−05		DBFGS		116	3	382	1.8e−05
	CBFGS		62	2	198	5.4e−05		CBFGS		116	3	382	1.8e−05
	BFGS	10	66	0	203	2.0e−05		BFGS	10	4547	1	18085	1.0e−04
	DBFGS		66	2	220	3.7e−05		DBFGS		119	3	405	4.1e−05
	CBFGS		66	2	217	9.0e−05		CBFGS		119	3	405	4.0e−05
	BFGS	−10	66	0	203	1.2e−05		BFGS	−10	4070	1	16177	1.0e−04
	DBFGS		66	2	221	1.8e−05		DBFGS		119	3	406	2.3e−06
	CBFGS		76	2	254	6.9e−05		CBFGS		119	3	407	2.3e−06

Continued

Dim	Method	Init	Iter	Inner	Numf	Fnorm	Dim	Method	Init	Iter	Inner	Numf	Fnorm
100	BFGS	10^2	69	0	213	6.7e−05	200	BFGS	10^2	6095	1	24277	1.0e−04
	DBFGS		67	2	220	5.7e−05		DBFGS		125	4	447	6.9e−05
	CBFGS		66	2	217	9.0e−05		CBFGS		120	4	424	9.7e−05
	BFGS	-10^2	69	0	213	6.2e−05		BFGS	-10^2	6049	1	24093	1.0e−04
	DBFGS		65	2	211	9.8e−05		DBFGS		125	3	433	6.9e−05
	CBFGS		65	2	211	8.5e−05		CBFGS		124	3	439	5.6e−05
	BFGS	10^3	72	0	221	9.2e−05		BFGS	10^3	8330	1	33217	1.0e−04
	DBFGS		70	2	236	5.8e−05		DBFGS		150	3	543	7.3e−05
	CBFGS		67	2	222	2.3e−05		CBFGS		540	3	3704	8.5e−05
	BFGS	-10^3	71	0	220	4.3e−05		BFGS	-10^3	8296	1	33081	1.0e−04
	DBFGS		69	2	232	8.5e−05		DBFGS		145	3	511	9.8e−05
	CBFGS		67	2	222	2.7e−05		CBFGS		533	3	3662	8.3e−05
	BFGS	10^4	73	0	227	6.7e−05		BFGS	10^4	9962	1	39745	1.0e−04
	DBFGS		72	2	246	4.9e−05		DBFGS		185	3	704	8.9e−05
	CBFGS		69	2	233	8.3e−05		CBFGS		1299	2	10043	5.0e−05
	BFGS	-10^4	73	0	227	6.3e−05		BFGS	-10^4	9915	1	39557	1.0e−04
	DBFGS		72	2	242	4.3e−05		DBFGS		590	3	3123	8.5e−05
	CBFGS		69	2	232	5.6e−05		CBFGS		1268	2	9837	4.8e−05

Table 2 Test results for unconstrained optimization problems $B_0 = I$

Method	Prob	Dim	Iter	Inner	Numf	Fnorm	Method	Prob	Dim	Iter	Inner	Numf	Fnorm
BFGS	rose	2	103	0	415	6.3e−005	BFGS	froth	2	—	—	—	—
DBFGS			—	—	—	—	DBFGS			—	—	—	—
CBFGS			668	7	6301	9.1e−05	CBFGS			282	7	3155	9.1e−06
BFGS	beale	2	347	0	1331	9.4e−05	BFGS	jensam	—	—	—	—	—
DBFGS			—	—	—	—	DBFGS			—	—	—	—
CBFGS			155	4	1330	2.6e−05	CBFGS			12	5	65	8.3e−05
BFGS	helix	3	279	0	1205	8.9e−05	BFGS	gulf	3	1	1	4	5.6e−086
DBFGS			—	—	—	—	DBFGS			1	1	4	1.9e−10
CBFGS			156	6	1413	3.1e−05	CBFGS			1	1	4	1.0e−10
BFGS	gauss	3	2	0	8	5.9e−006	BFGS	meyer	3	—	—	—	—
DBFGS			2	2	10	6.0e−06	DBFGS			1	4	14	4.2e−07
CBFGS			2	2	10	6.0e−06	CBFGS			1	4	14	4.2e−07
BFGS	sing	4	218	1	875	8.6e−05	BFGS	wood	4	—	—	—	—
DBFGS			214	9	1847	9.9e−05	DBFGS			—	—	—	—
CBFGS			97	6	650	9.7e−05	CBFGS			617	8	8971	6.9e−05
BFGS	kowosb	5	—	—	—	—	BFGS	biggs	6	59	0	211	4.4e−05
DBFGS			661	4	7031	1.0e−04	DBFGS			101	5	589	6.9e−05
CBFGS			661	4	7028	1.0e−04	CBFGS			101	5	589	6.9e−05
BFGS	osb2	11	225	1	775	4.0e−05	BFGS	watson	2	24	0	90	2.8e−06
DBFGS			—	—	—	—	DBFGS			18	5	124	1.1e−05
CBFGS			—	—	—	—	CBFGS			18	5	124	1.1e−05
BFGS	trid	10	152	0	609	1.8e−05	BFGS	singx	40	—	—	—	—
DBFGS			115	5	682	6.2e−05	DBFGS			—	—	—	—
CBFGS			115	5	682	6.2e−05	CBFGS			741	6	6375	1.0e−04

Continued

Method	Prob	Dim	Iter	Inner	Numf	Fnorm	Method	Prob	Dim	Iter	Inner	Numf	Fnorm
BFGS	pen1	10	248	0	1048	2.9e−05	BFGS	pen2	10	320	0	1499	5.0e−05
DBFGS			148	7	1235	4.5e−05	DBFGS			—	—	—	—
CBFGS			148	7	1235	4.5e−05	CBFGS			—	—	—	—
BFGS	bv	10	30	0	104	1.0e−05	BFGS	ie	10	5	0	17	2.6e−05
DBFGS			31	3	135	1.8e−05	DBFGS			4	2	18	2.6e−05
CBFGS			31	3	135	1.8e−05	CBFGS			4	2	18	2.6e−05
BFGS	lin	10	1	0	4	1.0e−13	BFGS	lin1	10	2	0	17	7.7e−06
DBFGS			1	1	4	8.9e−16	DBFGS			2	11	28	1.1e−10
CBFGS			1	1	4	8.9e−16	CBFGC			2	11	28	1.1e−10
BFGS	lin0	10	2	0	17	7.7e−07							
DBFGS			2	11	30	1.3e−11							
CBFGS			2	11	30	1.3e−11							

Table 3 Average performance for nonlinear equation problems

Dim	Method	Iter	Inner	Numf	Dim	Method	Iter	Inner	Numf
10	BFGS	13.4	0	42	50	BFGS	46	0	123.6
	DBFGS	11.8	1.9	46.1		DBFGS	38	1.9	129.6
	CBFGS	11.8	1.9	46.1		CBFGS	38	1.9	129.6
100	BFGS	66.8	0	205.6	200	BFGS	5629.7	0.9	22446
	DBFGS	65.5	1.9	215.9		DBFGS	176.8	3.1	770.8
	CBFGS	65.2	1.9	214.1		CBFGS	409.5	2.6	2799.1

All the three methods terminate at solutions of nonlinear equation problems for all tested starting points. However, for the 33 unconstrained optimization problems, all the three methods fail to converge to a solution for at least 10 problems. The numbers of problems for which the NBFGS method, the DBFGS method, and the CBFGS method fail to converge are 16, 19, and 12, respectively.

The numerical results show that for low dimensional problems, the performance of these three methods is not different very much. For most of the test problems, the DBFGS method and the CBFGS method perform better than the NBFGS method in the iteration number, but worse in the number of the function evaluation. However, for high dimensional problems ($n = 200$ in Tables 1 and 3), both the DBFGS and the CBFGS methods perform much better than the NBFGS method in the iteration number as well as the number of the function evaluation. The maximum numbers of the inner iteration to generate a descent direction of a DBFGS method are generally very small. We also note that the performance of the DBFGS and CBFGS methods is almost the same if the both methods terminate regularly. For unconstrained optimization problems, the DBFGS method fails more frequently than the CBFGS method does.

In summary, the presented numerical results reveal that the DBFGS and CBFGS methods, compared with the NBFGS method, have potential advantages when applied to solve symmetric nonlinear equation whose function is not difficult to compute. In Tables 1—3, we simply denote the NBFGS method as the BFGS method.

References

[1] D. H. Li and M. Fukushima, A globally and superlinearly convergent Gauss-Newton-based BFGS method for symmetric nonlinear equations, *SIAM Journal on Numerical Analysis*, 37 (1999), 152-172.

[2] D. H. Li and M. Fukushima, A modified BFGS method and its global convergence in nonconvex minimization, *Journal of Compututational and Applied Mathematics*, 129 (2001), 15-35.

[3] D. H. Li and M. Fukushima, On the global convergence of the BFGS method for nonconvex unconstrained optimization problems, *SIAM Journal on Optimization*, 11 (2001), 1054-1064.

[4] R. H. Byrd and J. Nocedal, A tool for the analysis of quasi-Newton methods with application to unconstrained minimization, *SIAM Journal on Numerical Analysis*, 26 (1989), 727-739.

[5] R. H. Byrd, J. Nocedal, and Y.-X. Yuan, Global convergence of a class of quasi-Newton methods on convex problems, *SIAM Journal on Numerical Analysis*, 24 (1987), 1171-1190.

[6] M. J. D. Powell, Some global convergence properties of a variable metric algorithm for minimization without exact line searches, In: *Nonlinear Programming, SIAM-AMS Proceedings*, Vol. IX, R. W. Cottle and C. E. Lemke, eds., AMS, Providence, RI, 1976, pp. 55-92.

[7] Y. Dai, *Convergence Properties of the BFGS Algorithm*, Technical report, The State Key Laboratory of Scientific and Engineering Computing, Chinese Academy of Sciences, Beijing, China, 2001.

[8] J. E. Dennis, Jr., and J. J. Moré, A characterization of superlinear convergence and its application to quasi-Newton methods, *Mathematics of Computation*, 28 (1974), 1171-1190.

[9] J. E. Dennis, Jr., and J. J. Moré, Quasi-Newton methods, motivation and theory, *SIAM Review*, 19 (1977), 46-89.

[10] J. M. Ortega and W. C. Rheinboldt, *Iterative Solution of Nonlinear Equations in Several Variables*, Academic Press, New York, 1970.

A New Exceptional Family of Elements for a Variational Inequality Problem on Hilbert Space*†

S.Z. Zhou (周叔子)† M.R. Bai (白敏茹)

Abstract This paper introduces a new exceptional family for a variational inequality problem in the Hilbert space and related existence theorems for the solution to the variational inequality problem.

Keywords Variational inequality, exceptional family, existence of solution.

1 Introduction

Let V be a Hilbert space whose inner product and norm are denoted by $\langle\cdot,\cdot\rangle$ and $\|\cdot\|$, respectively. Let K be an unbounded closed convex subset of V. Given a mapping $F: V\to V$, we consider the following variational inequality: find $x^*\in K$ such that

$$\langle F(x^*), x-x^*\rangle \geqslant 0,\quad \forall x\in K. \tag{1.1}$$

Inequality (1.1) arises in a lot of practical problems of science, engineering and economics, and is studied by many authors, see [1–5] and the references therein. The solvability of (1.1) is a basic problem and has been studied by several methods. In the case of $V=R^n$ and $K=R^n_+$, i.e., Smith [6] and Isac et al. [7] have proposed, respectively, the concepts of exceptional sequence and exceptional family to study the solvability of it, which opens a new research direction for the solvability of (1.1). See [8–12] and the references therein.

Isac and Zhao [10] have first extended the concept of exceptional family in R^n to general Hilbert space V and proved the basic theorem: If F is without exceptional family of elements with respect to K, then (1.1) is solvable provided that F and $I-F$ are completely continuous, which is a generalization of the existing basic theorem in the case of $V=R^n$.

* 发表于：Applied Mathematics Letters, Vol. 17, 2004, pp. 423-428.
† Institute of Applied Mathematics, Hunan University, Changsha, Hunan 410082, China.

In this paper, we will propose a new definition of exceptional family in the general Hilbert space V, and prove the corresponding basic theorem, an existence theorem, and several corollaries.

2 Exceptional Family

Without loss of generality, we assume $x = 0 \in K$ from now on.

For a given $x \in K$, the normal cone of K at x is defined as follows (see [13, 14]):

$$N_K(x) = \{z \in V : \langle z, y - x\rangle \leqslant 0, \quad \forall y \in K\}.$$

It is easy to see that $\alpha N_K(x^r) = N_K(x^r)$ for any $\alpha > 0$.

The basic definition of exceptional family, which is given in [10], is the following.

Definition A *$\{x^r\}_{r>0} \subset V$ is called an exceptional family of elements for $F(x) = x - T(x)$ defined on V with respect to K if*

(1) $\|x^r\| \to \infty$ as $r \to \infty$;

(2) for any $r > 0$ there exists $\mu_r > 1$ such that $\mu_r x^r \in K$ and $T(x^r) - \mu_r x^r \in N_K(\mu_r x^r)$.

Now we propose another definition of exceptional family.

Definition 2.1. *$\{x^r\}_{r>0} \subset K$ is called an exceptional family of (1.1) if*

(1) $\|x^r\| \to \infty$ as $r \to \infty$;

(2) for any $r > 0$ there exists $t_r \in (0,1)$ such that

$$-t_r x^r - (1 - t_r)F(x^r) \in N_K(x^r). \tag{2.1}$$

The following lemma formulates an important property of exceptional family.

Lemma 2.2. *Assume $\{x^r\}$ is an exceptional family of (1.1). Then for any $x^r \neq 0$ holds*

$$\langle F(x^r), x^r\rangle < 0.$$

Proof. By the definition of $N_K(x)$ and (2.1) we have

$$\langle -t_r x^r - (1 - t_r)F(x^r), y - x^r\rangle \leqslant 0, \ \ \forall y \in K, \ r > 0,$$

i.e.,

$$\langle (1 - t_r)F(x^r), y - x^r\rangle \geqslant -t_r\langle x^r, y - x^r\rangle, \ \ \forall y \in K, \ r > 0,$$

in which we take $y = 0$ and obtain

$$\langle F(x^r), x^r\rangle \leqslant -(1 - t_r)^{-1} t_r \|x^r\|^2 < 0, \ \ \forall x^r \neq 0.$$

The proof is complete. □

Assume $\{x^r\}$ is an exceptional family defined by Definition A. Then there exists $\mu_r > 1$ such that $\mu_r x^r \in K$ and

$$x^r - F(x^r) - \mu_r x^r \in N_K(\mu_r x^r),$$

i.e.,

$$-(1-\mu_r^{-1})x^r - \mu_r^{-1}F(x^r) \in \mu_r^{-1}N_K(\mu_r x^r). \tag{2.2}$$

If $N_K(\mu_r x^r) = \mu_r N_K(x^r)$ then (2.2) can be rewritten as

$$-t_r x^r - (1-t_r)F(x^r) \in N_K(x^r), \tag{2.3}$$

where $t_r = 1-\mu_r^{-1}$. Equation (2.3) means $\{x^r\}$ is an exceptional family of (1.1). Hence we have the following lemma.

Lemma 2.3. *Assume $\mu K \subset K$ and $N_K(\mu x) = \mu N_K(x)$ for any $\mu > 1$, $x \in K$. Then Definition A is equivalent to Definition 2.1.*

Remark 2.1. Obviously, $V = R^n$, $K = R^n_+$ satisfies the conditions of Lemma 2.3.

Now consider a special case: $V = R^n$ and K is defined as follows:

$$K = \{x \in R^n : g_i(x) \leqslant 0, i = 1, \cdots, I; h_j(x) = 0, j = 1, \cdots, J\}, \tag{2.4}$$

where $g_i : R^n \to R$ is a continuously differentiable convex function and h_j is an affine function. Furthermore, we assume K satisfies Slater condition, i.e., there is a point $x^0 \in K$ such that $g_i(x^0) < 0$, $i = 1, \cdots, I$. Zhao et al. [9]has defined exceptional family as follows.

Definition B. *$\{x^r\}_{r>0} \subset K$ is called an exceptional family of (1.1) if $\|x^r\| \to \infty$ $(r \to \infty)$ and for any $r > 0$ there exist $\alpha_r > 0$, $\lambda_r \in R^I_+$ and $\mu_r \in R^J$ such that*

$$\begin{cases} F(x^r) = -\alpha_r x^r - \dfrac{1}{2}\left[\displaystyle\sum_{i=1}^{I}\lambda_{r_i}\nabla g_i(x^r) + \sum_{j=1}^{J}\mu_{r_j}\nabla h_j(x^r)\right], \\ \displaystyle\sum_{i=1}^{I}\lambda_{r_i} g_i(x^r) = 0. \end{cases} \tag{2.5}$$

It is well known that for K defined by (2.4) we have that (see [13, 14])

$$N_K(x) = \left\{ z = \sum_{i=1}^{I}\lambda_i \nabla g_i(x) + \sum_{j=1}^{J}\mu_j \nabla h_j(x) : \lambda \in R^I_+, \mu \in R^J, \sum_{i=1}^{I}\lambda_i g_i(x) = 0 \right\}. \tag{2.6}$$

It follows from (2.5) and (2.6) that

$$-2\alpha_r x^r - 2F(x^r) \in N_K(x^r). \tag{2.7}$$

With $\alpha = 2(\alpha_r + 1)$ and $t_r = 2\alpha^{-1}\alpha_r$, we have $t_r \in (0, 1)$ and $1 - t_r = 2\alpha^{-1}$. Hence, (2.7) can be rewritten as

$$-t_r x^r - (1-t_r)F(x^r) \in \alpha^{-1}N_K(x^r) = N_K(x^r).$$

Therefore, we have proved the following result.

Lemma 2.4. *If K is given by (2.4), then Definition 2.1 is equivalent to Definition B.*

Remark 2.2. Equation (2.1) means x^r is a solution of a variational inequality. Hence, the existence of exceptional family is a problem as difficult as the existence of (1.1). However, we discuss not the existence of exceptional family, but the nonexistence of exceptional family which implies the existence of the solution of (1.1). Refer to Theorems 3.4, 3.6, and the corollaries below.

3 Existence of Solution for (1.1)

At first we formulate two lemmas on Leray-Schauder degree $\deg(f, D, y)$ (see [15, 16]), where D is a bounded open subset of V.

Lemma 3.1 (Poincaré-Bohl). *Assume $H : \bar{D} \times [0,1] \to V$ is a completely continuous field, $h_t(x) = x - H(x,t)$. If $y \bar{\in} h_t(\partial D)$ then $\deg(h_t, D, y)$ is a constant for $0 \leqslant t \leqslant 1$.*

Lemma 3.2 (Kronecker). *If $\deg(f, D, y) \neq 0$, then equation $f(x) = y$ has at least one solution in D.*

We define a homotopy related to (1.1) as follows:

$$H(x,t) = P_K((1-t)(x - F(x))), \quad 0 \leqslant t \leqslant 1. \tag{3.1}$$

Let $\varphi(x) = x - P_K(x - F(x))$. Then $H(x,1) = 0$, $H(x,0) = x - \varphi(x)$. It is well known that equation $\varphi(x) = 0$ is equivalent to (1.1).

Lemma 3.3. *Let $T(x) = x - F(x)$. If $T(x)$ is completely continuous then so is $H(x,t)$.*

Proof. Assume $(x^{(m)}, t^{(m)}) \to (x^{(0)}, t^{(0)})$ weakly. Then we have by complete continuity of $T(x)$ that

$$\|T(x^{(m)}) - T(x^{(0)})\| \to 0, \quad (m \to \infty).$$

On the other hand, it follows from the nonexpansivity of P_K that

$$\begin{aligned}&\| P_K[(1-t^{(m)})T(x^{(m)})] - P_K[(1-t^{(0)})T(x^{(0)})]\| \\ \leqslant &\|(1-t^{(m)})T(x^{(m)}) - (1-t^{(0)})T(x^{(0)})\|,\end{aligned}$$

Hence we obtain

$$\begin{aligned}&\| H(x^{(m)}, t^{(m)}) - H(x^{(0)}, t^{(0)})\| \\ \leqslant &|(1-t^{(m)})| \cdot \|T(x^{(m)}) - T(x^{(0)})\| + |(t^{(m)} - t^{(0)})| \cdot \|T(x^{(0)})\| \to 0 \ \ (m \to \infty),\end{aligned}$$

which means $H(x,t)$ is complete continuous. □

Now we can prove the basic theorem as follows.

Theorem 3.4. *Assume $F(x)$ and $T(x) = x - F(x)$ are complete continuous. Then (1.1) has either*

(a) at least one solution; or

(b) at least one exceptional family.

Proof. It is sufficient to prove that if (a) is not true, then (b) is true. Assume (a) is not true. Then we declare that for any $r > 0$ and $D_r = \{x \in V : \|x\| < r\}$ there exist $z^r \in \partial D_r$ and $t_r \in [0,1]$ such that $z^r - H(z^r, t_r) = 0$. Otherwise, there exists $r > 0$ such that

$$0 \bar{\in} \{h_t(x) = x - H(x,t) : x \in \partial D_r, t \in [0,1]\}. \tag{3.2}$$

It follows from Lemma 3.1, 3.3 and (3.2) that $\deg(x - H(x,t), D_r, 0)$ is constant for $t \in [0,1]$ and

$$\deg(\varphi, D_r, 0) = \deg(x, D_r, 0) = 1,$$

which implies by Lemma 3.2 that equation $\varphi(x) = 0$ has a solution in D_r, i.e., (1.1) has a solution, which contradicts with the assumption of the beginning of the proof. Therefore, we have

$$z^r - H(z^r, t_r) = z^r - P_K((1-t_r)(z^r - F(z^r)) = 0, \quad r > 0. \tag{3.3}$$

Now we prove $\{z^r\}$ is an exceptional family of (1.1). Since $z^r \in \partial D_r$, we know $\|x^r\| \to \infty (r \to \infty)$. If $t_r = 0$ in (3.3), then $z^r = P_K(z^r - F(z^r))$ and z^r is a solution of (1.1). Contradiction again! If $t_r = 1$ in (3.3), then $z^r = P_K(0) = 0$, which contradicts with $z^r \in \partial D_r$. Hence, $t_r \in (0,1)$. At last, by (3.3) we have $z^r = P_K((1-t_r)(z^r - F(z^r)) \in K$ which means (see [1]):

$$\langle (1-t_r)(z^r - F(z^r)) - z^r, y - z^r \rangle \leqslant 0, \quad \forall y \in K,$$

i.e.,

$$-t_r z^r - (1-t_r)F(z^r) \in N_K(z^r).$$

The proof is complete. □

Remark 3.1. By Lemma 2.3 and Lemma 2.4 we know Theorem 3.4 is a generalization of the basic theorem in [7, 9].

Corollary 3.5. *Assume the conditions of Theorem 3.4 are fulfilled. If (1.1) has no exceptional family, then (1.1) has at least one solution.*

Based on Corollary 3.5, the following theorem gives a condition for the solvability of (1.1).

Theorem 3.6. *Assume $F(x)$ and $T(x) = x - F(x)$ are completely continuous. If for any sequence $\{x^r\}_{r>0}$ with $\|x^r\| \to \infty (r \to \infty)$ there exists $x^r \neq 0$ such that*

$$\langle F(x^r), x^r \rangle \geqslant 0, \tag{3.4}$$

then (1.1) has at least one solution.

Proof. If (1.1) has an exceptional family $\{x^r\}$, then it follows from Lemma 2.2 that $\langle F(x^r), x^r \rangle < 0$ for any $r > 0$ with $x_r \neq 0$, which contradicts (3.4). Therefore, (3.4) implies that (1.1) has no exceptional family and has at least one solution by Corollary 3.5. □

Remark 3.2. Theorem 3.6 is a generalization of Theorem 2.1 in [11].

We say that F satisfies the Karamadian condition on K if there exist convex compact $D \subset K$ and $y \in D$ such that $\langle F(x), x-y\rangle \geqslant 0$, $\forall x \in K \setminus D$ (see [8] for example).

Corollary 3.7. *Assume that $F(x)$ and $T(x)$ are completely continuous and that $F(x)$ satisfies the Karamadian condition. Then (1.1) has at least one solution.*

Proof. Let $\tilde{x} = x - y$, $\tilde{F}(\tilde{x}) = F(\tilde{x}+y)$, $\tilde{K} = K - y$. Then (1.1) is equivalent to the following problem:

$$\tilde{x} \in \tilde{K}, \quad \langle \tilde{F}(\tilde{x}), \tilde{y} - \tilde{x}\rangle \geqslant 0, \quad \forall \tilde{y} \in \tilde{K}. \tag{3.5}$$

It is obvious that $\tilde{F}$ satisfies the Karamadian condition on $\tilde{K}$ and (3.5) satisfies the conditions of Theorem 3.2. Hence, (3.5) has at least one solution, and so does (1.1) . □

We say that F is p-order coercive on K if there exists $p \in (-\infty, 1]$ and $y \in K$ such that

$$\lim_{x\in K,\ \|x\|\to\infty} \|x\|^{-p}\langle F(x), x-y\rangle = +\infty. \tag{3.6}$$

Corollary 3.8. *Assume that $F(x)$ and $T(x)$ are completely continuous and that $F(x)$ is p-order coercive. Then (1.1) has at least one solution.*

Proof. It is similar to that of Corollary 3.7. □

References

[1] D. Kinderlehrer and G. Stampacchia, *An Introduction to Variational Inequualities and Their Applications*, Academic Press, New York, 1980.

[2] R. Glowinski, J.L. Lions and R. Tremolieres, *Numerical Analysis of Variational Inequalities*, North-Holland, Amsterdam, 1981.

[3] R.W. Cottle, J.S. Pang and R.E. Stone, *The Linear Complementarity Problems*, Academic Press, New York, 1992.

[4] G. Isac, *Complementarity Problems*, Springer-Verlag, Berlin, 1992.

[5] R.T. Harker and J.S. Pang, Finite-dimensional variational inequality and nonlinear complementarity problems: A survey of theory, algorithms and applications, *Mathematical Programm*, 48 (1990), 161-220.

[6] T.E. Smith, A solution condition for complementarity problems, with an application to spatial price equilibrium, *Applied Mathematics and Computation*, 15 (1984), 61-69.

[7] G. Isac, V. Bulavaski and V. Kalashnikov, Exceptional families, topological degree and complementarity problems, *Journal of Global Optimization*, 10 (1997), 207-225.

[8] G. Isac and W.T. Obuchowska, Functions without exceptional family of elements and complementarity problems, *Journal of Optimization, Theory and Applicaitons*, 99 (1998), 147-163.

[9] Y.B. Zhao, J.Y. Han and H.D. Qi, Exceptional families and existence theorems for variational inequality problems, *Journal of Optimization, Theory and Applicaitons*, 101 (1999), 475-495.

[10] G. Isac and Y.B. Zhao, Exceptional family of elements and the solvability of variatioal inequalities for unboounded sets in infinite dimensional Hilbert spaces, *Journal of Mathematical Analysis and Applications*, 246 (2000), 544-556.

[11] L.P. Zhang, J.Y. Han and D.C. Xu, The existence of solution for variational inequality problems, *Science in China, Series A*, 44 (2001), 212-219.

[12] W.T. Obuchowska, Exceptional families and existence results for nonlinear complementarity problem, *Journal of Global Optimization*, 19 (2001), 183-198.

[13] J. Outrata, M. Kocvara and J. Eowe, *Nonsmooth Approach to Optimization Problems with Equilibrium Constraints*, Kluwer Acad. Publishier, Dordrecht, 1998.

[14] F.F. Clarke, *Optimization and Nonsmooth Analysis*, John Wiley & Sons, New York, 1983.

[15] E.H. Rothe, *Introduction to Various Aspects of Degree Theory in Banach Spaces*, AMS, Providence, 1986.

[16] D.J. Guo, *Nonlinear Functional Analysis*, Shandon Publisher of Science & Technology, Jinan, 1985.

A Smoothing Projected Newton-Type Method for Semismooth Equations with Bound Constraints*

X.J. Tong (童小娇)† S.Z. Zhou (周叔子)‡

Abstract This paper develops a smoothing algorithm to solve a system of constrained equations. Compared with the traditional methods, the new method does not need the continuous differentiability assumption for the corresponding merit function. By using some perturbing technique and suitable strategy of the chosen search direction, we find that the new method not only keeps the strict positivity of the smoothing variable at any non-stationary point of the corresponding optimization problem, but also enjoys global convergence and locally superliner convergence. The former character is the key requirement for smoothing methods. Some numerical examples arising from semi-infinite programming (SIP) show that the new algorithm is promising.

Keywords Constrained nonsmooth equations, smoothing method, global convergence, locally superlinear convergence.

1 Introduction

We consider the following system of nonlinear equations:

$$F(x) = 0, \quad x \in X, \tag{1.1}$$

where$X = \{x \in R^n | l \leqslant x \leqslant u\}, l < u, l \in \{R \cup \{-\infty\}\}^n$, $u \in \{R \cup \{\infty\}\}^n, F : R^n \supset u \to R^n$ is defined on an open set$U$containing the feasible set$X$and is locally Lipschitzian continuous.

The common methods for (1.1) are optimization-based ones in which the global minimum is zero and the minimizer is the solution of (1.1). The typical optimization problem in these

* 发表于: Journal of Industrial & Management Optimization, Vol. 1, 2005, pp. 235-250.
† Institute of Mathematics, Changsha University of Science and Technology, Changsha, China.
‡ Institute of Applied Mathematics, Hunan University, Changsha, China.

methods is of the form

$$\begin{cases} \min & \theta(x) = \frac{1}{2}||F(x)||^2 \\ \text{s.t} & l \leqslant x \leqslant u \end{cases} \tag{1.2}$$

From this viewpoint, various numerical methods have been developed, see [1–7] and references therein. Almost all of these methods require $\theta(x)$ being continuously differentiable, and indeed, many systems of equations such as equations arising from nonlinear complementarity problems, the box constrained variational inequality problems, and the KKT system of variational inequality problems, satisfy this requirement. However, we also note that in some cases, the continuously differentiable condition of $\theta(x)$ can not be satisfied even when $F(x)$ is semismooth, for example, the system contains maximum function

$$g(x) = \max_{1 \leqslant i \leqslant m} g_i(x)$$

with continuously differentiable functions $g_i(x)$ $(i = 1, \cdots, m)$. The maximum function can be used to model operating limit in engineering, such as Available Transfer Capability in power system [8], vertical nonlinear complementarity problems [9], and generalized linear complementarity problems [10]. So there is a necessity to design a numerical method for a general system of equations without the continuous differentiability condition.

The aim of this paper is to design a new algorithm to solve (1.1) without the assumption of continuous differentiability for $\theta(x)$. There are two key strategies in our new method. One is a smoothing technique with a perturbing approach, which can ensure the strictly positivity of the smoothing parameter at any non-stationary point of the equivalent optimization problem. This character is a necessary requirement for a smoothing algorithm. The other is to design an optimal search direction in a similar way as [4, 5]. By using this technique, we can keep the feasible property of iterations and ensure the global and locally superlinear convergence of the algorithm.

Our method belongs to the so-called smoothing methods. In many papers (see [11–14] and reference therein), smoothing methods are used to solve unconstrained nonsmooth equations. This paper focuses on a system of constrained nonsmooth equations.

The paper is organized as follows. In Section 2, we give some mathematical preliminaries which will be used later. Section 3 describes the smoothing projected Newton-type algorithm. In Section 4, we analyze the convergence of the new method. The nice global and locally superrlinear convergence is established. Section 5 chooses the approximate KKT system of semi-infinite programming as numerical examples to test the new algorithm. Some comments are made in the last section.

Some words about the notation. For a smooth (continuously differentiable) function $G : R^m \to R^m$, we denote the Jacobian of G at $x \in R^m$ by $G'(X)$, whereas the transposed Jacobian as $\nabla G(x)$. For the function $f : R^n \times R^m \to R$, $\nabla_x f(x, y)$ indicates the gradient of f at (x, y) with respect to x. Similarly, the Jacobian of a continuously differentiable vector function $H(X, Y)$

is denoted by

$$H'(x,y) = (\partial_x H(x,y), \partial_y H(x,y))$$

with $\partial_x H(x,y) = (\nabla_y H(x,y))^{\mathrm{T}}$, $\partial_y H(x,y) = (\nabla_y H(x,y))^{\mathrm{T}}$. For a nonsmooth function $F(x)$, $\partial F(x)$ means the generalized Jacobian in the sense of Clarke [15]. $\|\cdot\|$ denotes the Euclidean norm. If δ is a small quantity, $O(\delta)$ and $o(\delta)$ mean the same order and the higher order of it, respectively. $\prod_W(\cdot)$ represents the orthogonal projection on the set W. R_{++} and R_+ mean $R_{++} = \{t : t > 0, t \in R\}$ and $R_+ = \{t : t \geqslant 0, t \in R\}$, respectively.

2 Preliminaries

In this section, we recall some concepts and properties related to semismooth functions and smoothing functions which will be used later. We prefer to [10–14, 16–19] for details.

2.1 Semismoothness

Semismoothness was originally introduced by Mifflin [17] for functionals. In [19], Qi and Sun extended the definition of semismooth functions to $H : R^n \to R^n$.

F is said to be semismooth at $x \in R^n$, if

$$\lim_{\substack{V \in \partial_F(x+th') \\ h' \to h, t \downarrow 0}} \{Vh'\}$$

exists for any $h \in R^n$.

F is said to be strongly semismooth at x if F is semismooth at x and for any $v \in \partial F(x+h)$ and $h \to 0$,

$$F(x+h) - F(x) - Vh = O(||h||^2).$$

F is said to be a (strongly) semismooth function if it is (strongly) semismooth everywhere on R^n.

F is said to be BD-regular at x if every element in$V \in \partial F(x)$is nonsingular.

Furthermore, semismooth functions enjoy some typical properties of smooth functions, for example, the composition of semismooth functions is still semismooth.

The following lemma is used extensively in locally convergent analysis for solving semismooth equations.

Lemma 2.1 ([19]). *Suppose that $F : R^n \to R^n$is a locally Lipschitz function and semismooth at x. Then, for any $h \to 0$ and $V \in \partial F(x+h)$, it holds*

$$F(x+h) - F(x) - Vh = O(||h||)$$

2.2 Smoothing Functions and Smoothing Methods.

We give some concepts of a smoothing function as follows.

Definition 2.2 ([13]). *Let F be a Lipschitz continuous function in R^n.*

(i) We call $G(t,x) : R \times R^n \to R^n$ a smoothing approximation function of F if it satisfies: (a) $G(0,x) = F(x)$; (b) for any $t > 0$, $G(t,x)$ being smooth (continuously differentiable) with respect to the second variable $x \in D$; (c)

$$\lim_{t\downarrow 0, z\to x} G(t,x) = F(x)$$

(ii) $G(t,x)$ is called a regular smoothing function of F if for any $t > 0, G(t,x)$ is smooth and for any compact set $D \subseteq R^n$ and $\bar{t} > 0$, there exists a constant $C > 0$ such that for any $x \in D$ and $t \in (0,\bar{t}]$

$$||G(t,x) - F(x)|| \leqslant Ct$$

(iii) Furthermore, if for any $x \in R^n$,

$$\lim_{t\to 0+} dist(\partial_x G(t,x), \partial F(x)) = 0$$

then we say $G(t,x)$ satisfies the Jacobian consistency property.

For some simple nonsmooth functions such as the one-dimensional plus function $p(t) = \max\{0,x\}, (x \in R)$, the absolute value function $p(t) = |x|$, and the projected function $p(x) = \prod_D(x)$ with $D = [l,u]$, the corresponding smoothing function (denoted by $P(t,x)$) can be constructed via convolution, and some of the useful properties of $P(t,x)$, for example, the continuously differentiable property for the variable x in the case $t > 0$ and the semismooth property for variable t and x jointly, are analyzed. The continuously differentiable property is a necessary condition for smoothing methods, and the the semismooth property is the crucial requirement to ensure the superlinear convergence for some smoothing methods (see [13,14]).

There are two types of smoothing algorithm for a system of equations in literature. One views t as a parameter (see [11,16,20] and reference therein), and at each iteration this smoothing method reduces t by a suitable strategy. Under Jacobian consistency condition, the algorithm has global and locally superlinear convergence. The other way of smoothing methods views t as a variable, that is, t and x share the same role in $G(t,x)$ (see [13,14]). The central idea of the second smoothing method is to construct the sequence $\omega^k = (t^k, x^k)$ such that the mapping $G(\cdot)$ is continuously differentiable at each point ω^k, and may be non-differentiable at the limiting point of $\{\omega^k\}$. Under the semismooth condition of $G(t,x)$ for variable t and x jointly at the limiting point, this method also enjoys the locally superlinear convergence.

3 A Smoothing Projected Newton-type Algorithm

We will use the second smoothing idea, i.e., t is viewed as a variable to design a new algorithm in this section. Some assumptions are given as follows.

AS.1 $F(x)$is semismooth in D;

AS.2 $G(t,x)$ is the smoothing function of $F(x)$, and it is semismooth for a joint mapping of t and x on $R_+ \times X$.

We need to give some explanations for AS.2. On the one hand, some typical smoothing functions satisfy this assumption (see [13, 14]). On the other hand, we will see later that, this assumption can guarantee $||G(t,x)||^2$ to be continuously differentiable at any iterations generated by our algorithm. The semismooth assumption will be used for locally superlinear analysis of numerical methods as usual.

According to the chosen smoothing idea, the corresponding smoothing system of (1.1) is defined by

$$\begin{cases} \Phi(t,x) = \begin{pmatrix} t \\ G(t,x) \end{pmatrix} = 0, \\ l \leqslant x \leqslant u. \end{cases} \tag{3.1}$$

It is obvious that the system of (3.1) is equivalent to one of (1.1). Moreover, for $t > 0$, we have

$$\Phi'(t,x) = \begin{pmatrix} 1 & 0 \\ \partial_t G(t,x) & \partial_x G(t,x) \end{pmatrix}. \tag{3.2}$$

Denote $\omega = (t,x)$ and define a merit function of (3.1) as

$$\Psi(\omega) = \frac{1}{2}||\phi(t,x)||^2 \tag{3.3}$$

Then the optimization-based problem of (3.1) is defined by

$$\begin{cases} \min & \Psi(\omega) \\ \text{s.t.} & l \leqslant x \leqslant u. \end{cases} \tag{3.4}$$

Under the assumption of AS.2, it is not difficult to see that $\Psi(\omega)$ is continuously differentiable on $R_{++} \times X$ and

$$\nabla\Psi(\omega) = \nabla\Phi(\omega)\Phi(\omega).$$

Then based on the equivalent problem (3.4), the common stopping-rule of a numerical method for solving (3.4) is specialized by a projected gradient direction defined below. Denote $W = R \times X$ and

$$\bar{d}_G(1) = \prod_W(\omega - \gamma\nabla\Psi(\omega)) - \omega = \begin{pmatrix} -\gamma\nabla_t\Psi(w) \\ \prod_X(x - \gamma\nabla_x\Psi(w)) - x, \end{pmatrix} \tag{3.5}$$

where $\gamma > 0$ is a constant related to W, $\prod_W(\cdot)$ is an orthogonal projection operator onto W. Then a stationary point of (3.4) is characterized by (see [5])

$$||\bar{d}_G(1)|| = 0. \tag{3.6}$$

(3.4) and $\bar{d}_G(1)$ are basis that we construct a descent algorithm for the system (1.1).

(3.1) is a smoothing system for $t>0$, and the smoothing parameter t is set a variable. By using this smoothing method, we can see that the crucial technique is to keep $t>0$ during the iterative process, which is done by using a perturbation strategy in our new algorithm. So a perturbation parameter is introduced. Let $\alpha \in (0,1)$ be a constant and $\beta_0 = \alpha \ \min\{1, ||\bar{d}_G^0(1)||^2\}$. For $k=1,2,\cdots$, we define a sequence $\{\beta_k\}$ by

$$\beta_k = \begin{cases} \beta_{k-1}, & \text{if } \ \alpha \ \min\{1, ||\bar{d}_G^k(1)||^2\} > \beta_{k-1}, \\ \alpha \min\{1, ||\bar{d}_G^k(1)||^2\}, & \text{otherwise,} \end{cases} \tag{3.7}$$

where $\bar{d}_G(1)$ expresses the projected gradient direction at point ω^k. In the convergent analysis later, we will see that β_k is a crucial quantity, and by using this parameter, we can keep $t>0$ at non-stationary points of (3.4) during iterations.

3.1 Search direction

We will describe how to design a search direction during the iterative process in this subsection, which is done by using a similar way proposed in [4–6], but with a perturbation strategy. The aim of constructing the search direction is to guarantee the global and fast local convergence of the algorithm.

Let $\bar{t}>0$ be a given constant, $\bar{w}=(\bar{t},0)\in R\times R^n$. At current iteration point$w^k=(t^k,x^k)$ satisfying $t^k>0$, the search direction is computed by three steps with $\lambda>0$.

Step 1. Compute the projected perturbed-gradient direction.

Let d_G^k be a solution of

$$\gamma_k \nabla \bar{\Psi}(w^k) + d_G^k = \beta_k \bar{w} \tag{3.8}$$

where $\gamma_k>0$ is a scalar dependent on w^k. d_G^k is called a perturbed gradient direction. Then the so-called *projected perturbed-gradient direction* is computed by

$$\bar{d}_G^k(\lambda) := \prod_W (w^k + \lambda d_G^k) - w^k. \tag{3.9}$$

Step 2. Compute the projected perturbed-Newton direction.

Let d_N^k be a solution of

$$\Phi(w^k) + \bar{\Phi}'(w^k) d_N^k = \beta_k \bar{w}. \tag{3.10}$$

Similarly, d_N^k is called a perturbed Newton direction. The so-called *projected perturbed-Newton direction* then is computed by

$$\bar{d}_N^k(\lambda) := \prod_W (w^k + \lambda d_N^k) - w^k. \tag{3.11}$$

Step 3. Compute an optimal combination direction.

According to the descent properties of $d_G^k(\lambda)$ and $d_N^k(\lambda)$, i.e., global descent and local fast descent properties respectively, the search direction used in this paper is computed by

$$\bar{d}^k(\lambda) = \tau_k^*(\lambda)\bar{d}_G^k(\lambda) + (1-\tau_k^*(\lambda))\bar{d}_N^k(\lambda), \tag{3.12}$$

where $\tau_k^*(\lambda)$ is a solution of the following optimization problem

$$\min_{\tau\in[0,1]} \frac{1}{2}\left\|\bar{\Phi}(w^k) + \Phi'(w^k)[\tau\widetilde{d}_G^k(\lambda) + (1-\tau)\widetilde{d}_N^k(\lambda)]\right\|^2.$$

We call $\bar{d}^k(\lambda)$ an optimal combination direction at point w^k.

By using a similar way in proof of [5, Lemma 3.1], we can obtain the conclusion of $\tau_k^*(\lambda)$ as follows.

Lemma 3.1. *Suppose that $w^k \in W$ and d_N^K exists. Then we have*

$$\tau_k^*(x) = \max\{0,\ \min\{1, \tau_k(\lambda)\}\} \tag{3.13}$$

where $\tau_k(\lambda)$ is defined by

$$\tau_k(\lambda) = \begin{cases} 0, & \text{if } \Phi'(w^k)[\widetilde{d}_G^k(\lambda) - \widetilde{d}_N^k(\lambda)] = 0, \\ -\dfrac{[\Phi(w^k) + \Phi'(w^k)\widetilde{d}_N^k(\lambda)]^T\phi'(w^k)[\widetilde{d}_G^k(\lambda) - \widetilde{d}_N^k(\lambda)]}{||\Phi'(w^k)[\widetilde{d}_G^k(\lambda) - \widetilde{d}_N^k(\lambda)]||^2}, & \text{otherwise.} \end{cases}$$

Remark 3.1. Since $W = R \times X$, this implies that the above projected approach acts just on the variable x at any point $w = (t, x)$.

3.2 Projected Newton algorithm

Now we state the smoothing projected Newton-type algorithm.

Algorithm 3.1.

Step 0. (Initialization)

Given some constants $\eta, \rho, \sigma \in (0,1)$, $p_1 > 0$, $p_2 > 0$ and $\alpha > 0$, $\bar{t} > 0$ with $\alpha\bar{t} < 1$. Let $t^0 = \bar{t}$ and $w^0 = (t^0, x^0)$. Set $k := 0$.

Step 1. (Stop Test)

Let

$$\gamma_k = \min\left\{1, \frac{t_k}{|t_k + \partial_t G(w^k)^T G(w^k)|}, \frac{\eta||\Phi(w^k)||}{||\nabla\psi(w^k)||}, \frac{\eta\Psi(w^k)}{||\nabla\Psi(w^k)||^2}\right\}. \tag{3.14}$$

Compute $\bar{d}_G^k(1)$ by (3.5) at point w^k. If $||\bar{d}_G^k(1) = 0||$, stop.

Step 2. (Compute Perturbed Direction)

Compute d_G^k by (3.8). If (3.10) has a solution d_N^k and satisfies

$$-\nabla\Psi(w^k)^T d_N^k \geqslant p, ||d_N^k||^{p_2},$$

then use the direction d_N^k. Otherwise, set $d_N^k := d_G^k$.

Step 3. (Line Search)

Let m_k be the smallest nonnegative integer m satisfying

$$\psi(w^k + \bar{d}^k(\rho^m) \leqslant \Psi(w^k) + \sigma\nabla\psi(w^k)^T\widetilde{d}_G^k(\rho^m), \tag{3.15}$$

where $\bar{d}^k(\lambda)$ is computed by (3.12) and (3.13). Let$\lambda_k = \rho^{mk}$ and $w^{k+1} = w^k + \bar{d}^k(\lambda_k)$

Step 4. Set $k := k+1$ and go to Step 1.

The following projection properties are used in our analysis (see [21]).

Lemma 3.2. *The projection operator $\prod_W(\cdot)$ with any convex set $W \subset R^n$ satisfies*

(i) for any $w \in W$,

$$\Big[\prod_W(w') - w'\Big]^T \Big[\prod_W(w') - w\Big] \leqslant 0, \quad \text{for all } w' \in R^n;$$

(ii)

$$\Big\|\prod_W(w') - \prod_W(w'')\Big\| \leqslant \|w' - w''\|, \quad \text{for all } w', w'' \in R^n;$$

(iii) Given $w, d \in R^n$, the function ζ defined by

$$\zeta(\lambda) = \Big\|\prod_W(w + \lambda d) - w\Big\|\Big/\lambda, \quad \lambda > 0$$

is non-increasing.

The following conclusion is obvious.

Proposition 3.3. *$\{\beta_k\}$ defined in (3.7) has the following properties:*

(i) $\{\beta_k\}$ is non-increasing sequence;

(ii) for all k, β_k satisfies

$$\beta_k \leqslant \alpha \min\{1, \|\bar{d}_G^k(1)\|^2\};$$

(iii) when $\|\bar{d}_G^k(1)\| > 0$, it holds $\beta_k > 0$ for all k.

The following result shows the descent property of the perturbed-gradient direction $\widetilde{d}_G^k$ at point w^k *for any* $\lambda > 0$.

Proposition 3.4. *Let $w^k \in W$ be not a stationary point of (3.4) and satisfy $t^k > 0$. Then for any $1 \geqslant \lambda > 0$, it holds*

$$\nabla\Psi(w^k)^T \widetilde{d}_G^k(\lambda) \leqslant -\frac{\lambda}{\gamma}(1 - \alpha\bar{t})\|\bar{d}_G^k(1)\|^2 < 0, \tag{3.16}$$

where $\bar{d}_G^k(1)$ is the projected gradient direction defined in (3.5).

Proof. We omit the iterative number k for simple expressions in the following proof. For $T > 0$, it holds

$$\nabla\Psi(w) = \nabla\Phi(w)\Phi(w) = \begin{pmatrix} t + \partial_t G(w)^T G(w) \\ \partial_x G(w)^T G(w) \end{pmatrix} \equiv \begin{pmatrix} \nabla_t \Psi(w) \\ \nabla_x \Psi(w) \end{pmatrix}.$$

Note that $W = R \times X$, from the computation of d_G, the projected perturbed gradient direction can be written as

$$\widetilde{d}_G(\lambda) \equiv \begin{pmatrix} (\widetilde{d}_G(\lambda))_t \\ (\widetilde{d}_G(\lambda))_x \end{pmatrix} = \begin{pmatrix} -\lambda\gamma(t + \partial_t G(w)^T G(w) + \lambda\beta\bar{t} \\ \prod_X x - \lambda\gamma\nabla_x \Psi(w) - x \end{pmatrix}.$$

We then have the following derivation:

$$
\begin{aligned}
&(t+\partial_t G(w)^T G(w)|^2+\lambda(t+\partial_t G(w)^T G(w))\beta\bar{t} \\
=&(\lambda\gamma|t+\partial_t G(w)^T G(w)|^2+\lambda(\partial_t G(w)^T G(w))\beta\bar{t} \\
\leqslant&-\frac{\lambda}{\gamma}|-\gamma\nabla_t\Psi(w)|^2+\frac{\lambda}{\gamma}|-\gamma\nabla_t\Psi(w)|\beta\bar{t} \\
\leqslant&-\frac{\lambda}{\gamma}|-\gamma\nabla_t\Psi(w)|^2+\frac{\lambda}{\gamma}|-\gamma\nabla_t\Psi(w)|(\alpha\bar{t})||\bar{d}G(1)|| \\
\leqslant&-\frac{\lambda}{\gamma}|-\gamma\nabla_t\Psi(w)|^2+\alpha\bar{t}\frac{\lambda}{\gamma}||\bar{d}_G(1)||^2,
\end{aligned}
\tag{3.17}
$$

where the second inequality comes from Proposition 3.3 (ii) and the fact that $\beta\leqslant\alpha||\bar{d}_G(1)||$ for any case; the last inequality is due to $|-\gamma\nabla_t\Psi(w)|\leqslant||\bar{d}_G(1)||$ (see (3.5)).

Furthermore, it holds

$$
\begin{aligned}
&\nabla_x\Psi(w)^T\left[\prod_X(x-\lambda\gamma\nabla_x\Psi(w))-x\right] \\
=&-\frac{1}{\lambda\gamma}\left[x-\lambda\gamma\nabla_x\Psi(w)-x\right]^T\left[\prod_X(x-\lambda\gamma\nabla_x\Psi(w))-x\right] \\
=&\frac{1}{\lambda\gamma}\left[\prod_X(x-\lambda\gamma\nabla_x\Psi(w))-(x-\lambda\gamma\nabla_x\Psi(w))\right]^T\left[\prod_X(x-\lambda\gamma\nabla_x\Psi(w))-x\right] \\
&-\frac{1}{\lambda\gamma}\left\|\prod_X(x-\lambda\gamma\nabla_x\Psi(w))-x\right\|^2 \\
\leqslant&-\frac{1}{\lambda\gamma}\left\|\prod_X(x-\lambda\gamma\nabla_x\Psi(w))-x\right\|^2 \\
\leqslant&-\frac{\lambda}{\gamma}\left\|\prod_X(x-\gamma\nabla_x\Psi(w))-x\right\|^2,
\end{aligned}
\tag{3.18}
$$

where the first inequality and the second inequality come from Lemma 3.2 (i) and (iii), respectively. From (3.17) and (3.18) we have

$$
\begin{aligned}
\nabla\Psi(w)^T\widetilde{d}_G(\lambda)=&(t+\partial_t G(w)^T G(w))[-\lambda\gamma(t+\partial_t G(w)^T G(w))+\lambda\beta\bar{t}] \\
&+\nabla_x\Psi(w)^T\left[\prod_X(x-\lambda\gamma\nabla_x\Psi(w))-x\right] \\
\leqslant&-\frac{\lambda}{\gamma}[|-\gamma\nabla_t\Psi(w)|^2+||\prod_X(x-\lambda\gamma\nabla_x\Psi(w))-x||^2]+\alpha\bar{t}\frac{\lambda}{\gamma}||\bar{d}_G(1)||^2 \\
=&-\frac{\lambda}{\gamma}(1-\alpha\bar{t})||\bar{d}_G(1)||^2\leqslant 0.
\end{aligned}
\tag{3.19}
$$

We then finish the proof of the proposition. □

Now we have the following conclusion, which shows Algorithm 3.1 to be well defined.

Theorem 3.5. *Suppose that $w^k\in W$ is not a stationary point of (3.4) with $t^k>0$. Then there exists a constant $\lambda'\in(0,1]$ such that for any $\lambda\in(0,\lambda']$, $\bar{d}^k(\lambda)$ is a descent direction of $\Psi(w)$*

at point w^k and

$$\Psi(w^k+\bar{d}^k(\lambda))\leqslant\Psi(w^k)+\sigma\nabla\Psi(w^k)^T\bar{d}_G^k(\lambda). \tag{3.20}$$

Proof. By using Proposition 3.4, we can prove the conclusion in a similar way of proof of [5, Theorem 3.1].

4 Convergence Analysis

In this section we analyze the global and local convergence of Algorithm 3.1.

4.1 Global convergence

The next proposition is a key result, which shows that Algorithm 3.1 can remain $t>0$ during the iteration process.

Proposition 4.1. *Let $w^k=(t^k,x^k)$ be generated by Algorithm 3.1 and not a stationary point of (3.4). Then for every $k\geqslant 0$ we have*

$$t^k\geqslant\beta_k\bar{t}>0 \tag{4.1}$$

Proof. We prove the proposition by induction.

For $k=0$, from Algorithm 3.1 and the choice of β_0 it holds

$$t^0=\bar{t}\geqslant\beta_0\bar{t}>0.$$

Suppose that (4.1) holds for k, we need to prove the conclusion for $k+1$.

We divide the search directions proposed in Section 3.1 into two parts respectively:

$$\bar{d}_G^k(\lambda_k)=\begin{pmatrix}(\bar{d}_G^k(\lambda(\lambda_k))_t\\(\bar{d}_G^k(\lambda_k))_x\end{pmatrix},\qquad \bar{d}_N^k(\lambda_k)=\begin{pmatrix}(\bar{d}_N^k(\lambda(\lambda_k))_t\\(\bar{d}_N^k(\lambda_k))_x\end{pmatrix},$$

$$\bar{d}^k(\lambda_k)=\tau_k^*(\lambda_k)\bar{d}_G^k(\lambda_k)+(1-\tau_k^*(\lambda_k))\bar{d}_N^k(\lambda_k)=\begin{pmatrix}(\bar{d}^k(\lambda_k))_t\\(\bar{d}^k(\lambda_k))_x\end{pmatrix},$$

where λ_k is the accepted step-length at the k-th iteration. We consider the part of variable t. From (3.2) and (3.8)–(3.11), we have

$$(\bar{d}_G^k(\lambda_k))_t=\lambda_k[-\lambda_k(t^k+\partial_t G_k^T G_k)+\beta_k t],\quad (\bar{d}_N^k(\lambda_k))_t=\lambda_k(-t^k+\beta_k\bar{t}),$$

where G_k indicates $G(w^k)$. Then According to the computation of the search direction $\bar{d}(\lambda_k)$, it follows that

$$\begin{aligned}(\bar{d}^k(\lambda_k))_t&=\tau_k^*(\lambda_k)\lambda_k[-\gamma k(t^k+\partial_t G_k^T G_k)+\beta_k\bar{t}]+(1-\tau_k^*(\lambda_k))\lambda_k[-t^k+\beta_k\bar{t}]\\&=-\lambda_k\gamma_k\tau_k^*(\lambda_k)(t^k+\partial_t G_k^T G_k)-(1-\tau_k^*(\lambda_k))\lambda_k t^k+\lambda_k\beta_k\bar{t}\\&\geqslant-\lambda_k\tau_k^*(\lambda_k)t^k-(1-\tau^*(\lambda_k))\lambda_k t^k+\lambda_k\beta_k\bar{t}\\&=-\lambda_k t^k+\lambda_k\beta_k\bar{t},\end{aligned}$$

where the inequality comes from the definition of γ_k (see (3.14)). Then we have

$$\begin{aligned} t^{k+1}-\beta_{k+1}\bar{t}&=t^k+(\bar{d}^k(\lambda_k))_t-\beta_{k+1}\bar{t}\\ &\geqslant(1-\lambda_k)t^k+\lambda_k\beta_k\bar{t}-\beta_{k+1}\bar{t}\\ &\geqslant(1-\lambda_k)t^k+\lambda_k\beta_k\bar{t}-\beta_k\bar{t}\\ &=(1-\lambda_k)t^k-(1-\lambda_k)\beta_k\bar{t}\geqslant 0, \end{aligned} \tag{4.2}$$

where the second inequality is due to the monotone property of β_k in Proposition 3.3, and the last inequality comes from the assumption of w^k. Therefore, it holds

$$t^{k+1}\geqslant\beta_{k+1}\bar{t}.$$

Furthermore, if $w^k$and w^{k+1} are not stationary points, from Proposition 3.3 we have $\beta_k\geqslant\beta_{k+1}>0$. So (4.1) holds at point w^{k+1}. We complete the proof. □

Let w^k be a sequence of points generated by Algorithm 3.1. Proposition 4.1 shows that if the algorithm does not stop at a stationary point (3.4) in any finite step, we have $t^k>0$ for every k. This result implies that $\Phi(w)$ and Ψw are continuously differentiable at any point generated by Algorithm 3.1. Hence, by using a similar way of proof of [5, Theorem 4.1], we have the globally convergent theorem for Algorithm 3.1. Here we omit the detailed proof.

Theorem 4.2. *Let $\{w^k\}\subset W$ be a sequence generated by Algorithm 3.1. Then any accumulation point of$\{w^k\}$is a stationary point of (3.4).*

Note that in our algorithm, the nonsingular property of $\Phi(w)$ does not need for the analysis of the global convergence.

4.2 Local convergence analysis

In the rest of this section, we analyze the local convergence of Algorithm 3.1 based on semismooth properties of $G(w)$ and $\Phi(w)$ in local analysis:

AS.3 Any stationary point of (3.4) is a solution of (3.1).

AS.4 Let $w^*=(t^*,x^*)$ be an accumulation point of $\{w^k\}$ generated by Algorithm 3.1 and a solution of the system of equations (1.1) (or equivalently, the solution of (3.1)). Suppose that $\Phi(w)$ is BD-regular at w^* and $\lim_{k\in K}w^k=w^*$ for some subset $K\subset\{1,2,\cdots\}$.

We give some explanations for the above assumptions. AS.3 is necessary in our local analysis. It ensures that any non-solution point w^k generated by Algorithm 3.1 satisfies $t^k>0$ due to Proposition 4.1. So $\Phi(w)$ is continuously differentiable at all iterations w^k. Some papers (for example, [22, 23] and references therein) consider the conditions where AS.3 holds for the system of equations arising from MCP and NCP problems. AS.4 is a common assumption used extensively in semismooth systems.

AS.3 and AS.4 together with Theorem 4.1 and the expression of (3.1) imply that the limit point $w^*=(t^*,x^*)$ of $\{w^k\}$ satisfies $w^*=(0,x^*)$ and $\Phi(0,x^*)=0$. Then from (3.1) and $G(t,x)$

being the smoothing function of $F(x)$, the nonsmooth point of $\Phi(w)$ may just happen at the point w^*in$R_+ \times X$. Furthermore, from the definition of $\Psi(w)$ (see (3.3)), we have

$$\partial\Psi(w) \subseteq V^T\Phi(w),$$

where $V \in \partial\Phi(w)$. This indicates that $\Phi(w)$ has the same smooth property of $\Phi(w)$ in $R_+ \times X$. However, the assumption of the solution point in AS.4 and the expression of $\partial\Psi(w)$ imply the possible nonsmooth point $w^* = (0, x^*)$ to be smoothed due to $\Phi(w^*) = 0$. Hence, under the assumptions of AS.1-AS.4, $\Psi(w)$ is continuously differentiable at all iterations and the accumulation point.

From the BD-regular condition and the smoothing properties of $G(t, x)$, we have the following well-known conclusions.

Lemma 4.3. *There exist positive constants* κ *and*ϵ *such that for every*$w \in R_+ \times X$ *satisfying* $||w - w^*|| \leqslant \epsilon$, *we have for each* $V \in \partial\Psi(w)$

(i) V *is nonsingular and satisfies*

$$V^{-1} \leqslant \kappa$$

(ii)

$$\|\Phi(w)\| = \sqrt{2}\Psi(w)^{\frac{1}{2}} = O(||w - w^*||).$$

The following lemma shows some properties of the search directions used in Algorithm 3.1.

Lemma 4.4. *For all* $k \in K$ *sufficiently large, we have*

(i)

$$\beta_k = O(\Psi(w^k)) = O(||w^k - w^*||^2).$$

(ii) For any $\lambda \in (0, 1]$

$$w^k + \lambda d_N^k = (1 - \lambda)w^k + \lambda w^* + \lambda o(\Psi(w^k)^{\frac{1}{2}}). \tag{4.3}$$

Proof. (i) From the definition of β_k, the choice ofγ_k, the projection property and Lemma , when$w^k$is sufficiently closed to w^*, it follows

$$\beta_k \leqslant \alpha||\bar{d}_G^k(1)||^2 \leqslant \alpha\gamma_k^2||\nabla\Psi(w^k)||^2 \leqslant \alpha\eta\Psi(w^k) = \frac{\alpha\eta}{2}||\Phi(w^k||^2 = O(||w^k - w^*||^2).$$

(ii) From conclusion (i) and Lemma 4.2, we derive that

$$\begin{aligned} w^k + \lambda d_N^k &= w^k + \lambda\Phi'(w^k)^{-1}[-\Phi(w^k) + \beta_k\bar{w}] \\ &= w^k - \lambda\Phi'(w^k)^{-1}[\Phi(w^k) - \Phi(w^*) - \Phi'(w^k)(w^k - w^*)] \\ &\quad -\lambda(w^k - w^*) + \lambda\Phi'(w^k)^{-1}\beta_k\bar{w} \\ &= (1 - \lambda)w^k + \lambda w^* + \lambda o(||w^k - w^*||) + \lambda O(\Psi(w^k)) \\ &= (1 - \lambda)w^k + \lambda w^* + \lambda o(\Psi(w^k)^{\frac{1}{2}}), \end{aligned}$$

where the third equality is due to the semismooth property and the conclusion (i). The proof is completed □

Lemma 4.5. *For $k \in K$ large enough, we have*

$$\widetilde{d}_N^k(\lambda) = -\lambda(w^k - w^*) + \lambda o(\Psi(w^k)^{\frac{1}{2}}) \tag{4.4}$$

and

$$\nabla\Psi(w^k)^T \widetilde{d}_N^k(\lambda) \leqslant -\mu\lambda\Psi(w^k), \tag{4.5}$$

where $\mu \in (0, 2)$ is any constant.

Proof. From Lemma 4.4 and the properties of a projector, we obtain that

$$\begin{aligned}
\widetilde{d}_N^k(\lambda) =& \prod -W(w^k + \lambda d_N^k) - w^k \\
=& \prod_W [(1-\lambda)w^k + \lambda w^* + \lambda o(\Psi(w^k)^{\frac{1}{2}})] - w^k \\
=& \prod_W [(1-\lambda)w^k + \lambda w^*] - w^k \\
& + \{\prod_W [(1-\lambda)w^k + \lambda w^k + \lambda o(\Psi(w^k)^{\frac{1}{2}})] - \prod_W [(1-\lambda)w^k + \lambda w^*]\} \\
=& -\lambda(w^k - w^*) + \lambda o(\Psi(w^k)^{\frac{1}{2}}),
\end{aligned}$$

where the last equality comes from $(1-\lambda)w^k + \lambda w^* \in W$ and the projected property (see Lemma 3.2 (ii)). Then the first conclusion is proved. Moreover, (4.4) follows

$$\begin{aligned}
\nabla\Psi(w^k)^T \widetilde{d}_N^k(\lambda) =& -\lambda\Phi(w^k)^T \Phi'(w^k)(w^k - w^*) + \lambda o(\Psi(w^k)) \\
=& -2\lambda\Psi(w^k) + \lambda\Phi(w^k)^T[\Phi(w^k) - \Phi(w^*) \\
& -\Phi'(w^k)(w^k - w^*)] + \lambda o(\Psi(w^k)) \\
\leqslant& -\mu\lambda\Psi(w^k),
\end{aligned}$$

where the last inequality comes from the semismooth property of $\Phi(w)$ at w^* and Lemma 4.2. So (4.5) is true. We complete the proof. □

Lemma 4.6. *We have that for $k \in K$ large enough,*

(i)

$$\tau^*(\lambda)_k \leqslant o(1) \tag{4.6}$$

where $\tau^*(\lambda)_k$ is defined in (3.13) at point w^k

(ii)

$$\bar{d}^k(\lambda) = -\lambda(w^k - w^*) + \lambda o(\Psi(w^k)^{\frac{1}{2}}). \tag{4.7}$$

(iii)

$$\nabla\Psi(w^k)^T \bar{d}^k(\lambda) = -2\lambda\Psi(w^k) + \lambda o(\Psi(w^k)) \tag{4.8}$$

Proof. Lemma 4.5 shows that for the projected disturbance-Newton direction used in our algorithm, the same conclusion of [5, Proposition 3.2] holds. Then the lemma is proved by using a similar way to the proof of [5, Theorem 3.2]. We omit the detailed proof. □

Now we prove the locally super linear property of Algorithm 3.1 under the BD regular condition.

Theorem 4.7. *Suppose that* $\{w^k\}$ *is a sequence generalized by Algorithm 3.1 and* w^* *is a point satisfying (AS.3). If the BD-regular condition holds at* w^*, *then the whole sequence* $\{w^k\}$ *converges to* w^* *superlinearly.*

Proof. From Lemma 4.6 we have that for sufficiently large $k \in K$,

$$\|w^k + \bar{d}^k(1) - w^*\| = o(\Psi(w^k)^{\frac{1}{2}}) = o(||\Phi(w^k)||) = o(||w^k - w^*||) \tag{4.9}$$

Furthermore, we derive that

$$\begin{aligned}\Psi(w^k + \bar{d}^k(1)) &= \frac{1}{2}||\Phi(w^k) + \bar{d}^k(1))||^2 \\ &= \frac{1}{2}||\Phi(w^k + \bar{d}^k(1)) - \Phi(w^*)||^2 \\ &= O(||w^k + \bar{d}^k(1)) - \Phi(w^*)||^2 \\ &= o(\Psi(w^k)),\end{aligned} \tag{4.10}$$

where the last equality is due to (4.9). On the other hand, it holds

$$\begin{aligned}-\nabla\Psi(w^k)^T \widetilde{d}_G^k(1) &\leqslant \|\nabla\Psi(w^k)\| \, \|\widetilde{d}_G^k(1)\| \\ &= \|\nabla\Psi(w^k)\| \left\| \prod_W [w^k - \gamma_k \nabla\Psi(w^k) + \beta_k \bar{w}] - w^k \right\| \\ &\leqslant \|\nabla\Psi(w^k)\| \left[||\gamma_k \nabla\Psi(w^k)|| + O(\Psi(w^k)) \right] \\ &\leqslant \eta\Psi(w^k) + o(\Psi(w^k)),\end{aligned} \tag{4.11}$$

where the second inequality is due to the property of β_k and the projection property, and the last inequality comes from the choice of γ_k (see (3.14)). From (4.10) and (4.11) we have

$$\Psi(w^k) + \sigma\nabla\Psi(w^k)^T \widetilde{d}_G^k(1) \geqslant (1 - \sigma\eta)\Psi(w^k) + o(\Psi(w^k)) \geqslant o(\Psi(w^k)) = \Psi(w^k + \bar{d}^k(1)). \tag{4.12}$$

Hence, from Algorithm 3.1 and (4.12) we obtain

$$w^{k+1} = w^k + \bar{d}^k(1).$$

Moreover, from (4.9) we conclude that w^k converges to w^* super linearly. We complete the proof. □

5 Numerical Examples

This section makes some numerical examples to illustrate the computational behavior of Algorithm 3.1. We consider the KKT system of a semi-infinite programming (denoted by SIP). The SIP problem is to find $x \in R^n$ such that

$$\min\{f(x) : x \in X\}, \tag{5.1}$$

where $X = \{x \in R^n : g(x,v) \leqslant 0, \forall v \in V\}, V = [a,b] \subset R$, is a nonempty compact subset, $f : R^n \to R, g : R^n \times R \to \times R$ are twice continuously differentiable functions.

Under some condition, the KKT system of (5.1) can be reformulated equivalently as:

$$\begin{cases} \nabla f(x) + \sum_{i=1}^{p} u_i \nabla_x g(x, v^i) = 0, \\ g(x,v) \leqslant 0, \forall v \in V, \\ u_i > 0,\ g(x, v^i) = 0,\ (i = 1, \cdots, p) \\ \Phi(x, v^i) = 0,\ (i = 1, \cdots, p), \end{cases} \tag{5.2}$$

where p is the number of active set at the solution of SIP, $\Phi(x,v) = v - \text{mid}\ (a, b, v + \nabla_v g(x,v))$ with

$$(\text{mid}\ (c,d,w))_j = \begin{cases} c_j, & \text{if}\ \ w_j < c_j, \\ w_j, & \text{if}\ \ c_j \leqslant w_j \leqslant d_j, \\ d_j, & \text{if}\ \ d_j < w_j. \end{cases}$$

See [20, 24] for details.

We use a finite set V_N to approximate V with

$$V_N = \left\{ v_j = a + \frac{j(b-a)}{N} : j = 0, 1, \cdots, N \right\}.$$

Denote $G_N(x) = \max_{v \in V_N} g(x,v)$. Then the approximate system of (5.2) can be written as

$$\begin{cases} H(z) = 0, \\ u \geqslant 0, y \geqslant 0, \end{cases} \tag{5.3}$$

where $z = (x, u, v, y) \in R^n \times R^p \times R^p \times R$, and

$$H(z) = \begin{pmatrix} \nabla f(x) + \sum_{i=1}^{p} u_i \nabla_x g(x, v^i) \\ g(x, v^i)(i = 1, \cdots, p) \\ G_N(x) + y \\ \Phi(x, v^i)(i = 1, \cdots, p) \end{pmatrix}.$$

$G_N(x)$ and $\Phi(x,v)$ are nonsmooth, but semismooth. In order to use Algorithm 3.1 for solving (5.3), we choose the following smoothing functions of $G_N(x)$ and $\Phi(x,v)$, respectively (see [9, 10, 13, 14]).

$$G_s(t,x) = \begin{cases} t\ \ln\left(\sum_{j=1}^{N} e^{g_j(x)/t} \right), & \text{if}\ \ t > 0, \\ G(x), & \text{if}\ \ t = 0, \end{cases}$$

$$
\Phi_s(t,x,v)=\begin{cases} v-\left[\dfrac{a+\sqrt{(a-v-(\nabla_v g(x,v)))^2+4t^2}}{2}\right.\\ \qquad\left.+\dfrac{b-\sqrt{(b-v-(\nabla_v g(x,v)))^2+4t^2}}{2}\right], & \text{if } t>0,\\ v-\ \operatorname{mid}\,(a,b,v+\nabla_v g(x,v))), & \text{if } t=0.\end{cases}
$$

It is not difficult to prove that $H(z)$ and the smoothing functions satisfy AS.1–AS.2 (see [10,20] for their properties in details). Then we can use Algorithm 3.1 to solve the approximate KKT system of SIP problems.

The stopping rule in Algorithm 3.1 is set $\epsilon = 1.0\mathrm{e}-6$. The parameters used in Algorithm 3.1 are specified as follows.

$$\eta = 0.9,\ \rho = 0.5,\ \sigma = 0.001,\ \alpha = 0.5,\ \bar{t} = 0.9.$$

The starting pointu^0 and y^0 for all examples are set $t^0 = \bar{t}$, $u^0 = 0.05\mathbf{e}(p)$, $y^0 = 0.5$, where $\mathbf{e(p)}$ represents p-order unity vector.

The examples are drawn from some references. Examples 1–3 are chosen from [25]; Example 4 comes from [26] with a revised region

Example 1.

$$f(x) = 1.21e^{x_1} + e^{x_2}, \quad g(x,v) = v - e^{x_1+x_2}, \quad V = [0,1].$$

Example 2.

$$f(x) = x_1^2 + x_2^2 + x_3^2, \quad g(x,v) = x_1 + x_2 e^{x_3 v} + e^{2v} = 2\sin(4v), \quad V = [0,1].$$

Example 3.

$$f(x) = \frac{1}{3}x_1^2 + \frac{1}{2}x_1 + x_2^2, \quad g(x,v) = (1 - x_1^2 v^2)^2 - x_1 v^2 - x_2^2 + x_2, \quad V = [0,1].$$

Example 4.

$$f(x) = x_1^2 + (x_2 - 3)^2, \quad g(x,v) = x_2 - 2 + x_1 \sin\left(\frac{v}{x_2 - 0.5}\right), \quad V = [0,10].$$

The computed results are reported in Table 1 and 2 below.

Table 1 Test Results for Algorithms 3.1

Ex.	p	NH	NdH	N_{dis}	CPU	$\|\|\bar{d}_G(w^k)\|\|$	$\|\|H(W^k)\|\|$	$F(x^k)$	$\max g(x^k, v_j)$
1	1	7	7	1280	0.46	9.9624e−13	1.1069e−12	2.2	−8.8418e−13
2	1	18	12	1280	0.3810	5.9545e−9	6.6161e−9	5.3347	−3.9968e−15
3	1	15	14	1600	0.7810	2.0427e−7	2.2697e−7	0.2675	−6.2390e−9
4	1	74	74	12800	5.7280	6.1885e−10	1.3737e−7	1.0	8.1425e−8

Table 2 Approximate Solution for Examples

Example	Initial Point (x^0, v^0)	Approximate Solution(x^k, v^k)
1	(0,0,0)	$(-0.0953, 0.0953, 1.0)$
2	(1,1,1,1)	$(-0.2133, -1.3615, 1.8353, 1.0)$
3	$(-1, -1, 1)$	$(-0.7500, -0.6745, 0.3595)$
4	(1,1,1)	(0,2,0)

where in Table 1, **p** is the guess of number in active set at solution point; **NH** and**NdH** represent the computing number of function and its derivative defined in constrained equations; N_{dis} indicates the dividing number for region V; **CPU**is the total cost time for solving SIP problems; $||\bar{d}_G(w^k)||$ and $||H(w^k)$ indicate the final values of the projected gradient and the norm of $H(w)$, respectively; $f(x^k)$ is the final value of the objective function in SIP; and $\max g(x^k, v_j)$ is the maximal value of constrained function in the approximate subset V_N.

From Table 1 and 2 we can see that Algorithm 3.1 performs very well.

6 Conclusions

This paper develops a smoothing algorithm for solving a general system of constrained equations. By choosing some suitable smoothing functions and a smoothing technique, we do not need the continuously differentiable assumption for the merit function, which is required in traditional methods. Moreover, the perturbed technique and strategy of the search direction not only ensure the process of the smoothing algorithm, but also guarantee the nice global and locally suberlinear convergence. Numerical examples arising from SIP problems show that this smoothing algorithm is promising.

Acknowledgments The authors would like to thank Dr. Houduo Qi and two anonymous referees for their detailed comments which considerably improved the presentation of the paper.

References

[1] S. Bellavia, M. Macconi, B. Morini, An affine scaling trust-region approach to boundedconstrained nonlinear systems, *Applied Numerical Mathematics*, 44 (2003), 257-280.

[2] C. Kanzow, An active-set type Newton method for constrained nonlinear equations, In: *Complementarity: Applications, Algorithms and Extensions*, M.C. Ferris, O.L. Mangasarian and J.-S. Pang, (eds), Kluwer Academic Press, 2001, 179-200.

[3] C. Kanzow, Strictly feasible equation-based method for mixed complementarity problems, *Numerische Mathematik*, 89 (2001), 135-160.

[4] L. Qi, X.J. Tong and D.H. Li, An active-set projected trust region algorithm for box constrained nonsmooth equations, Journal of Optimization, *Theory and Applicaitons*, 120 (2004), 601-625.

[5] D. Sun, R.S. Womersley and H. Qi, A feasible semismooth asymptotically Newton method for mixed complementarity problems, *Mathematical Programming*, 94 (2002), 167-187.

[6] X.J. Tong and L. Qi, On the convergence of a trust region method for solving constrained nonlinear equations with degenerate solution, *Journal of Optimization, Theory and Applicaitons*, (to appear).

[7] M. Ulbrich, Nonmonotone Trust-Region Method for Bound-Constrained Semismooth Equations with Applications to Nonlinear Mixed Complementarity Problem. *SIAM Journal on Optimization*, 11 (2001), 889-917.

[8] North American Electric Reliability Council, *Available transfer capability definitions and determination*, Reference Document Prepared by TTC Task Force, 1996.

[9] H.D. Qi, L.Z. Liao and Z.H. Lin, Regularized smoothing approximations to vertical nonlinear complementarity problems, *Journal of Mathematical Analysis and Applications*, 230 (1999), 261-276.

[10] J.M. Peng and Z. Lin, A non-interior continuation method for generalized linear complementarity problems, *Mathematical Programming*, 86 (1999), 533-563.

[11] X. Chen, L. Qi and D. Sun. Global and superlinear convergence of the smoothing Newton method and its application to general box constrained variational inequalities, *Mathematics of Computation*, 67 (1998), 519-540.

[12] S.A. Gabriel and J.J. Moré, Smoothing of mixed complementarity problems, in: *Complementarity and Variational Problems: State of the Art*, M.C. Ferris, J.S. Pang, eds., SIAM, Philadelphia, Pennsylvania, 1997, 105-116.

[13] L. Qi and D. Sun, Smoothing functions and smoothing Newton method for complementarity and variational inequality problems, *Journal of Optimization, Theory and Applicaitons*, 113 (2002), 121-147.

[14] L. Qi, D. Sun and G. Zhou, A new look at smoothing Newton methods for nonlinear complementarity problems and box constrained variational inequalities, *Mathematical Programming*, 87 (2000), 1-35.

[15] F.H. Clarke, *Optimization and nonsmooth analysis*, John Wiley and Sons. New York, 1983.

[16] C. Chen, O.L. Mangasarian, A class of smoothing functions for nonlinear and mixed complementarity problems, *Computational Optimization and Applications*, 5 (1996), 97-138.

[17] R. Mifflin, Semismooth and semiconvex functions in constrained optimization, *SIAM Journal on Control and Optimization*, 15 (1977), 957-972.

[18] L.Qi, Convergence analysis of some algorithms for solving nonsmooth equations, *Mathematics of Operations Research*, 18 (1993), 227-244.

[19] L. Qi and J. Sun, A nonsmooth version of Newton's method, *Mathematical Programming*, 58 (1993), 353-367.

[20] D.H. Li, L. Qi, J. Tam and S.-Y. Wu, A smoothing Newton method for semi-infinite programming, *Journal of Global Optimizaiton*, to appear.

[21] P.H. Calamai, J.J. More, Projected gradient methods for linear constrained problems, *Mathematical Programming*, 39 (1987), 93-116.

[22] M.C. Ferris, C. Kanzow and T.S. Munson, Feasible descent algorithm for mixed complementarity problems, *Mathematical Programming*, 86 (1999), 475-491.

[23] A.Fischer, A new constrained optimization reformulation for complementarity problems, Journal of Optimizaiton, *Theory and Applicaitons*, 97 (1998), 105-117.

[24] L. Qi, S.-Y. Wu and G. Zhou, Semismooth Newton methods for solving semi-infinite programming problems, *Journal of Global Optimizaiton*, 27 (2003), 215-232.

[25] G.A. Watson, Numerical experiments with globally convergent methods for semi-infinite programming problems, in: *Semi-infinite programming and applications*, A.V. Fiacco and K.O. Kortanek, eds., (Austin, Tex., 1981), 193-205, Lecture Notes in Econom. and Math. Systems, 215, Springer, Berlin-New York.

[26] K.L. Teo, X.Q. Yang and L.S. Jennings, Computational discretization algorithms for functional inequality constrained optimization, *Annals of Operations Research*, 98 (2000), 215-234.

Variational-like Inequalities with Relaxed $\eta-\alpha$ Pseudo-Monotone Mappings in Banach Spaces*

M.R. Bai (白敏茹)† S.Z. Zhou (周叔子)† G.Y. Ni (倪谷炎)‡

Abstract In this paper, we introduce a new concept of relaxed $\eta-\alpha$ pseudomonotonicity. By using the KKM technique, some existence results for variational-like inequalities with relaxed $\eta-\alpha$ pseudo-monotone mappings in reflexive Banach spaces are established.

Keywords Variational-like inequality, relaxed $\eta-\alpha$ pseudomonotone mapping, KKM maping.

1 Introduction

Variational inequality theory plays an important role in many fields of science, engineering and economics. Because of their wide applicability, variational inequality problems have been generalized in various directions for the past several years. For details, we refer to [1–9] and the references therein.

In recent years, a number of authors have proposed many important generalizations of monotonicity such as pseudomonotonicity, relaxed monotonicity, relaxed $\eta-\alpha$ monotonicity, quasimononicity, and semimonotonicity; see [1,6–19] and the references therein. In [6], Verma studied a class of variational inequalities with relaxed monotone operators. In [7], Fang and Huang introduced a new concept of relaxed $\eta-\alpha$ monotonicity and obtained the existence of solutions for variational-like inequalities with relaxed $\eta-\alpha$ monotone mappings in reflexive Banach spaces. In [10], Karamardian and Schaible introduced various kinds of generalized monotone mappings which in case of gradient mappings are related to generalized convex functions. The latter play an important role in certain applications of mathematical programming as well as in economic theory, see [11].

Inspired and motivated by [6, 7], in this paper we introduce a new concept of relaxed $\eta-\alpha$ pseudomonotone mappings. By using the KKM technique, some existence results for variational-like inequalities with relaxed $\eta-\alpha$ pseudo-monotone mappings in reflexive Banach

* 发表于: Applied Mathematics Letters, Vol. 19, 2006, pp. 547-554.

† College of Mathematics and Econometrics, Hunan University, Changsha 410082, China.

‡ Department of Mathematics, National University of Defense Technology, Changsha 410073, China.

spaces are established. Our results are the generalization of many existing works of [2-8] and [12].

2　Some Definition

Let E be a real reflexive Banach space with dual space E^*, and $\langle\cdot,\cdot\rangle$ denote the pairing between E and E^*. Let K be a nonempty subset of E, and 2^E denote the family of all the nonempty subset of E.

Definition 2.1. *A mapping $T : K \to E^*$ is said to be relaxed $\eta - \alpha$ pseudomonotone if there exist a mapping $\eta : K \times K \to E$ and a function $\alpha : E \to R$ with $\alpha(tz) = t^p\alpha(z)$ for all $t > 0$ and $z \in E$ such that, for any $x, y \in K$, we have*

$$\langle Ty, \eta(x,y)\rangle \geqslant 0 \quad \textit{implies} \quad \langle Tx, \eta(x,y)\rangle \geqslant \alpha(x-y), \tag{2.1}$$

where $p > 1$ is a constant.

Special case:

(1) If $\eta(x,y) = x - y\ \forall x, y \in K$, then (1) becomes that for any $x, y \in K$, we have

$$\langle Ty, x-y\rangle \geqslant 0 \text{ implies } \langle Tx, x-y\rangle \geqslant \alpha(x-y),$$

and T is said to be relaxed α pseudomonotone.

(2) If $\eta(x,y) = x - y$, $\forall x, y \in K$ and $\alpha(x,y) \equiv 0$, then (1) reduces to that: for any $x, y \in K$, we have

$$\langle Ty, x-y\rangle \geqslant 0 \ \text{ implies } \ \langle Tx, x-y\rangle \geqslant 0,$$

and T is said to be pseudomonotone; see [8, 10, 12, 13].

If a mapping $T : K \to E^*$ is relaxed $\eta - \alpha$ monotone (see [7]), i.e., there exist a mapping $\eta : K \times K \to E$ and a function $\alpha : E \to R$ with $\alpha(tz) = t^p\alpha(z)$ for all $t > 0$ and $z \in E$ such that

$$\langle Tx - Ty, \eta(x,y)\rangle \geqslant \alpha(x-y), \forall x, y \in K,$$

then it is easy to see that T is relaxed $\eta - \alpha$ pseudomonotone. However, the converse is not true in general, which is showed in the following example.

Example 1. Let $K = (-\infty, +\infty)$, $\eta(x-y) = x - y$, and

$$Tx = \begin{cases} (3/2)x, & x \geqslant 0, \\ -(1/2)x, & x < 0. \end{cases}$$

Then T is relaxed $\eta - \alpha$ pseudomonotone with $\alpha(z) = -(1/2)z^2$. It is easy to see that T is not a relaxed $\eta - \alpha$ monotone mapping.

Definition 2.2. *A mapping $T : K \to E^*$ is said to be B-pseudomonotone if $\{x_n\} \subset K$ is weakly convergent to $x \in K$, we have*

$$\lim_{n\to\infty}\sup\langle Tx_n, x_n - x\rangle \leqslant 0 \quad \textit{implies} \quad \langle Tx, x-y\rangle \leqslant \lim_{n\to\infty}\inf\langle Tx_n, x_n - y\rangle, \forall y \in K \tag{2.2}$$

The notion of B-pseudomonotone mappings was originally introduced by Brezis, see [14]. The following example shows that a relaxed $\eta-\alpha$ pseudomonotone mapping is different from a B-pseudomonotone mapping.

Example 2. Let l^2 be the space of all real functions x on $\{0,1,2,\cdots\}$ whose norm

$$\|x\|=\left\{\sum_{n=0}^{\infty}\|x(n)\|^2\right\}^{1/2}$$

is finite. Let $Tx=-x$ for all $x\in K$, $K=\{x\in l^2:\|x\|\leqslant 1\}$ and $e_n=(\underbrace{0,\cdots,0,1}_{n},0,\cdots)$, $n=1,2,\cdots$.

It is easy to see that e_n converges weakly to 0 as $n\to+\infty$. Now we prove that T is not B-pseudomonotone. Since

$$\langle Te_n,e_n-0\rangle=-1,$$

and so

$$\lim_{n\to\infty}\sup\langle Te_n,e_n-0\rangle=-1\leqslant 0.$$

However, for $y=0\in K$, we have

$$\langle T0,0-y\rangle=0,$$

$$\lim_{n\to\infty}\inf\langle Te_n,e_n-y\rangle=\lim_{n\to\infty}\inf\langle Te_n,e_n-0\rangle=-1,$$

so

$$\langle T0,0-y\rangle\nleqslant\lim_{n\to\infty}\inf\langle Te_n,e_n-y\rangle,$$

which means that the condition (2.2) is not satisfied for $y=0\in K$, i.e., T is not B-pseudomonotone. We can prove that T is relaxed $\eta-\alpha$ pseudomonotone, where $\eta(x,y)=x-y$ and $\alpha(z)=-\|z\|^2$.

We recall the following definition ([1]).

Definition 2.3. *Let $T:K\to E^*$ and $\eta:K\times K\to E$ be two mappings. T is said to be η-hemicontinuous if, for any fixed $x,y\in K$, the mapping $f:[0,1]\to(-\infty,+\infty)$ defined by $f(t)=\langle T(x+t(y-x)),\eta(y,x)\rangle$ is continuous at 0^+.*

The following definition and lemma (see [20]) will be needed in what follows.

Definition 2.4. *A mapping $F:K\to 2^E$ is said to be a KKM mapping if, for any $\{x_1,\cdots,x_n\}\subset K$, co$\{x_1,\cdots,x_n\}\subset\bigcup_{i=1}^{n}F(x_i)$, where co$\{x_1,\cdots,x_n\}$ denotes the convex hull of $x_1,\cdots,x_n$.*

Lemma 2.5. *Let K be a nonempty subset of a Hausdorff topological vector space X and let $F:K\to 2^X$ be a KKM mapping. If $F(x)$ is closed in X for every x in K and compact for some $x_0\in K$, then*

$$\bigcap_{x\in K}F(x)\neq\varnothing.$$

3 Existence Results

In this section, we discuss the existence of the following variational-like inequality: find $x \in K$ such that

$$\langle Tx, \eta(y,x)\rangle \geqslant 0, \quad \forall y \in K, \tag{3.1}$$

where K is a nonempty closed convex subset of a real reflexive Banach space E.

Theorem 3.1. *Let K be a nonempty closed convex subset of a real reflexive Banach space E. Let $T: K \to E^*$ be an η-hemicontinuous and relaxed $\eta-\alpha$ pseudomonotone. Assume that:*

(i) $\eta(x,x)=0$ for all x in K;

(ii) for any fixed y,z in K, the mapping $x \mapsto \langle Tz, \eta(x,y)\rangle$ is convex.

Then, $x \in K$ is a solution of (3.1) if and only if

$$\langle Ty, \eta(y,x)\rangle \geqslant \alpha(y-x), \quad \forall y \in K. \tag{3.2}$$

Proof. Suppose that $x \in K$ is a solution of (3.1). Since T is relaxed $\eta-\alpha$ pseudomonotone, we have

$$\langle Ty, \eta(y,x)\rangle \geqslant \alpha(y-x), \quad \forall y \in K.$$

Hence $x \in K$ is a solution of (3.2).

Conversely, suppose that $x \in K$ is a solution of (3.2) and $y \in K$ be any point. Letting

$$x_t = ty + (1-t)x, \quad t \in (0,1],$$

then $x_t \in K$. It follows from (3.2) that

$$\langle Tx_t, \eta(x_t,x)\rangle \geqslant \alpha(x_t - x) = \alpha(t(y-x)) = t^p\alpha(y-x). \tag{3.3}$$

By conditions (i) and (ii), we have

$$\langle Tx_t, \eta(x_t,x)\rangle \leqslant t\langle Tx_t, \eta(y,x)\rangle + (1-t)\langle Tx_t, \eta(x,x)\rangle = t\langle Tx_t, \eta(y,x)\rangle. \tag{3.4}$$

It follows from (3.3) and (3.4) that

$$\langle Tx_t, \eta(y,x)\rangle \geqslant t^{p-1}\alpha(y-x), \tag{3.5}$$

for all y in K. Since T is η-hemicontinuous and $p>1$, letting $t \to 0$ in (3.5), we get

$$\langle Tx, \eta(y,x)\rangle \geqslant 0, \quad \forall y \in K.$$

Therefore $x \in K$ is a solution of (3.1). □

Theorem 3.2. *Let K be a bounded closed convex subset of the reflexive Banach space E, and $T: K \to E^*$ be η-hemicontinuous relaxed $\eta-\alpha$ pseudomonotone. Assume that*

(i) $\eta(x,y)+\eta(y,x)=0$, for all x,y in K;

(ii) for any fixed y, z in K, the mapping $x \mapsto \langle Tz, \eta(x,y)\rangle$ is convex and lower semicontinuous;

(iii) $\alpha : E \to R$ is weakly lower semicontinuous, i.e., for any net $\{x_\beta\} \subset E$, x_β converges weakly to x_0 implies that

$$\alpha(x_0) \leqslant \liminf_{\beta} \alpha(x_\beta).$$

Then there exists at least one solution for (3.1).

Proof. For any $y \in K$, define two set-valued mappings $F, G : K \to 2^K$ as follows:

$$F(y)=\{x \in K : \langle Tx, \eta(y,x)\rangle \geqslant 0\},$$
$$G(y)=\{x \in K : \langle Ty, \eta(y,x)\rangle \geqslant \alpha(y-x)\}.$$

We claim that F is a KKM mapping. If F is not a KKM mapping, then there exists $\{y_1, \cdots, y_n\} \subset K$ such that

$$co\{y_1, \cdots, y_n\} \not\subset \bigcup_{i=1}^{n} F(y_i),$$

i.e., there exists a $y_0 \in co\{y_1, \cdots, y_n\}$, $y_0 = \sum_{i=1}^{n} t_i y_i$, where $t_i \geqslant 0$, $i = 1, \cdots, n$, $\sum_{i=1}^{n} t_i = 1$, but $y_0 \notin \bigcup_{i=1}^{n} F(y_i)$. By the definition of F, we have

$$\langle Ty_0, \eta(y_i, y_0)\rangle < 0,$$

for $i = 1, \cdots, n$. It follows from conditions (i) and (ii) that

$$0 = \langle Ty_0, \eta(y_0, y_0)\rangle = \langle Ty_0, \eta\left(\sum_{i=1}^{n} t_i y_i, y_0\right)\rangle \leqslant \sum_{i=1}^{n} t_i \langle Ty_0, \eta(y_i, y_0)\rangle < 0,$$

which is a contradiction. This implies that F is a KKM mapping. Now we prove that

$$F(y) \subset G(y), \quad \forall y \in K.$$

For any given $y \in K$, letting $x \in F(y)$, then

$$\langle Tx, \eta(y,x)\rangle \geqslant 0.$$

Since T is relaxed $\eta - \alpha$ pseudomonotone, we have

$$\langle Ty, \eta(y,x)\rangle \geqslant \alpha(y-x).$$

It follows that $x \in G(y)$ and so

$$F(y) \subset G(y), \ \ \forall\, y \in K. \tag{3.6}$$

This implies that G is also a KKM mapping.

Since $x \mapsto \langle Tz, \eta(x,y)\rangle$ is convex lower semicontinuous, we know that it is weakly lower semicontinuous. From the definition of G and the weakly lower semicontinuity of α, it is easy to see that $G(y)$ is weakly closed for all $y \in K$. Since K is bounded closed and convex, we known that K is weakly compact, and so $G(y)$ is weakly compact in K for each $y \in K$. Therefore, the conditions of Lemma 2.5 are satisfied in the weak topology. It follows from Lemma 2.5 and Theorem 3.1 that

$$\bigcap_{y \in K} F(y) = \bigcap_{y \in K} G(y) \neq \varnothing.$$

Hence, there exists $x \in K$ such that

$$\langle Tx, \eta(y, x) \rangle \geqslant 0, \quad \forall\, y \in K.$$

□

Theorem 3.3. *Let K be an unbounded closed convex subset of the reflexive Banach space E, and $T : K \to E^*$ be η-hemicontinuous relaxed $\eta - \alpha$ pseudomonotone. Assume that*

(i) $\eta(x, y) + \eta(y, x) = 0$, *for all x, y in K;*

(ii) for any fixed y, z in K, the mapping $x \mapsto \langle Tz, \eta(x, y) \rangle$ is convex and lower semicontinuous;

(iii) $\alpha : E \to R$ is weakly lower semicontinuous, i.e., for any net $\{x_\beta\} \subset E$, x_β converges weakly to x_0 implies that

$$\alpha(x_0) \leqslant \liminf_{\beta} \alpha(x_\beta);$$

(iv) T is weakly η-coercive, i.e., there exists $x_0 \in K$ such that

$$\langle Tx, \eta(x, x_0) > 0,$$

whenever $\|x\| \to +\infty$ and $x \in K$.

Then there exists at least one solution for (3.1).

Proof. For $r > 0$, letting

$$B_r = \{y \in X : \|y\| \leqslant r\}.$$

Consider the following problem: find $x_r \in K \cap B_r$ such that

$$\langle Tx_r, \eta(y, x_r) \rangle \geqslant 0, \quad \forall\, y \in K \cap B_r. \tag{3.7}$$

By Theorem 3.2, we known that problem (3.7) has one solution $x_r \in K \cap B_r$. Choose $r > \|x_0\|$ with x_0 as in the coercive condition. Then $x_0 \in K \cap B_r$ and

$$\langle Tx_r, \eta(x_0, x_r) \rangle \geqslant 0. \tag{3.8}$$

By assumption (i), we get

$$\langle Tx_r, \eta(x_0, x_r) \rangle = -\langle Tx_r, \eta(x_r, x_0) \rangle$$

If $\|x_r\| = r$ for all r, we may choose r large enough so that the above equality and the weak η-coercivity of T imply that

$$\langle Tx_r, \eta(x_0, x_r) \rangle < 0,$$

which contradicts (3.8). So there exists r such that $\|x_r\| < r$. For any $y \in K$, we can choose $0 < \varepsilon < 1$ small enough such that

$$x_r + \varepsilon(y - x_r) \in K \cap B_r.$$

It follows from (3.7) that

$$\begin{aligned} 0 &\leqslant \langle Tx_r, \eta(x_r + \varepsilon(y - x_r), x_r)\rangle \\ &\leqslant (1-\varepsilon)\langle Tx_r, \eta(x_r, x_r)\rangle + \varepsilon\langle Tx_r, \eta(y, x_r)\rangle \\ &= \varepsilon\langle Tx_r, \eta(y, x_r)\rangle, \end{aligned}$$

which implies x_r such that

$$\langle Tx_r, \eta(y, x_r)\rangle \geqslant 0, \quad \forall\, y \in K.$$

This completes the proof. □

Remark 3.1.

(1) If instead T is continuous on finite dimensional subspaces, the conclusions of Theorem 3.1, Theorem 3.2 and Theorem 3.3 are also true.

(2) Theorem 3.2 and Theorem 3.3 improve and generalize the known results of Hartman-Stampacchia ([4]) and the corresponding results of [2, 3, 5, 6, 8, 12]. Theorem 3.2 and Theorem 3.3 also improve and generalize the corresponding results of [7] when $f \equiv 0$.

References

[1] Y.Q. Chen, On the semimonotone operator theory and applications, *Journal of Mathematical Analysis and Applications*, 231 (1999), 177-192.

[2] D. Goeleven and D. Motreanu, Eigenvalue and dynamic problems for variational and hemivariational inequalities, *Communications on Applied Nonlinear Analysis*, 3 (1996), 1-21.

[3] A.H. Siddqi, Q.H. Ansari and K.R. Kazmi, On nonlinear variational inequalities, *Indian Journal of Pure and Applied Mathematics*, 25 (1994), 969-973.

[4] G.J. Hartman and G. Stampacchia, On some nonlinear elliptic differential functional equations, *Acta mathematica*, 115 (1966), 271-310.

[5] R.U. Verma, On monotone nonlinear variational inequalities problems, *Commentationes Mathematicae Universitatis Carolinae*, 39 (1998), 91-98.

[6] R.U. Verma, On generalized variational inequalities involving relaxed Lipschitz and relaxed monotone operators, *Journal of Mathematical Analysis and Applications*, 213 (1997), 387-392.

[7] Y.P. Fang and N.J. Huang, Variational-like inequalities with generalized monotone mappings in Banach spaces, *Journal of Optimization, Theory and Applications*, 118 (2003), 327-338.

[8] R.W. Cottle and J.C. Yao, Pseudomonotone complementarity problems in Hilbert spaces, *Journal of Optimization, Theory and Applications*, 78 (1992), 281-295.

[9] N.J. Huang, M.R. Bai, Y.J. Cho and S.M. Kang, Generalized nonlinear mixed quasi-Variational inequalities, *Computers and Mathematics with Applicaitons*, 40 (2000), 205-215.

[10] S. Karamardian and S. Schaible, Seven kinds of monotone maps, *Journal of Optimization, Theory and Applications*, 66 (1990), 37-46.

[11] M. Avriel, W.E. Diewert, S. Schaible and I. Zang, *Generalized Concavity*, Plenum Publishing Corporation, New York, (1988).

[12] S. Karamardian, Complementarity over cones with monotone and pseudomonotone maps, *Journal of Optimization, Theory and Applications*, 18 (1976), 445-454.

[13] S. Karamardian, S. Schaible and J.P. Crouzeix, Characterizations of generalized monotone maps, *Journal of Optimization, Theory and Applications*, 76 (1993), 399-413.

[14] H. Brezis, Équations et inéquations non linéaires dans les espaces vectoriels en dualité, *Annales-Institut Fourier*, 18 (1968), 115-175.

[15] I.V. Konnov and S. Schaible, Duality for equalibrium problems under generalized monotonicity, *Journal of Optimization, Theory and Applications*, 104 (2000), 395-408.

[16] I.V. Konnov, Application of the proximal point method to nonmonotone equilibrium problems, *Journal of Optimization, Theory and Applications*, 119 (2003), 317-333.

[17] N. El Farouq, Pseudomonotone variational inequalities: convergence of proximal method, *Journal of Optimization, Theory and Applications*, 109 (2001), 311-326.

[18] Y.J. Wang, N.H. Xiu and J.Z. Zhang, Modified extragradient method for variational inequalities and verification of solution existence, *Journal of Optimization, Theory and Applications*, 119 (2003), 167-183.

[19] D.T. Luc, Existence results for densely pseudomonotone variational inequalities, *Journal of Mathematical Analysis and Applications*, 254 (2001), 309-320.

[20] K. Fan, Some properties of convex sets related to fixed-point theorems, *Mathematische Annalen*, 266 (1984), 519-537.

非线性互补约束优化问题的可行性条件 *

万 中 † 周叔子 †

摘要: 本文研究了非线性互补约束优化问题的可行性条件, 其中约束条件除含有带参数 (决策变量) 的互补问题外, 还包括第一水平 (设计) 变量和第二水平 (状态) 变量同时出现的其他非线性约束. 它是线性互补约束优化问题的可行性条件的推广.

关键词: 带均衡约束的数学规划; 非线性互补; 序列二次规划; 可行性.

1 引言

带均衡约束的优化问题 (简称 MPEC) 是指约束条件中除通常的等式或不等式约束条件外, 还包含带参数 (模型的决策变量) 的互补条件或带参数的变分不等式, 后者称为均衡约束条件. 至今该类问题的大部分主要分析结果局限于仅仅允许第一水平 (设计) 变量与第二水平 (状态) 变量同时出现于均衡约束条件中, 在其他的等式或不等式约束中不能包含第二水平变量 (参看文献 [1-4]). 其根本原因在于无法解决 MPEC 本身或数值方法 (如磨光 SQP 方法) 中构造的子问题的可行性问题, 寻找可行域非空的条件.

文献 [5] 第一次给出了线性互补约束的数学规划问题的如下可行性充分条件:

$$
\begin{aligned}
\text{minimize} \quad & f(x,y) \\
\text{subject to} \quad & Ax+By=b, \\
& w = Nx+My-q, \\
& w^T y=0, x\in \mathcal{C}, y\geqslant 0, w\geqslant 0,
\end{aligned}
\tag{1.1}
$$

其中 $f: R^{n+m}\to R$ 连续可微, $A\in R^{p\times n}$, $B\in R^{p\times m}$, $N\in R^{m\times n}$, $M\in R^{m\times m}$ 为已知矩阵, $b\in R^p$ 和 $q\in R^n$ 为已知向量, $\mathcal{C}$ 为 R^n 中的凸多面锥, 均衡约束条件是线性互补问题. 本文在文献 [5] 的结果的基础上进一步讨论带非线性互补问题的 MPEC 的可行性条件.

2 问题及其可行性条件

我们考虑如下均衡约束优化问题:

* 发表于: 应用数学学报, Vol. 26, 2003, pp. 646-651.

† 湖南大学数学与计量经济学院, 湖南长沙, 410082.

$$
\begin{aligned}
\text{minimize} \quad & f(x,y) \\
\text{subject to} \quad & g(x,y)=0, \\
& h(x,y)\geqslant 0, \\
& 0\leqslant y\perp F(x,y)\geqslant 0,
\end{aligned} \tag{2.1}
$$

其中 $f:R^{n+m}\to R$, $g:R^{n+m}\to R^l$, $h:R^{n+m}\to R^p$, $F:R^{n+m}\to R^m$ 连续可微, x 为第一水平 (设计) 变量, y 为第二水平 (状态) 变量, 均衡约束条件为带决策变量 x 的非线性互补问题:

$$0\leqslant y\perp F(x,y)\geqslant 0.$$

针对问题 (2.1), 我们引入两个假设:

(A_1): 蕴含式

$$
\left.\begin{array}{r}
\left[\nabla_x g^T, -\nabla_x h^T\right]u+\nabla_x F^T v=0 \\
v\circ([\nabla_y g^T, -\nabla_y h^T]u+\nabla_y F^T v)\leqslant 0
\end{array}\right\}\Rightarrow v\circ\left(\left(\nabla_y g^T, -\nabla_y h^T\right)u+\nabla_y F^T v\right)=0 \tag{2.2}
$$

成立, 其中 $\circ$ 表示 Hadamard 乘积.

(A_2): 下列非线性系统的解集 $\mathcal{X}$ 非空,

$$
\begin{cases}
g(x,y)=0, \\
h(x,y)\geqslant 0, \\
y\geqslant 0, \\
F(x,y)\geqslant 0,
\end{cases} \tag{2.3}
$$

且

$$SFD(x,\mathcal{X})=LFD(x,\mathcal{X}), \qquad \forall x\in\mathcal{X}. \tag{2.4}$$

这里, SFD 和 LFD 分别表示序列可行方向集和线性化可行方向集 (见 [6]). 实际上, (2.4) 可减弱到仅要求在下列非线性规划问题的全局极小点处成立:

$$
\begin{aligned}
\text{minimize} \quad & y^T F(x,y) \\
\text{subject to} \quad & (2.2).
\end{aligned} \tag{2.5}
$$

文献 [6] 给出了许多使 (2.4) 式成立的充分性条件.

注. 可以验证当 (2.1) 中的 $g(x,y)$, $h(x,y)$ 和 $F(x,y)$ 为仿射函数时, 文献 [5] 中两个可行性充分条件蕴含假设 (A_1) 和 (A_2) 成立.

对给定的 (x,y), 用 $\mathcal{S}$ 表示下列线性方程组的解集:

$$\left[\nabla_x g^T, -\nabla_x h^T\right]u+\nabla_x F^T v=0,$$

则 $\mathcal{S}$ 为闭凸锥. 下面进一步给出 (A_1) 成立的两个充分性条件.

定理2.1. *对给定的 (x,y), 如果矩阵*

$$
\begin{pmatrix}
0 & 0 & \nabla_y g \\
0 & 0 & -\nabla_y h \\
\nabla_y g^T & -\nabla_y h^T & 2\nabla_y F
\end{pmatrix}\equiv\mathcal{P}
$$

在 $\mathcal{S}$ 上协正定, 则假设 (A_1) 成立.

证明: 设 (u,v) 满足蕴含式 (2.2) 的左边, 则 $(u,v)\in\mathcal{S}$. 因此, 根据 $\mathcal{P}$ 在 $\mathcal{S}$ 上协正定, 我们有

$$\begin{aligned}(u^T,v^T)\mathcal{P}\begin{pmatrix}u\\v\end{pmatrix}&=(u^T,v^T)\begin{pmatrix}0&0&\nabla_y g\\0&0&-\nabla_y h\\\nabla_y g^T&-\nabla_y h^T&2\nabla_y F\end{pmatrix}\begin{pmatrix}u\\v\end{pmatrix}\\&=(u^T,v^T)\left(\begin{pmatrix}0&0\\0&0\\\nabla_y g^T&-\nabla_y h^T\end{pmatrix}u+\begin{pmatrix}\nabla_y g\\-\nabla_y h\\2\nabla_y F\end{pmatrix}v\right)\\&=u^T\begin{pmatrix}\nabla_y g\\-\nabla_y h\end{pmatrix}v+v^T\left[\nabla_y g^T,-\nabla_y h^T\right]u+2v^T\nabla_y Fv\\&=2v^T[\nabla_y g^T,-\nabla_y h^T]u+2v^T\nabla_y Fv\\&=2v^T\left(\left[\nabla_y g^T,-\nabla_y h^T\right]u+\nabla_y F^T v\right)\geqslant 0.\end{aligned}$$

考虑到 (u,v) 满足条件:

$$v\circ([\nabla_y g^T,-\nabla_y h^T]u+\nabla_y F^T v)\leqslant 0,$$

可以断言 $v\circ([\nabla_y g^T,-\nabla_y h^T]u+\nabla_y F^T v)=0$. 否则, 存在 $i_0\in\{1,2,\cdots,m\}$, 使得

$$v_{i_0}\circ\left(\left[\nabla_y g^T,-\nabla_y h^T\right]_{i_0}u+\left(\nabla_y F^T\right)_{i_0}v\right)<0.$$

从而与下式矛盾:

$$v^T\left(\left[\nabla_y g^T,-\nabla_y h^T\right]u+\nabla_y F^T\right)v\geqslant 0,$$

故蕴含式 (2.2) 得证. ■

定理2.2. *假设 $\nabla_y F$ 为半正定阵, 对给定的 $(x,y)\in R^{n+m}$, 下列关于变量 $z\in R^n$ 的线性方程组:*

$$\begin{pmatrix}\nabla_x g\\-\nabla_x h\\\nabla_x F\end{pmatrix}z=\begin{pmatrix}0\\{[\nabla_y g^T,-\nabla_y h^T]u}\end{pmatrix}\tag{2.6}$$

对一切 $u\in R^{p+l}$ 恒有解, 则 (A_1) 成立.

证明: 假设 (u,v) 满足蕴含式 (2.2) 的左边, 则有

$$\left[\nabla_x g^T,-\nabla_x h^T\right]u+\nabla_x F^T v=0,\tag{2.7}$$

与

$$v\circ\left(\left[\nabla_y g^T,-\nabla_y h^T\right]u+\nabla_y F^T v\right)\leqslant 0,\tag{2.8}$$

成立. 由 (2.8) 式可得

$$v^T\left(\left[\nabla_y g^T,-\nabla_y h^T\right]u+\nabla_y F^T v\right)\leqslant 0,$$

即

$$v\circ\left[\nabla_y g^T,-\nabla_y h^T\right]u\leqslant -v^T\nabla_y F^T v\leqslant 0,\tag{2.9}$$

其中第二个不等式用到 $\nabla_y F$ 的半正定性.

对上述 u, 由于线性方程组 (2.6) 式有解, 不妨记之为 $\bar{z}$. 则有

$$\nabla_x F\bar{z} = [\nabla_y g^T, -\nabla_y h^T]u, \quad \begin{pmatrix}\nabla_x g\\ -\nabla_x h\end{pmatrix}\bar{z} = 0.$$

因此,

$$\begin{aligned} v^T\left[\nabla_y g^T, -\nabla_y h^T\right]u &= v^T\nabla_x F\bar{z} + u^T\begin{pmatrix}\nabla_x g\\ -\nabla_x h\end{pmatrix}\bar{z} \\ &= (\bar{z})^T\left(\left[\nabla_x g^T, -\nabla_x h^T\right]u + \nabla_x F^T v\right) = 0. \end{aligned} \tag{2.10}$$

利用 (2.9), (2.10) 可得:

$$v^T\nabla_x F^T v = 0.$$

因而

$$v^T([\nabla_y g^T, -\nabla_y h^T]u + \nabla_y F^T v = 0.$$

从而不难证明:

$$v\circ\left(\left[\nabla_y g^T, -\nabla_y h^T\right]u + \nabla_y F^T v\right) = 0.$$

■

定理2.3. *在假设 (A_1) 和 (A_2) 成立时, 且 (2.3) 定义的解集 $\mathcal{X}$ 与 $y^TF(x,y)$ 的水平集的交集有界. MPEC (2.1) 具有非空可行域.*

证明: 由于 $y\geqslant 0$, $F(x,y)\geqslant 0$, 所以函数 $y^TF(x,y)$ 下有界. 因此 (2.5) 必存在最优解 $(\bar{x},\bar{y})$ 和相应乘子向量 $(\lambda,\mu,\theta,\zeta)\in R^{l+p+2m}$ 使得:

$$\begin{cases} lr\nabla_x F^T y + \nabla_x g^T\lambda - \nabla_x h^T\mu - \nabla_x F^T\zeta = 0, & \\ F(x,y) + \nabla_y F^T y + \nabla_y g^T\lambda - \nabla_y h^T\mu - \theta - \nabla_y F^T\zeta = 0, & \\ \mu_i\geqslant 0, h_i(x,y)\geqslant 0, \mu_i h_i(x,y) = 0, & i = 1,2,\cdots,p, \\ \theta_j\geqslant 0, y_j\geqslant 0, \theta_j y_j = 0, & j = 1,2,\cdots,m, \\ \zeta_k\geqslant 0, F_k(x,y)\geqslant 0, \zeta_k F_k(x,y) = 0, & k = 1,2,\cdots,m. \end{cases}$$

即

$$\nabla_x F^T(y-\zeta) + \left[\nabla_x g^T, -\nabla_x h^T\right]\begin{pmatrix}\lambda\\ \mu\end{pmatrix} \tag{2.11}$$

$$\nabla_y F^T(y-\zeta) + \left[\nabla_y g^T, -\nabla_y h^T\right]\begin{pmatrix}\lambda\\ \mu\end{pmatrix} = \theta - F(x,y), \tag{2.12}$$

$$\mu\geqslant 0, h(x,y)\geqslant 0, \mu\circ h(x,y) = 0, \tag{2.13}$$

$$\theta\geqslant 0, y\geqslant 0, \theta\circ y = 0, \tag{2.14}$$

以及

$$\zeta\geqslant 0, F(x,y)\geqslant 0, \zeta\circ F(x,y) = 0. \tag{2.15}$$

记

$$\phi \equiv \theta - F(x,y), \psi \equiv \nabla_y F^T(y-\zeta) + \left[\nabla_y g^T, -\nabla_y h^T\right] \begin{pmatrix} \lambda \\ \mu \end{pmatrix}.$$

在 (2.12) 式两边同乘以 y, 则

$$y \circ F(x,y) = y \circ \theta - y \circ \psi.$$

根据 (2.14) 式有

$$y \circ F(x,y) = -y \circ \psi.$$

又因为

$$\begin{aligned}(y-\zeta)\circ\psi &= (y-\zeta)\circ(\theta - F(x,y))\\ &= y\circ\theta - y\circ F(x,y) - \zeta\circ\theta + \zeta\circ F(x,y)\\ &= -y\circ F(x,y) - \zeta\circ\theta \leqslant 0,\end{aligned}$$

其中第三个等式和最后的不等式由 (2.14) 和 (2.15) 式得到. 令

$$u = \begin{pmatrix} \lambda \\ \mu \end{pmatrix}, v = y - \zeta,$$

则由假设 (A_1) 和 (2.11) 式可得

$$(y-\zeta)\circ\psi = 0.$$

从而, $y \circ F(x,y) = -y\circ\psi = -\zeta\circ\psi = -\zeta\circ(\theta - F(x,y)) = -\zeta\circ\theta \leqslant 0.$

此外, $y \geqslant 0, F(x,y) \geqslant 0$, 从而, $y \circ F(x,y) = 0$. 由此可知,

$$y^T F(x,y) = 0. \tag{2.16}$$

否则, 存在 $i_0 \in \{1,2,\cdots,m\}$, 使得 $y_{i_0}F_{i_0} \neq 0$. 从而, $y_{i_0}F_{i_0}(x,y) > 0$. 从而与 (2.16) 矛盾. ■

3　一类 MPEC 的 SQP 算法中子问题的可行性

文献 [7] 研究了求解下述带非线性互补问题约束优化问题的磨光 SQP 算法:

$$\begin{aligned}\text{minimize}\quad & f(x,y)\\ \text{subject to}\quad & h(x,y) \geqslant 0,\\ & 0 \leqslant y \perp F(x,y) \geqslant 0.\end{aligned}$$

它是问题 (2.1) 略去等式约束时的特殊情形. 具体地说, 该算法将上述 MPEC 磨光为通常的带参数的非线性规划问题:

$$\begin{aligned}\text{minimize}\quad & f(x,y)\\ \text{subject to}\quad & h(x,y) \leqslant 0,\\ & F(x,y) - w = 0,\\ & \Phi(y,w,\mu) = 0,\end{aligned} \tag{3.1}$$

其中 μ 为非负磨光参数, 函数 $\Phi: R^{2m}\times[0,\infty)\to R^m$ 定义为

$$\Phi(y,w,\mu)\equiv\begin{pmatrix} y_1+w_1-\left(y_1^2+w_1^2+\mu\right)^{1/2} \\ \vdots \\ y_m+w_m-\left(y_m^2+w_m^2+\mu\right)^{1/2}\end{pmatrix}.$$

该算法的所有迭代步均需求解 (3.1) 的如下二次逼近问题:

$$\begin{aligned}\text{minimize}\quad & \nabla f(x^k,y^k)^T\begin{pmatrix} dx \\ dy\end{pmatrix}+\frac{1}{2}(dx^T,dy^T,dw^T)Q_k\begin{pmatrix} dx \\ dy \\ dw\end{pmatrix} \\ \text{subject to}\quad & \nabla_x g dx+\nabla_y g dy+g(x^k,y^k)\leqslant 0, \\ & \nabla_x F dx+\nabla_y F dy-dw+\left(F(x^k,y^k)-w^k\right)=0, \\ & D_y^k dy+D_w^k dw+\Phi\left(y^k,w^k,\mu_k\right)=0.\end{aligned}\tag{3.2}$$

这里, $Q_k\in R^{(n+2m)\times(n+2m)}$ 为对称正定阵, 而

$$\begin{aligned} &\nabla_x g\equiv\nabla_x g(x^k,y^k), && \nabla_y g\equiv\nabla_y g(x^k,y^k), \\ &\nabla_x F\equiv\nabla_x F(x^k,y^k), && \nabla_y F\equiv\nabla_y F(x^k,y^k), \\ &D_y^k\equiv\nabla_y\Phi(y^k,w^k,\mu_k), && D_w^k\equiv\nabla_w\Phi(y^k,w^k,\mu_k).\end{aligned}$$

我们将证明假设 (A_1) 能够保证子问题 (3.2) 的可行性.

消去 (3.2) 约束条件中的 dw, 则约束条件可写成:

$$\begin{cases}\nabla_x g dx+\nabla_y g dy\leqslant -g(x^k,y^k), \\ (D_w^k\nabla_x F)dx+(D_y^k+D_w^k\nabla_y F)dy=s^k,\end{cases}$$

其中,

$$s^k\equiv-\Phi(y^k,w^k,\mu_k)+D_w^k\left(F(x^k,y^k)-w^k\right).$$

或写成:

$$\begin{cases}\nabla_x g dx+\nabla_y g dy\leqslant -g(x^k,y^k) \\ \nabla_x F dx+((D_w^k)^{-1}D_y^k+\nabla_y F)dy=(D_w^k)^{-1}s^k\end{cases}\tag{3.3}$$

由文献 [8, 定理 2.7.8] 可知, (3.3) 式关于变量 dx, dy 有解等价于

$$\left.\begin{aligned}\nabla_x g^T u+\nabla_x F^T v=0 \\ \nabla_y g^T u+((D_w^k)^{-1}D_y^k+\nabla_y F)^T v=0 \\ (u,v)\geqslant 0\end{aligned}\right\}\Rightarrow -g(x^k,y^k)^T u+(D_w^k)^{-1}s^k v\geqslant 0.\tag{3.4}$$

只要 $\nabla_y g^T u+\left(\left(D_w^k\right)^{-1}D_y^k+\nabla_y F\right)^T v=0$, 由于 D_y^k, D_w^k 均为正的对角阵, 则必有

$$v\circ\left[\nabla_y g^T u+\nabla_y F^T v\right]=-v\circ\left(\left(D_y^k\right)^T\left(D_w^k\right)^{-T}v\right)\leqslant 0.\tag{3.5}$$

为了证明蕴含式 (3.4) 成立, 设存在 (u,v) 使 (3.4) 式的左边成立, 从而有 (3.5) 式成立. 利用假设 (A_1), 可得 $v^T[\nabla_y g^T u+\nabla_y F^T]=0$. 即 $-v^T(D_y^k)(D_w^k)^{-1}v=0$. 因此 $v=0$. 如果我们要求对所有的 k, 恒有 $g(x^k,y^k)\leqslant 0$, 则由 $u\geqslant 0$ 可知蕴含式 (3.4) 成立.

综上所述, 我们得到如下结论:

定理3.1. 如果假设 (A_1) 成立，且对所有的 k，恒有 $g(x^k, y^k) \leqslant 0$. 则 QP 子问题 (3.2) 可行.

参考文献

[1] F. Facchinei, H. Jiang and L. Qi, A smoothing method for mathematical programs with equilibrium constraints, Mathematical Programming, 85 (1999), 107-134.

[2] M. Fukushima, Z.Q. Luo and J.S. Pang, A globally convergent sequential quadratic programming algorithm for mathematical programs with linear complementarity constraints, Computational Optimization and Applicaitons, 10 (1998), 5-34.

[3] Z.Q. Luo, J.S. Pang and D. Ralph, Mathematical Progroms With Equilibrium Constraints, Cambridge University Press, London, 1996.

[4] J.V. Outrata, M.Kočvar, J. Zowe, Nonsmooth Approach To Optimization Problems With Equilibrium Constraints, Kluwer Academic Publishers, 1998.

[5] M. Fukushima and J.S. Pang, Some feasibility issues in mathematical programs with equilibrium constraints, SIAM Journal on Optimizaiton, 8 (1998), 673-681.

[6] 袁亚湘, 孙文瑜, 最优化理论与算法, 科学出版社, 北京, 1997.

[7] H.Y. Jiang and D. Ralph, Smooth SQP methods for mathematical programs with nonlinear complementarity constraints, SIAM Journal on Optimization, 10 (2000), 779-808.

[8] R.W. Cottle, J.S. Pang and R.E. Stone, The Linear Complementarity Problems, Academic Press, New York, 1992.

On the Feasibility of Mathematical Programs with Nonlinear Complementarity Constraints

Z. Wan, S.Z. Zhou

Abstract This paper studies the feasibility of mathematical programs with nonlinear complementarity constraints(MPEC), where additional joint constraints exist such that they must be satisfied by the first-level (design) and the second-level (state) variables of the problem. It is an extension of the feasibility conditions for the mathematical programs with linear complementarity constraints.

Keywords mathematical programs with equilibrium constraints, nonlinear complementarity, sequential quadratic programming, feasibility.

第三篇 指导的部分学生论文

Translating Solitons of Mean Curvature Flow of Noncompact Spacelike Hypersurfaces in Minkowski Space*

H.Y. Jian (简怀玉)†

我于 1985 年 9 月到 1988 年 7 月在湖南大学念硕士, 期间周叔子教授给我们讲授二阶椭圆型偏微分方程、Sobolev 空间和变分不等式理论及应用等课程, 正是这三门课程为我从事偏微分方程的研究打下了坚实的基础. 周老师当时就跟我们说: "L. Caffarelli 虽然年轻, 但他做的自由边值问题的工作代表当今偏微研究的最高水平." 他还指导我们办起了读 Caffarelli 论文的讨论班, 推荐我去北大听 Caffarelli 的学术报告. 我的第一篇论文就是关于自由边值问题的, 是周老师给我修改后才得以投稿发表的. 三十多年过去了, 回想起来当初周老师的眼光是何等的独到, 因为最近三十年 Caffarelli 一直是国际上椭圆和抛物型偏微分方程研究领域最顶尖和最有成就的学者之一. 谨以此文纪念我敬爱的老师周叔子教授!

Abstract In this paper, we study the existence, uniqueness and asymptotic behavior of rotationally symmetric translating solitons of the mean curvature flow in Minkowski space. We also study the asymptotic behavior and the strict convexity of general solitons of such flows.

Keywords Spacelike hypersurface, elliptic equation, singular ODE, asymptotic behavior, strictly convex solution.

---

1 Introduction

Minkowski space $R^{n,1}$ is the linear space R^{n+1} endowed with the Lorentz metric

$$ds^2 = \sum_{i=1}^{n} dx_i^2 - dx_{n+1}^2.$$

Spacelike hypersurfaces in $R^{n,1}$ are Riemanian n-manifolds, having an everywhere lightlike normal field ν which we assume to be future directed and thus satisfy the condition $<\nu,\nu>=$

---

* 发表于: Journal of Differential Equations, Vol. 220, 2006, pp. 147-162.

† Department of Mathematics, Tsinghua University, Beijing 100084, China.

-1. Locally, such surfaces can be expressed as graphs of functions $x_{n+1} = u(x_1, \cdots, x_n)$: $R^n \longmapsto R$ satisfying the spacelike conditions $|\nabla u(x)| < 1$ for all $x \in R^n$.

If a family of spacelike embeddings $X_t = X(\cdot, t) : R^n \mapsto R^{n,1}$ with corresponding hypersurfaces $M_t = X(R^n, t)$ satisfy the evolution equation

$$\frac{\partial X}{\partial t} = H\nu \tag{1.1}$$

on some time interval, we say that the surfaces $\{M_t\}$ are evolved by *Mean Curvature Flow* (MCF). Here $H = \text{div}\,_{M_t}\nu$ denotes the mean curvature of the hypersurface M_t. Let $V(\cdot, t)$ be the graph expression of M_t. Then $|\nabla V(\cdot, t)| < 1$ and MCF equation (1.1) is equivalent, up to a diffeomorphism in R^n, to the equation

$$\frac{\partial V}{\partial t} = \sqrt{1 - |\nabla V|^2}\, \text{div} \left(\frac{\nabla V}{\sqrt{1 - |\nabla V|^2}} \right) \quad \text{in} \quad R^n. \tag{1.2}$$

MCF has been extensively studied in Euclidean space; see [1] and the references therein, while in Minkowski space, MCF was studied in [2, 3] for compact hypersurfaces and in [4, 5] for noncompact hypersurfaces. The method of MCF was used in [2, 3] to constructed spacelike hypersurfaces with prescribed mean curvature, which, as it is well-known, have played important roles in studying Lorentzian manifolds. In particular, maximal hypersurfaces, i.e., the ones with zero mean curvature, were used by Schoen and Yau in the first proof of the famous positive mass theorem [6].

The solutions of MCF (1.1) (or (1.2), equivalently) which move by vertical translation are called *Translating Solitons*. Therefore, a translating soliton of MCF (1.2) is characterized by $V(x, t) = u(x) + t$, where $u : R^n \mapsto R$ is an initial spacelike hypersurface satisfying

$$\text{div} \left(\frac{\nabla u(x)}{\sqrt{1 - |\nabla u(x)|^2}} \right) = \frac{1}{\sqrt{1 - |\nabla u(x)|^2}}, \forall x \in R^n. \tag{1.3}$$

The spacelike condition reads as

$$|\nabla u(x)| < 1, \forall x \in R^n. \tag{1.4}$$

Translating solitons can be regarded as a natural way of foliating spacetimes by almost null like hypersurfaces. It may be expected that this kind of translating solitons would have applications in general relativity [5]. For this purpose, it is useful to understand their geometric structure sufficiently. In [5], the existence of smooth solutions of (1.3)–(1.4) was proved by a PDE method. However, using ODE techniques we can find strictly convex radially symmetric solutions of (1.3)–(1.4).

Theorem 1.1. *There exists exact one solution $r \in C^2[0, \infty)$ to initial value problem*

$$\frac{r''(t)}{1 - (r'(t))^2} + \frac{n-1}{t} r'(t) = 1, \quad t \in (0, \infty) \tag{1.5}$$

and

$$r(0) = r'(0) = 0 \tag{1.6}$$

such that $u(x) = r(|x - x_0|) + u(x_0)$ *in* R^n *for any radially symmetric* C^2 *solution* u *of (1.3)–(1.4), where* x_0 *is the vertex of* u*. Moreover, the function* $r \in C^\infty[0,\infty)$ *satisfies*

$$\frac{t}{\sqrt{n^2+t^2}} \leqslant r'(t) < 1, \forall t \in [0,\infty) \tag{1.7}$$

and

$$0 < r''(t) \leqslant 1, \forall t \in [0,\infty). \tag{1.8}$$

Therefore, all rotationally symmetric spacelike translating solitons of MCF (1.2) is smooth, strictly convex, unique up to a translation in R^{n+1}*, and of linear growth.*

To describe the asymptotic behavior of general solitons as $|x| \to \infty$, we use the tangent cones methods in [7, 8] for entire spacelike convex hypersurfaces of constant mean curvature and in [9] for constant Gauss curvature. Define the *blow down* of F at infinity by

$$V_F(x) = \lim_{\rho\to\infty} \frac{F(\rho x)}{\rho}. \tag{1.9}$$

Since $\dfrac{d}{d\rho}\left(\dfrac{F(\rho x)}{\rho} - \dfrac{F(0)}{\rho}\right) \geqslant 0$ if F is convex, and $\dfrac{F(\rho x)}{\rho} - \dfrac{F(0)}{\rho} \leqslant |x|$ if F is spacelike. V_F is well-defined over R^n and the limit in (1.9) is uniform on any compact set in R^n if F is a convex function satisfying (1.4). Using Theorem 1.1 and the methods in [7, 8], we will prove

Theorem 1.2. *Suppose that* u *is a convex solution to (1.3)–(1.4). Then the blowdown function* V_u *is a positive homogeneous degree one convex function satisfying the 1-Lipschitz condition*

$$|V_u(x) - V_u(y)| \leqslant |x - y|, \forall x, y \in R^n \tag{1.10}$$

and the null condition, i.e., for any $x \in R^n$ *and any* $\delta > 0$*, there is* $y \in R^n$ *such that*

$$|V_u(x) - V_u(y)| = |x - y| = \delta. \tag{1.11}$$

Furthermore, one has

$$V_u(y) = \lim_{\rho\to\infty} \frac{u(\rho y)}{\rho} = 1 \quad \textit{uniformly} \quad \textit{for} \quad y \in \overline{\nabla u(R^n)} \bigcap S^{n-1} \tag{1.12}$$

and

$$V_u(x) = \lim_{\rho\to\infty} \frac{u(\rho x)}{\rho} = |x| \quad \textit{for} \quad x \in \overline{\nabla u(R^n)}, \tag{1.13}$$

where $\overline{\nabla u(R^n)}$ *is the smallest closed set containing* $\{y : y = \nabla u(x), x \in R^n\}$ *in* R^n*.*

A natural question is whether any solution to (1.3)–(1.4) is convex. This question seems very difficult to the author. However, we obtain the following result which is related to this question in some way.

Theorem 1.3. *Let u be a convex solution of equation (1.3)–(1.4). If the set $\Omega_0 = \{x \in R^n : (u_{ij}(x)) > 0\}$ is nonempty, then $\Omega_0 = R^n$.*

similar result was obtained for the equation $\Delta u = f(u, \nabla u)$ in [10], for the equation of entire spacelike hypersurfaces of constant mean curvature in [8] and for the mean curvature flow in Eclidean space in [11–13].

This paper is organized as follows. In Section 2, we will use ODE theory and a priori estimate techniques to prove Theorem 1.1. In Section 3, we will give the proof of Theorem 1.2. In Section 4, we will prove Theorem 1.3.

2 Radially Symmetric Solutions

We start with some simple facts which will be used throughout this section.

If $u(x) = r(|x - x_0|) + u(x_0)$ and $u \in C^{k,\alpha}(R^n)$ for some $k \geqslant 1, 0 \leqslant \alpha \leqslant 1$ with $k + \alpha \geqslant 2$, then $r \in C^{k,\alpha}[0, \infty)$ since $r(t) = u((t, 0) + x_0) = u((-t, 0) + x_0)$ for all $t \geqslant 0$. Thus $r'(0) = 0$ and equation (1.3) is equivalent to

$$\frac{r''(t)}{1 - (r'(t))^2} + \frac{n-1}{t} r'(t) = 1, \forall t \in (0, \infty) \tag{2.1}$$

and

$$r(0) = r'(0) = 0; \tag{2.2}$$

the spacelike condition (1.4) is equivalent to

$$0 < r'(t) < 1, \forall t \in (0, \infty) \tag{2.3}$$

and the strict convexity to

$$1 \geqslant r''(t) > 0, \forall t \in [0, \infty). \tag{2.4}$$

Conversely, if $r \in C^2[0, \infty)$ is a solution to (2.1)–(2.2), then it follows from a direct computation that $u(x) = r(|x|) \in C^{1,1}(R^n)$ is a solution to (1.3)–(1.4). By the standard regularity theory of elliptic equations in [14] we see that $r(|x|) \in C^\infty(R^n)$ and thus $r \in C^\infty[0, \infty)$.

Lemma 2.1. *If $r \in C^2[0, \infty)$ is a solution to (2.1)–(2.4), then it satisfies (1.7).*

Proof. If $r'(t) < 1 - \delta$ for all $t \in [0, \infty)$ and some $\delta \in (0, 1)$, then $r''(t) \geqslant \frac{\delta}{2}$ for all $t \geqslant t_0$ and for some large $t_0 > 0$ by (2.1). Integrating this inequality over $[t_0, t)$ we obtain

$$1 - \delta > r'(t) \geqslant \frac{\delta}{2}(t - t_0) - r'(t_0)$$

for all $t \geqslant t_0$, a contradiction. Therefore, there is a sequence $t_k \to \infty$ such that $r'(t_k) \to 1$. Using (2.4), we get

$$\lim_{t \to +\infty} r'(t) = 1. \tag{2.5}$$

Note that the inequality on the right sides of (1.7) follows directly from (2.3) . We want only to prove

$$r'(t) \geqslant \frac{t}{\sqrt{n^2+t^2}}, \forall t \geqslant 0. \tag{2.6}$$

On the contrary that (2.6) is false. Then we have a $t_0 > 0$ such that

$$r'(t_0) < \frac{t_0}{\sqrt{n^2+t_0^2}}.$$

Observing that $r'(0) = 0$ and

$$\lim_{t\to+\infty}\left(r'(t) - \frac{t}{\sqrt{n^2+t^2}}\right) = 0$$

by (2.5), we see that the function $r'(t) - \dfrac{t}{\sqrt{n^2+t^2}}$ attains its negative minimum at a point $t_1 > 0$. Hence

$$r''(t_1) = \left(\frac{t_1}{\sqrt{n^2+t_1^2}}\right)' = n^2(n^2+t_1^2)^{-\frac{3}{2}}$$

and

$$r'(t_1) < \frac{t_1}{\sqrt{n^2+t_1^2}}.$$

This, together with (2.1), imply

$$\begin{aligned} 1 &= \frac{r''(t_1)}{1-(r'(t_1))^2} + \frac{n-1}{t_1} r'(t_1) \\ &< \frac{n^2(n^2+t_1^2)^{-\frac{3}{2}}}{1-\dfrac{t_1^2}{n^2+t_1^2}} + \frac{n-1}{t_1}\cdot\frac{t_1}{\sqrt{n^2+t_1^2}} \\ &= \frac{n}{\sqrt{t_1^2+n^2}} < 1, \end{aligned}$$

a contradiction! □

Lemma 2.2. *There exists a* $r \in C^\infty[0,\infty)$ *to (2.1)–(2.4).*

Proof. Since equation (2.1) is singular at $t = 0$, we consider the approximation problem

$$\frac{r''(t)}{1-(r'(t))^2} + \frac{n-1}{t+\varepsilon} r'(t) = 1, \forall t \in (0,\infty) \tag{2.7}$$

$$|r'(t)| < 1, \forall t \in (0,\infty) \tag{2.8}$$

and

$$r(0) = 0, \quad r'(0) = \frac{\varepsilon}{n}. \tag{2.9}$$

Integrating (2.7) over $[0,t)$ we have

$$\frac{1}{2}\left[\ln\frac{1+r'(t)}{1-r'(t)} - \ln\frac{1+r'(0)}{1-r'(0)}\right] + (n-1)\int_0^t \frac{r'(s)}{s+\varepsilon}\mathrm{d}s = t,$$

which implies that for any $R > 0$, there exist a constant $0 < C(R) < 1$ depending on R such that

$$|r'(t)| < 1 - C(R), \forall t \in [0, R).$$

Therefore, by local existence result of ODE, we see that for any $\varepsilon \in (0,1)$ there is a unique smooth solution to (2.7)–(2.9). Denote this solution by r_ε. Obviously,

$$r''_\varepsilon(0) = \left[1 - \frac{n-1}{\varepsilon} r'(0)\right] [1 - (r'(0))^2] = \frac{n^2 - \varepsilon^2}{n^3}. \tag{2.10}$$

This leads us to conclude that

$$r''_\varepsilon(t) \geqslant 0, \quad \forall t \in [0, \infty). \tag{2.11}$$

Otherwise, there is a $t_1 \in (0, \infty)$ such that $r''_\varepsilon(t_1) < 0$. Then we may choose $t_0 > 0$ and $\delta > 0$ such that

$$r''_\varepsilon(t_0) = 0, r''_\varepsilon(t) < 0, \forall t \in (t_0, t_0 + \delta). \tag{2.12}$$

By (2.9) and (2.10), we may further assume

$$r'_\varepsilon(t) > 0, \forall t \in [t_0, t_0 + \delta). \tag{2.13}$$

Hence, $0 < r'_\varepsilon(t) < r'_\varepsilon(t_0)$ for all $t \in (t_0, t_0 + \delta)$. But this, together with (2.7), (2.8), (2.12) and (2.13), implies

$$1 = \frac{n-1}{t_0 + \varepsilon} r'_\varepsilon(t_0) > \frac{n-1}{t + \varepsilon} r'_\varepsilon(t) > \frac{r''_\varepsilon(t)}{1 - (r'_\varepsilon(t))^2} + \frac{n-1}{t + \varepsilon} r'_\varepsilon(t) = 1$$

for all $t \in (t_0, t_0 + \delta)$, a contradiction! This proves (2.11).

It follows from (2.11) and (2.9) that

$$r'_\varepsilon(t) \geqslant \frac{\varepsilon}{n}, \forall t \in (0, \infty). \tag{2.14}$$

Using this, (2.7) and (2.11) again, we claim that

$$r''_\varepsilon(t) > 0, \quad \forall t \in [0, \infty). \tag{2.15}$$

In fact, on the contrary that there is a $t_2 > 0$ such that $r''_\varepsilon(t_2) = 0$. Then the function

$$y(t) := \frac{n-1}{t + \varepsilon} r'_\varepsilon(t) = 1 - \frac{r''_\varepsilon(t)}{1 - (r'_\varepsilon(t))^2}$$

attains a maximum at t_2. Hence, $y'(t_2) = 0$ and therefore, $r'_\varepsilon(t_2) = 0$, contradicting with (2.14). This proves (2.15).

Now we use (2.8), (2.9), (2.14), (2.15)and (2.7) to see that

$$\frac{\varepsilon}{n} \leqslant r'_\varepsilon(t) < 1, \forall t \in [0, \infty) \tag{2.16}$$

$$\frac{t\varepsilon}{n} \leqslant r_\varepsilon(t) \leqslant t, \forall t \in [0, \infty) \tag{2.17}$$

$$0<r_\varepsilon''(t)=\left(1-\frac{n-1}{t+\varepsilon}r_\varepsilon'(t)\right)(1-(r_\varepsilon'(t))^2)\leqslant 1-\frac{(n-1)\varepsilon}{n(t+\varepsilon)},\forall t\in[0,\infty). \tag{2.18}$$

By estimates (2.16)–(2.18) we can choose a subsequence $\varepsilon_k\to 0$ $(k\to\infty)$ and a function $r\in C^{1,\alpha}[0,\infty)$ $(\alpha\in(0,1)$ fixed) such that

$$r_{\varepsilon_k}\to r \text{ in } C^{1,\alpha}[0,\infty) \text{ as } k\to\infty. \tag{2.19}$$

Obviously,

$$r(0)=0=r'(0), \tag{2.20}$$

$$0\leqslant r'(t)\leqslant 1 \text{ and } 0\leqslant r''(t)\leqslant 1,\forall t\in[0,\infty). \tag{2.21}$$

Furthermore, we can conclude that

$$0\leqslant r'(t)<1,\forall t\in[0,\infty) \tag{2.22}$$

Otherwise, there is a $t_3>0$ such that $r'(t_3)=1$ and $0\leqslant r'(t)<1$ for all $t\in[0,t_3)$. Integrating (2.7) for r_{ε_k} over $\left[\frac{t_3}{2},t\right)$ we have

$$\frac{1}{2}\left[\ln\frac{1+r_{\varepsilon_k}'(t)}{1-r_{\varepsilon_k}'(t)}-\ln\frac{1+r_{\varepsilon_k}'\left(\frac{t_3}{2}\right)}{1-r_{\varepsilon_k}'\left(\frac{t_3}{2}\right)}\right]+(n-1)\int_{\frac{t_3}{2}}^{t}\frac{r_{\varepsilon_k}'(s)}{s+\varepsilon_k}\mathrm{d}s=\frac{t_3}{2},\forall t\in\left(\frac{t_3}{2},t_3\right).$$

Letting $k\to\infty$ and $t\to t_3^-$ then, we obtain

$$+\infty-\ln\frac{1+r'\left(\frac{t_3}{2}\right)}{1-r'\left(\frac{t_3}{2}\right)}+2(n-1)\int_{\frac{t_3}{2}}^{t_3}\frac{r'(s)}{s}\mathrm{d}s=\frac{t_3}{2},$$

a contradiction! This shows (2.22). Observing that r_ε satisfies equation (2.7), we use (2.19)–(2.22) to see that $r\in C^2[0,\infty)$ satisfies (2.1) and (2.2) , which implies $r\in C^\infty[0,\infty)$ as we have said in the beginning of this section.

Therefore, in order to finish the proof of Lemma 2.2, we want only to prove

$$r'(t)>0,\forall t\in(0,\infty) \tag{2.23}$$

and

$$r''(t)>0,\forall t\in[0,\infty). \tag{2.24}$$

In fact, by (2.10) we have $r''(0)=\frac{1}{n}$. Then (2.23) follows from the fact that $r'(0)=0$ and $r''(t)\geqslant 0$ for all $t>0$ as in (2.20) and (2.21).

If there were a $t_4\in(0,\infty)$ such that $r''(t_4)=0$, then it follows from (2.21) that the function

$$Z(t):=\frac{n-1}{t}r'(t)=1-\frac{r''}{1-(r')^2}$$

attains a maximum at the point t_4. Thus

$$Z'(t_4) = 0 \quad \text{and} \quad \text{therefore} \quad r'(t_4) = 0,$$

contradicting (2.23). This proves (2.24) and thus Lemma 2.2. □

Lemma 2.3. *If $r_1, r_2 \in C^2[0,\infty)$ are both solutions to initial problem (2.1), (2.2) and (2.3) , then $r_1(t) = r_2(t)$ for all $r \geqslant 0$.*

Proof. Let $u_i(x) = r_i(|x|)$ $(i = 1, 2)$. As we have seen, $u_i \in C^\infty(R^n)$ are solutions of (1.3)–(1.4). Fix $t > 0$, arbitrarily. We see that both $u_1(x)$ and $u_2(x) + r_1(t) - r_2(t)$ are solutions of the Dirichlet problem of equation (1.3)–(1.4) over the ball $B_t(0)$ with the same boundary value $r_1(t)$. Thus $u_1(x) = u_2(x) + r_1(t) - r_2(t)$ for all $x \in B_t(0)$ by the uniqueness theorem [14, Theorem 10.2]. Taking $x = 0$, we obtain $r_1(t) = r_2(t)$. □

Proof of Theorem 1.1: observing the simple facts at the beginning of this section and using Lemmas 2.1, 2.2 and 2.3, we immediately obtain Theorem 1.1.

3 Proof of Theorem 1.2

In this section, we use the concept of tangent cones at infinity to describe the asymptotic behavior of the solitons as $|x| \to \infty$. This method was used in [7, 8] for entire spacelike convex hypersurfaces of constant mean curvature and in [9] for constant Gauss curvature.

Recall that the <u>*blowdown*</u> function

$$V_F(x) = \lim_{\rho \to 0\infty} \frac{F(\rho x)}{\rho} \tag{3.1}$$

is well defined over R^n and the limit is uniform on any compact set in R^n if F is a convex function satisfying (1.4).

Lemma 3.1. *If u is a convex function satisfying (1.4), then V_u is a positively homogeneous degree one convex function satisfying the 1-Lipschitz condition*

$$|V_u(x) - V_u(y)| \leqslant |x - y|, \forall x, y \in R^n; \tag{3.2}$$

while if u is a convex solution to (1.3)–(1.4), then V_u satisfies the null condition, i.e., for any $x \in R^n$ and any $\delta > 0$, there is $y \in R^n$ such that

$$|V_u(x) - V_u(y)| = |x - y| = \delta. \tag{3.3}$$

Proof. The convexity and the positive homogeneity are obviously from the definition of V_u and the convexity of u.

For any $x, y \in R^n$, by (1.4) we have

$$|V_u(x) - V_u(y)| \leqslant \limsup_{\rho \to \infty} \frac{|u(\rho x) - u(\rho y)|}{\rho} \leqslant |x - y|.$$

Hence, it is sufficient to prove the null condition. On the contrary that there would exist an $x \in R^n$, $\delta > 0$ and $\theta > 0$ such that

$$V_u(y) \leqslant V_u(x) + (1 - 2\theta)\delta$$

for all $y \in R^n$ with $|x - y| = \delta$. Observing that the limit in (3.1) is uniform on any compact set, we may choose a $\rho_0 > 0$ so that

$$u_\rho(y) \leqslant V_u(x) + (1 - \theta)\delta \tag{3.4}$$

for all $\rho > \rho_0$ and all $y \in B(x, \delta)$, where we have used the notation

$$B(x, \delta) = \{y \in R^n : |y - x| < \delta\} \text{ and } u_\rho(x) = \frac{u(\rho x)}{\rho}.$$

It follows from (1.3)–(1.4) that u_ρ satisfies

$$(\delta_{ij} + \frac{(u_\rho)_i(x)(u_\rho)_j(x)}{1 - |\nabla u_\rho(x)|^2})(u_\rho)_{ij} = \rho, \quad \forall x \in R^n \tag{3.5}$$

and

$$|\nabla u_\rho(x)| < 1, \quad \forall x \in R^n. \tag{3.6}$$

Let $r(|x|)$ be the same solution to (1.3)–(1.4) as in Theorem 1.1, where r is the unique solution of (1.5) and (1.6). Then the function

$$W(y) = W(y; \rho) := V_u(x) + \left(\delta - \frac{r(\rho\delta)}{\rho} + \frac{r(\rho|y - x|)}{\rho}\right) - \theta\delta$$

also satisfies the same (3.5)–(3.6) as u_ρ for any $\rho > 0$ and any $x \in R^n$. Note that

$$W(y) = V_u(x) + (1 - \theta)\delta, \forall y \in \partial B(x, \delta).$$

We use (3.4) and maximum principle on the domain $B(x, \delta)$ to obtain

$$u_\rho(y) \leqslant W(y; \rho), \forall y \in B(x, \delta).$$

Letting $\rho \to \infty$, we have

$$\begin{aligned} V_u(x) &\leqslant V_u(x) + (\delta - \theta\delta) + \lim_{\rho\to\infty} \frac{r(\rho|y - x|)}{\rho} - \lim_{\rho\to\infty} \frac{r(\rho\delta)}{\rho} \\ &= V_u(x) + (\delta - \theta\delta) + (|y - x| - \delta) \\ &= V_u(x) + |y - x| - \theta\delta. \end{aligned}$$

Here, in order to determine the limit, we have used the estimate

$$\sqrt{n^2 + t^2} - n \leqslant r(t) \leqslant t,$$

which follows directly from (1.6) and (1.7) in Theorem 1.1. Taking $y = x$ yields

$$V_u(x) \leqslant V_u(x) - \theta\delta,$$

a contradiction. In this way, we have shown the desired lemma. □

Recall that the tangential mapping of convex function V_u at a point $x_0 \in R^n$ is defined by

$$T_{V_u}(x_0) = \{\alpha \in R^n : V_u(x) \geqslant \alpha \cdot (x - x_0) + V_u(x_0), \forall x \in R^n\}.$$

Obviously, it is a closed, convex set and equals to $\nabla V_u(x_0)$ if V_u is differential at x_0. The tangent cone of u is defined by

$$T_{V_u}(R^n) = \bigcup_{x \in R^n} T_{V_u}(x).$$

Lemma 3.2. *If u is a convex function satisfying (1.4), then its tangent cone satisfies*

$$\overline{T_{V_u}(R^n)} = T_{V_u}(0) = \overline{\nabla u(R^n)}.$$

Proof. To show $T_{V_u}(0) \subset \overline{\nabla u(R^n)}$, we choose $\xi \in T_{V_u}(0)$. Since $V_u(0) = 0$, $V_u(y) \geqslant \xi \cdot y$ for all $y \in R^n$. Given a $\delta > 0$. Observing that the limit

$$V_u(y) = \lim_{\rho \to 0} \frac{u(\rho y)}{\rho} = \lim_{\rho \to 0} \frac{u(\rho x) - u(0)}{\rho}$$

holds uniformly on any compact set in R^n, we see that

$$\phi(y) := \frac{u(\rho_\delta y) - u(0)}{\rho_\delta} - \xi \cdot y + |y|^2 \geqslant \frac{\delta^2}{2}$$

for all $y \in \partial B(0, \delta)$ and some large $\rho_\delta > 1$. But $\phi(0) = 0$, so ϕ attains its minimum at a point $x_\delta \in B(0, \delta)$. Thus

$$\nabla \phi(x_\delta) = \nabla u(\rho_\delta x_\delta) - \xi + 2x_\delta = 0.$$

Letting $\delta \to 0$ we get $\xi \in \overline{\nabla u(R^n)}$. Therefore, $T_{V_u}(0) \subset \overline{\nabla u(R^n)}$.

To finish the proof, we follow the arguments in [7, p.793]. Let $\xi \in T_{V_u}(R^n)$. Then there is an $x \in R^n$ such that

$$V_u(\rho y) \geqslant \xi \cdot (\rho y - x) + V_u(x), \forall y \in R^n, \forall \rho > 0.$$

Dividing this inequality by ρ, using the homogeneity of V_u and then letting $\rho \to \infty$, we get

$$V_u(y) \geqslant \xi \cdot y, \forall y \in R^n.$$

This means $\xi \in T_{V_u}(0)$. Thus, $\overline{T_{V_u}(R^n)} = T_{V_u}(0)$ Since $T_{V_u}(0)$ is closed.

Now for any $x \in R^n$, the convexity implies

$$u(\rho y) \geqslant \nabla u(x) \cdot (\rho y - x) + u(x), \forall y \in R^n, \forall \rho > 0.$$

Dividing this by ρ, and letting $\rho \to \infty$, we see that

$$V_u(y) \geqslant \nabla u(x) \cdot y, \forall y \in R^n,$$

which implies $\nabla u(x) \in T_{V_u}(0)$. Since x is arbitrary and $T_{V_u}(0)$ is closed, we conclude that

$$\overline{\nabla u(R^n)} \subset T_{V_u}(0) = \overline{T_{V_u}(R^n)}.$$

This proves the lemma. □

Proof of Theorem 1.2: Since we have Lemma 3.1, it is enough to prove (1.12) and (1.13).

Choose $y \in \overline{\nabla u(R^n)}$. By Lemma 3.2, $y \in T_{V_u}(0)$. Because of $V_u(0) = 0$, we have

$$V_u(y) \geqslant y \cdot y = |y|^2.$$

On the other hand, Lemma 3.1 yields

$$V_u(y) \leqslant |V_u(y) - V_u(0)| \leqslant |y|.$$

Thus, we have

$$|y|^2 \leqslant V_u(y) \leqslant |y|, \forall y \in \overline{\nabla u(R^n)}. \tag{3.7}$$

Hence (1.12) follows. Note that

$$\frac{u(\rho x)}{\rho} = |x| \frac{u\left(\rho|x| \cdot \dfrac{x}{|x|}\right)}{\rho|x|}$$

and $\dfrac{x}{|x|} \in S^{n-1}$ for $x \neq 0$. This, together with (1.12), yields (1.13).

4　Proof of Theorem 1.3

On the contrary that there exists a $x_1 \in R^n \backslash \Omega_0$. We will derive a contradiction. We may assume Ω_0 is nonempty and connected. (Otherwise, we replace it by one of its connected components). Then there exists a short segment $l \subset \Omega_0$ such that $\bar{l} \cap \partial\Omega_0 = \{x_1\}$ and $\varepsilon_1 =$ dist $(l, \partial\Omega) > 0$. Take $x_2 \in l$ such that $B_\varepsilon(x_2) \subset \Omega_0$ for some $\varepsilon \in (0, \varepsilon_1)$. Translating the ball $B_\varepsilon(x_2)$ along the line l we come to a point $\bar{x}$ where the ball and $\partial\Omega_0$ are touched at the first time. It follows that

$$\bar{x} \in R^n \backslash \Omega_0, \quad B_\varepsilon(x_0) \subset \Omega_0 \quad \text{and} \quad \overline{B_\varepsilon(x_0)} \cap \partial\Omega_0 = \{\bar{x}\} \tag{4.1}$$

for some $x_0 \in \Omega_0$. Moreover, the minimum eigenvalue $\lambda(x)$ of the Hessian $(u_{ij}(x))$ satisfies $\lambda(\bar{x}) = 0$. By a coordinate translation and rotation we may arrange that

$$\bar{x} = 0, \quad u(0) = 0, \quad \nabla u(0) = 0 \ \text{ and } \ u_{11}(0) = \lambda(0) = 0. \tag{4.2}$$

Thus, the origin $0 \in \partial B_\varepsilon(x_0)$ and

$$(u_{ij}(x)) > 0 \ \text{ in } \ B_\varepsilon(x_0). \tag{4.3}$$

Rewrite equation (1.3) as

$$\Delta u = 1 + A(|\nabla u|^2) u_i u_j u_{ij} \ \text{ in } \ R^n, \tag{4.4}$$

where $A(t)=\dfrac{1}{t-1}$ is analytic for $t\in(-1,1)$. Differentiating (4.4) twice with respect to $\dfrac{\partial}{\partial x_1}$, we have

$$\begin{aligned}\Delta u_{11}=&4A''u_lu_{l1}u_mu_{m1}u_iu_ju_{ij}+2A'u_{m1}u_{m1}u_iu_ju_{ij}\\&+2A'u_mu_{m11}u_iu_ju_{ij}+8A'u_mu_{m1}u_{i1}u_ju_{ij}\\&+4A'u_mu_{m1}u_iu_ju_{ij1}+2Au_{i11}u_ju_{ij}\\&+2Au_{i1}u_{j1}u_{ij}+4Au_{i1}u_ju_{ij1}\\&+Au_iu_ju_{ij11}\quad\text{in}\quad R^n.\end{aligned}\tag{4.5}$$

Since u is analytic in R^n, we expand u_{11} at $x=0$ as a power series to obtain $u_{11}(x)=P_k(x)+R(x)$ for all $x\in\overline{B_\varepsilon(x_0)}$ (one can choose a smaller ε in advance if necessary), where $P_k(x)$ is the lowest order term, which, by (4.2) and (4.3), is a nonzero homogeneous polynomial of degree k, and $R(x)$ is the rest. The convexity of u yields $k\geqslant 2$. It follows from (4.3) that $u_{ii}u_{11}-(u_{i1})^2>0$ in $B_\varepsilon(x_0)$. Summing over i we have

$$\Delta uu_{11}>\sum_{j=1}^{n}u_{j1}^2\geqslant u_{i1}^2\tag{4.6}$$

for each $i=1,2,\cdots,n$.

We claim that each u_{i1} is of order at least $\dfrac{k}{2}$. Otherwise, we expand u_{i1} at $x=0$ as a power series so that the lowest order term $h(x)$ must be a nonzero homogeneous polynomial. Choose

$$a=(a_1,a_2,\cdots,a_n)\in B_\varepsilon(x_0)\backslash\{x\in B_\varepsilon(x_0):h(x)=0\}$$

so that the segment

$$L=\{ta:t\in(0,1)\}\subset B_\varepsilon(x_0).$$

Now restricting (4.6) on L, multiplying the both sides by t^{-k} and then letting $t\to 0^+$, we see the limit of the left-hand side of (4.6) is a nonzero constant multiplied by $\Delta u(0)$ which equals to 1 by (4.4), but the limit of the right-hand side is positive infinite. This is a contradiction.

Therefore, each u_{i1} is of order at least $\dfrac{k}{2}$. Hence u_{ij1}, u_{11i} and u_{11ij} are of order at least $\dfrac{k}{2}-1$, $k-1$ and $k-2$ respectively. Also note that each u_i is of order at least 1 by (4.2). With these facts one can check that the right-hand side of equation (4.5) is of order at least of k; while the left-hand side, Δu_{11}, is either of order $k-2$, or $\Delta P_k=0$ for all $x\in B_\varepsilon(x_0)$. Since the first case is impossible by comparing the orders of the two sides, we obtain that P_k is a harmonic polynomial in $B_\varepsilon(x_0)$.

We claim that $P_k\geqslant 0$ for all $x\in B_\varepsilon(x_0)$. Otherwise, there exists $a=(a_1,a_2,\cdots,a_n)\in B_\varepsilon(x_0)$ such that $P_k(a)<0$. Then

$$\frac{u_{11}(ta)}{t^k}=P_k(a)+\frac{R(ta)}{t^k},\quad\forall t\in(0,1),$$

which implies $\lim_{t\to 0^+} \frac{u_{11}(ta)}{t^k} = P_k(a) < 0$ contradicting the fact that $u_{11} > 0$ in $B_\varepsilon(x_0)$ (see (4.3)).

Now we use the strong maximum principle to see that $P_k > 0$ for all $x \in B_\varepsilon(x_0)$. But $P_k(0) = 0$, and it follows from Hopf's lemma that $\dfrac{\partial P_k}{\partial \nu}(0) < 0$, where ν is the unit outward normal to the sphere $\partial B_\varepsilon(x_0)$. This means that the degree of P_k is only one, contradicting the fact $k \geqslant 2$. This contradiction proves the theorem.

References

[1] G. Huisken, Local and global behavior of hypersurfaces moving by mean curvature, *Proceeding of Symposia in Pure Mathematics*, 54 (1993), 175-191.

[2] K. Ecker and G. Huisken, Parabolic methods for the construction of spacelike slices of prescribed mean curvature in cosmological spacetimes, *Communications in Mathematical Physics*, 135 (1991), 595-613.

[3] G. Huisken and S. T. Yau, Definition of center of mass for isolated physical system and unique foliations by stable spheres with constant curvature, *Inventiones Mathematicae*, 124 (1996), 281-311.

[4] K. Ecker, On mean curvature flow of spacelike hypersurfaces in asymptotical flat spacetimes, *Journal of the Australian Mathematical Society, Ser. A*, 55 (1993), 41-59.

[5] K. Ecker, Interior estimates and longtime solutions for mean curvature flow of noncompact spaceike hypersurfaces in Minkowski space, *Journal of Differential Geometry*, 45 (1997), 481-498.

[6] R. Schoen and S. T.Yau, On the proof of the positive mass conjecture in general relativity, *Communications in Mathematical Physics*, 65 (1979), 45-76.

[7] H. I. Choi and A. E. Treibergs, Gauss maps of spacelike constant mean curvature hypersurfaces of Minkowski space, *Journal of Differential Geometry*, 32 (1990), 775-817.

[8] A. E. Treibergs, Entire spacelike hypersurfaces of constant mean curvature in Minkowski space, *Inventiones Mathematicae*, 66 (1982), 39-56.

[9] B. Guan, H. Jian and R. Schoen, Entire spacelike hypersurfaces of Prscribed Gauss curvature in Minkowski space, *Journal für die reine und angewandte Mathematik*, 595 (2006), 167-188.

[10] N. Korevaar and J. Lewis, convex solutions to certain equations have constant rank Hessian, *Archive for Rational Mechanics and Analysis*, 97 (1987), 19-32.

[11] H. Jian, Q. Liu and X. Chen, Convexity and symmetry of translating solitons in mean curvature flow, *Chinese Annals of Mathematics*, 26B (2005), 413-422.

[12] C. Gui and H. Jian, Properties of translating solutions to mean curvature flow, *Discrete & Continuous Dynamical Systems*, 28B (2010), 441-453.

[13] X. Wang, Convexity solutions to the mean curvature flow, *Annals of Mathematics* 173 (2011), 1185-1239.

[14] D. Gilbarg and N. S. Trudinger, *Elliptic partial differential equations of second order*, 2nd Edition, Springer-Verlag, 1983.

自由边界问题的非协调有限元逼近及其离散问题迭代解 *

Z.H. Xie (谢正辉)†

Abstract In this paper, we discuss a non-conforming finite element approximation of a kind of fixed supportplate obstacle problem. A density theorem is proved. We prove the convergence of the discrete problem to the continuous problem, discuss iterative methods for the discrete problem and show the convergence of the iterative solution to the discrete solution.

1 连续问题稠密性定理

来源于力学、物理学及工程中的薄板弯曲问题可归结为求解下面的变分不等式:

$$\begin{cases} \text{求} \ \ u \in K, \text{使} \\ a(u, v-u) \geqslant (f, v-u), \forall v \in K, \end{cases} \tag{1.1}$$

其中,

$$a(u,v) = \int_\Omega \Delta u \Delta v \mathrm{d}x, \forall u, v \in H^2(\Omega), \tag{1.2}$$

$$(f,v) = \int_\Omega f v \mathrm{d}x, \forall v, f \in L^2(\Omega), \tag{1.3}$$

K 为 $H^2(\Omega)$ 的非空闭凸集.

R. Glowinski 等 [1, 2] 用混合法研究了固支障碍问题和带平均曲率约束问题, A. Fusciardi 等 [3] 混合法研究了简支障碍问题, R. Glowinski 等 [2] 用非协调元法研究了固支带平均曲率约束问题, 沈树民用 C^1-Clough Tocher 元研究了简支带平均率约束问题的有限元法. 本文利用 Morley 元研究了边界固支障碍问题的有限元逼近证明了一个稠密性定理和离散解的收敛性. 讨论了 Morley 元离散问题迭代解并证明了迭代解对离散解的收敛性. 以下恒设 K 为下面的形式:

$$K = \{v \in H_0^2(\Omega); v \geqslant \varphi \text{ a.e } \text{于 } \Omega\},$$

其中 $\varphi \in H^2(\Omega)$.

我们有下面的命题:

命题1.1. 如果 $f \in H^{-2}(\Omega)$, $\varphi \in H^2(\Omega)$ 且 $\varphi \leqslant 0$ 于 $\partial\Omega$ 的一个邻域, 则问题 (1.1) 存在唯一解.

* 发表于: 数值计算与计算机应用, Vol. 11, 1990, 42-53.

† 湖南怀化师专数学系.

注. 仅仅设 $\varphi \leqslant 0$ 于 $\partial\Omega$ 不能保证解存在. 如取 $\Omega \subset R^2, \Omega = \{(x,y) \in R^2; x^2+y^2 \leqslant 1\}$, 令 $\varphi = 1-x^2-y^2 \in H^2(\Omega)$, 此时 $K=\phi$. 另外, 若 $\varphi \in H_0^2(\Omega)$ 则 K 非空, 因而此问题 (1.1) 的解是存在唯一的.

稠密性定理如下:

定理1.1. 若 Ω 是星形域, $\varphi \in H^2(\Omega)$ 且 $\varphi \leqslant 0$ 于 $\partial\Omega$ 的某邻域, 则

$$\overline{C_0^\infty(\Omega) \cap K} = K. \tag{1.4}$$

证明: 假设 Ω 是关于 $(0,0)$ 的星形域, 设 $v \in K$ 和 $\epsilon > 0$, 定义 $\tilde{v} \in H^2(R^2)$ 如下:

$$\tilde{v}(x) = \begin{cases} v(x), & \text{如果 } x \in \Omega, \\ 0, & \text{如果 } x \in R^2 - \Omega. \end{cases} \tag{1.5}$$

易证由 (1.5) 式定义的 $\tilde{v}(x) \in H^2(R^2)$ 且 $\tilde{v}(x) \geqslant \tilde{\varphi}(x)$ a.e 于 R^2, 其中 $\tilde{\varphi}(x)$ 是 $\varphi(x)$ 的延拓, $\tilde{\varphi} \leqslant 0$ 于 $R^2 - \Omega$ 且 $\tilde{\varphi}(x) \in H^2(R^2)$.

定义 $\tilde{v}_\epsilon(x) = \tilde{v}((1+\epsilon)x), \forall x \in R^2$, 则容易看出 $\tilde{v}_\epsilon(x) \in H^2(R^2)$ 且 $\operatorname{supp} \tilde{v}_\epsilon(x) \subset\subset \Omega$. 定义 v_ϵ 为

$$v_\epsilon(x) = \tilde{v}_\epsilon(x)|_\Omega$$

那么有 $\tilde{v}_\epsilon(x) \in H_0^2(\Omega)$, 且 $v_\epsilon \Rightarrow v$ 于 $H_0^2(\Omega)$ ($\epsilon \to 0$ 时). 从而

$$\mathfrak{X}_1 = \{v \in K, v \text{ 有紧支集于 } \Omega\}$$

在 K 中稠密.

设 $v \in \mathfrak{X}_1$ 且 $\{\rho_n\}_n$ 是一串正则化光滑子, $n \to \infty$ 时 ρ_n 的支集收缩到一点. 定义 $\tilde{v} \in H^2(R^2)$ 和 $\{\tilde{v}_n\}_n, \tilde{v}_n \in C_0^\infty(R^2), \forall n$

$$\tilde{v}_n = \tilde{v} * \rho_n,$$

那么易证得

$$\tilde{v}_n \Rightarrow \tilde{v} \text{ 于 } H^2(R^2).$$

进一步有 $\tilde{v}_n(x) = \int_{R^2} \tilde{v}(y)\rho_n(x-y)\mathrm{d}y \geqslant \int_{R^2} \tilde{\varphi}(y)\rho_n(x-y)\mathrm{d}y$.

不难证明 $\tilde{v}_i \in C_0^\infty(R^2)$, 且 i 充分大时

$$\operatorname{supp} \tilde{v}_i \subset\subset \Omega,$$

$$\tilde{v}_i \Rightarrow v \text{ 于 } H^2(\Omega)\ (i \to \infty \text{ 时}),$$

$$\tilde{v}_i \Rightarrow v \text{ 于 } C^0(\overline{\Omega})\ (i \to \infty \text{ 时}),$$

故当 i 充分大时 $\tilde{v}_i \in C_0^\infty(\Omega)$.

另外, 由定理的假设和 $v \in \mathfrak{X}_1$ 推知, 有常数 $\delta > 0$, 使 $v = 0, \varphi \leqslant 0$ 于 Ω_δ, 其中 Ω_δ 定义为

$$\Omega_\delta = \{p \in \Omega, \operatorname{dist}(p, \partial\Omega) < \delta\},$$

故当 $i \geqslant i_0$ 时, $v(p)-\epsilon \leqslant \tilde{v}_i(p) \leqslant v(p)+\epsilon$, 当 $p \in \Omega - \Omega_{\delta/2}$ 时, $v_i(p)=v(p)=0$, 当 $p \in \Omega_{\delta/2}$ 时因 $\Omega - \Omega_{\delta/2} \subset\subset \Omega$, 故有函数 $\theta(x)$ 满足下面的条件:

$$\theta(x) \in C_0^\infty(\Omega), \theta \geqslant 0 \text{ 于 } \Omega, \theta = 1 \text{ 于 } \Omega - \Omega_{\delta/2}.$$

我们定义 $w_{i,\epsilon} = \tilde{v}_i + \epsilon\theta(x)$, 则当 i 充分大时, $w_{i,\epsilon} \in C_0^\infty(\Omega)$ 且

$$w_{i,\epsilon} = \tilde{v}_i + \epsilon\theta(x) \geqslant \varphi \text{ 于 } \Omega,$$

故 $w_{i,\epsilon} \in K$, 而 $\epsilon \to 0$ 时

$$w_{i,\epsilon} \Rightarrow \tilde{v}_i \text{ 于 } H^2(\Omega).$$

因此 $i \to \infty, \epsilon \to 0$ 时, $w_{i,\epsilon} \Rightarrow v$ 于 $H^2(\Omega)$. 从而定理证毕. ■

假设 Ω 适当光滑, 如果 $f \in L^q(\Omega), q > 2$, 则由文献 [4] 知 $u \in W^{3,p}(\Omega)$.

2　非协调有限元逼近

假设 $f \in L^2(\Omega)$ 且用 $(\cdot\ ,\cdot)$ 表示通常的 $L^2(\Omega)$ 空间的内积. 在引入问题 (1.1) 的非协调有限元之前, 我们重新写出连续问题如下:

$$\begin{cases} \text{求 } u \in K, \text{使} \\ \tilde{a}(u, v-u) \geqslant \displaystyle\int_\Omega f(v-u)\mathrm{d}x, \ \forall v \in K, \end{cases} \tag{2.1}$$

这里 $K = \{v \in H_0^2(\Omega); v \geqslant \varphi \ \text{ a.e. 于 } \Omega\}$, 而 $\tilde{a}(\cdot\ ,\cdot)$ 定义为

$$\tilde{a}(v,w) = \int_\Omega \sum_{|a|=2} D^a v D^a w \mathrm{d}x.$$

我们利用 Green's 公式容易算出 (参见文献 [2])

$$\tilde{a}(v,w) = \int_\Omega \Delta v \Delta w \mathrm{d}x, \ \forall v, w \in H_0^2(\Omega). \tag{2.2}$$

从 (2.2) 知半范数 $|v|_{2,\Omega}$ 满足

$$|v|_{2,\Omega}^2 = \tilde{a}(v,v) = |\Delta v|_{0,\Omega}^2.$$

它是 $H_0^2(\Omega)$ 上的一个范数且等价于通常的 $H_0^2(\Omega)$ 范数 $\|v\|_{H_0^2(\Omega)}$, 其中

$$\|v\|_{H_0^2(\Omega)} = \left(\sum_{|a|=2} \|D^a v\|^2 \right)^{1/2}.$$

下面定义非协调有限元空间及其对应的逼近问题.

为简单起见我们假设 Ω 是 R^2 的一个有界多边形区域. 设 $\{\mathcal{T}_h\}_h$ 是 Q 的一个三角形剖分族, h 为剖分 $\mathcal{T}_h$ 的单元的最大边长, 我们使用 Morley 有限元空间 V_h, 其定义见文献 [2]. 用下面定义的 K_h 和 $\tilde{a}_h(v_h, w_h)$ 来逼近 K 和 $\tilde{a}(\cdot\,,\cdot)$.

$$K_h = \{v_h \in V_h; v_h(p) \geqslant \varphi(p),\ \forall p \text{ 为 } \mathcal{T}_h \text{ 的单位顶点}\},$$

$$\tilde{a}_h(v_h, w_h) = \sum_{T\in\mathcal{T}_h} \int_T \sum_{|a|=2} D^a v_h D^a w_h dx, \forall v_h, w_h \in V_h.$$

易证泛函: $v_h \in V_h \to [\tilde{a}_h(v_h, v_h)]^{\frac{1}{2}}$ 是 V_h 上的一个范数, 用 $||\cdot||_h$ 表示, 即

$$||v_h||_h = \{\tilde{a}_h(v_h, v_h)\}^{\frac{1}{2}},\ \forall v_h \in V_h.$$

我们考虑问题 (2.1) 的逼近问题

$$\begin{cases} \text{求 } u_h \in K_h, \text{使} \\ \tilde{a}_h(u_h, v_h - u_h) \geqslant \displaystyle\int_\Omega f(v_h - u_h)\mathrm{d}x, \forall v_h \in K_h, \end{cases} \tag{2.3}$$

命题 2.1 问题 (2.3) 存在唯一解.

现在讨论 $\{u_h\}_h$ 当 $h \to 0$ 时的收敛性, 建立两个引理.

引理2.1. 设 Ω 是凸的, $\{v_h\}_h$ 满足

$$\forall h,\ v_h \in K_h, \tag{2.4}$$

$$\forall h, ||v_h||_h \leqslant C, \tag{2.5}$$

那么我们可以从 $\{v_h\}_h$ 中选一个子列, 使得

$$v_h \to v^* \text{ 于 } L^2(\Omega)\ (h \to 0 \text{ 时}),$$

$$v* \in K. \tag{2.6}$$

证明: 从文献 [5] 知, 存在一个与 h 无关的常数 C, 使 $||v_h||_{L^2(\Omega)} \leqslant C||v_h||_h,\ \forall v_k \in V_h$. 这实际上是对于 Morley 有限元空间的离散 Poincaré-Friedrichs 不等式.

如果 $\{v_h\}_h$ 满足 (2.5), 那么由上式可知 $\{v_h\}_h$ 是 $L^2(\Omega)$ 中的有届序列, 必存在 $\{v_h\}_h$ 的子列以及 $v^* \in L^2(\Omega)$, 使 $v_h \to v^*$ 于 $L^2(\Omega)$.

下面只须证明 $v^* \in K$.

我们引入双线性型 G_h:

$$G_h(v_h, v) = \sum_{T\in\mathcal{T}_h} \int_T \sum_{|a|=2} D^a v_h D^a v \mathrm{d}x,\ \forall\{v_h, v\} \in V_h \times H_0^2(\Omega).$$

因线性泛函 $v \to G_h(v_h, v)$ 是连续的, 故存在唯一的 $g_h \in H_0^2(\Omega)$, 使

$$\int_\Omega \Delta g_h \Delta v \mathrm{d}x = G_h(v_h, v),\ \forall v \in H_0^2(\Omega). \tag{2.7}$$

在 (2.7) 中取 $v = g_h$, 那么我们有 (注意 (2.2))

$$|g_h|_{2,\Omega} \leqslant ||v_h||_h,$$

故有子列使

$$g_h \to g^* \text{ 于 } H_0^2(\Omega). \tag{2.8}$$

为了证明 $g^* = v^*$, 我们引入 $\phi \in C_0^\infty(\Omega)$. 考虑下面的双调和问题:

$$\begin{cases} \Delta^2 w = \phi, \\ w \in H_0^2(\Omega). \end{cases} \tag{2.9}$$

其变分形式为

$$\begin{cases} w \in H_0^2(\Omega), \\ \displaystyle\int_\Omega \Delta w \Delta v \mathrm{d}x = \int_\Omega \phi v \mathrm{d}x, \ \forall v \in H_0^2(\Omega). \end{cases} \tag{2.10}$$

从 Q 的凸性及文献 [6] 我们有

$$w \in H^3(\Omega) \bigcap H_0^2(\Omega). \tag{2.11}$$

在 (2.10) 中取 $v = g_h$, 得

$$\int_\Omega \phi g_h \mathrm{d}x = \int_\Omega \Delta w \Delta g_h \mathrm{d}x. \tag{2.12}$$

从 (2.12) 和 (2.7) 知

$$\int_\Omega \phi g_h \mathrm{d}x = G_h(v_h, w). \tag{2.13}$$

利用 Green's 公式有 (参见文献 [2])

$$\begin{aligned} G_h(v_h, w) = \sum_{T \in \mathcal{T}_h} \Big\{ & \int_T v_h \Delta^2 w \mathrm{d}x - \int_{\partial T} \frac{\partial \Delta w}{\partial n} v_h \mathrm{d}s + \int_{\partial T} \Delta w \frac{\partial v_h}{\partial n} \mathrm{d}s \\ & + \int_{\partial T} \left(-\frac{\partial^2 w}{\partial s^2} \frac{\partial v_h}{\partial n} + \frac{\partial^2 w}{\partial n \partial s} \frac{\partial v_h}{\partial s} \right) \mathrm{d}s \Big\}, \end{aligned} \tag{2.14}$$

其中 s 和 n 分别为边界 ∂T 的单位切向量和单位外法向量.

令

$$F_h(v_h, w) = \sum_{T \in \mathcal{T}_h} \int_{\partial T} \left(\Delta w \frac{\partial v_h}{\partial n} - \frac{\partial \Delta w}{\partial n} v_h - \frac{\partial^2 w}{\partial s^2} \frac{\partial v_h}{\partial n} + \frac{\partial^2 w}{\partial n \partial s} \frac{\partial v_h}{\partial s} \right) \mathrm{d}s,$$

利用 (2.13) 和 (2.14) 有

$$\int_\Omega (g_h - v_h) \phi \mathrm{d}x = F_h(v_h, w). \tag{2.15}$$

利用 w 的 $H^3(\Omega)$ 正则性并参见文献 [5] 易证得

$$|F_h(v_h, w)| \leqslant Ch(|w|_{3,\Omega} + h||\Delta^2 w||_{L^2(\Omega)})||v_h||_h, \tag{2.16}$$

故由 (2.15) 和 (2.16) 令 $h \to 0$ 可得

$$\int_{\Omega}(g^* - v^*)\phi \mathrm{d}x = 0,\ \forall \phi \in C_0^{\infty}(\Omega). \tag{2.17}$$

因此, $v^* = g^*$ a.e 于 Ω, 有 $v* \in H_0^2(\Omega)$.

剩下须证

$$v^* \geqslant \varphi \text{ a.e 于 } \Omega. \tag{2.18}$$

它等价于

$$\forall \phi \in C_0^{\infty}(\Omega), \phi \geqslant 0,\ 有\ \int_{\Omega}(v^* - \varphi)\phi \mathrm{d}x \geqslant 0. \tag{2.19}$$

我们知道:

$$\int_{\Omega}(v_h - \varphi)\phi \mathrm{d}x = \sum_{T \in \mathcal{T}_h}\int_T (v_h - \varphi)\phi \mathrm{d}x, \tag{2.20}$$

$$\int_{\Omega}(v_h - \varphi)\phi \mathrm{d}x = \int_T (v_h - v_{hl} + v_{hl} - \varphi_l + \varphi_l - \varphi)\phi \mathrm{d}x, \tag{2.21}$$

其中 v_{hl} 和 φ_l 分别为 v_h 和 φ 得线性插值. 不难证明 $v_{hl} \geqslant \varphi_l$ 于 T, 故由 (2.21) 有

$$\begin{aligned}\int_T (v_h - \varphi)\phi \mathrm{d}x &= \int_T (v_h - v_{hl})\phi \mathrm{d}x + \int_T (\varphi_l - \varphi)\phi \mathrm{d}x,\\ &\geqslant -Ch^2||v_h||_{2,T}||\phi||_{0,T} - Ch^2||\varphi||_{2,T}||\phi||_{0,T},\end{aligned} \tag{2.22}$$

亦即:

$$\int_{\Omega}(v_h - \varphi)\phi \mathrm{d}x \geqslant -Ch^2||v_h||_{h,\Omega}||\phi||_{0,\Omega} - Ch^2||\varphi||_{2,\Omega}||\phi||_{0,\Omega}. \tag{2.23}$$

在 (2.23) 中令 $h \to 0$, 由 v_h 得弱收敛性可得到要证明的 (2.19) 式. 引理证毕. ■

我们构造算子 $r_h : H_0^2(\Omega) \to V_h$ 如下:

$$\begin{cases} r_h v \in V_h, \forall v \in H_0^2(\Omega), r_h v(p) = v(p), \forall p \ 为\ \mathcal{T}_h\ 的顶点\\ \displaystyle\int_{LiT}\frac{\partial}{\partial n}r_n v \mathrm{d}s = \int_{LiT}\frac{\partial v}{\partial n}\mathrm{d}s, LiT\ 是\ T \in \mathcal{T}_h\ 的边\ (i = 1, 2, 3). \end{cases} \tag{2.24}$$

引理2.2. $r_h \in \mathcal{L}(H_0^2(\Omega), v_h)$, *更精确地*

$$||r_h v||_h \leqslant |v|_{2,\Omega}, \forall v \in H_0^2(\Omega). \tag{2.25}$$

由算子 r_n 的定义和 Green's 公式容易证明.

引理2.3. *假设 Q 是 R^2 中的星形多边形区域, 那么有*

$$r_h v \in K_h,\quad \forall h, \tag{2.26}$$

$$\lim_{h \to 0}||r_h v - v||_h,\quad \forall v \in K. \tag{2.27}$$

证明: 因 $r_hv \in V_h$, 而 $r_hv(p) = v(p) \geqslant \phi(p), \forall p$ 为 $\mathcal{T}_h$ 的顶点. 故 $r_hv \in K_h$, 即 (2.26) 式成立.

由稠密性定理 1.1, $\forall v \in K, \exists v_n \in C_0^\infty(\Omega) \bigcap K$, 使

$$\lim_{n\to\infty} ||v_n - v||_{H^2(\Omega)=0}, \tag{2.28}$$

故任给 $\epsilon > 0, \exists n_0$, 使

$$|v - v_{n_0}| \leqslant \frac{\epsilon}{4}. \tag{2.29}$$

对 v_{n_0} 的插值 $r_hv_{n_0}$, 当 k 充分小时有 [5, 7]

$$||v_{n_0} - r_hv_{n_0}||_h \leqslant \frac{\epsilon}{2}. \tag{2.30}$$

故由 (2.25), (2.29), (2.30) 式可知当 h 充分小时,

$$\begin{aligned}||v - r_hv||_h &\leqslant |v - v_{n_0}|_{2,\Omega} + ||v_{n_0} - r_hv_{n_0}||_h + ||r_hv_{n_0} - r_hv_{n_0}||_h\\ &\leqslant 2|v - v_{n_0}|_{2,\Omega} + ||v_{n_0} - r_hv_{n_0}||_h < \epsilon,\end{aligned}$$

从而 (2.27) 式成立, 引理证毕. ■

为了讨论逼近问题 (2.3) 对问题 (1.1) 的解 u 当 $h \to 0$ 时的收敛性, 我们定义空间 $\tilde{V}_h$ 如下:

$$\tilde{V}_h = \{v \in L^2(\Omega); v_h|_T \in H^2(T), \forall T \in \mathcal{T}_h\}.$$

易知 $H^2(\Omega)$、$V_h \subset \tilde{V}_h$, 定义双线性泛函 $\tilde{a}_h$

$$\tilde{a}_h(v, w) = \sum_{T\in\mathcal{T}_h} \int_T \sum_{|a|=2} D^a v D^a w \mathrm{d}x, \forall v, w \in \tilde{V}_h.$$

令 $||v_h||_h = [\tilde{a}_h(v, v)]^{\frac{1}{2}}, \forall v \in \tilde{V}_h$,

$$G_h(v_h, v) = \tilde{a}_h(v_h, v)(= \tilde{a}_h(v, v_h)), \forall v \in V_h.$$

现在证明 $h \to 0$ 时逼近解的收敛性.

定理2.1. 设 u 是连续问题 (1.1) 的解, u_h 是离散问题 (2.3) 的解, 再假设 Q 是 R^2 的有界凸多边形区域 (从而是星形区域), 那么有

$$\lim_{h\to 0} ||u - u_h||_h = 0 \tag{2.31}$$

证明: 我们分两步进行证明, 以下的 C 表示与 h 无关的常数, 在各处取值可不同.

$$\tilde{a}_h(u_h, u_h) \leqslant \tilde{a}_h(u_h, v_h) - \int_\Omega f(v_h - u_h)\mathrm{d}x, \forall v_h \in K_h.$$

故

$$||u_h||_h^2 \leqslant C||u_h||_h||v_h||_h + ||f||_{0,\Omega}(||v_h||_h + ||u_h||_h), \forall v_h \in K_h. \tag{2.32}$$

取 $v_0 \in C_0^2(\Omega)\bigcap K$, 则由引理 2.3 知 $v_h = r_h v_0 \in K_h$, 且

$$\lim_{h\to 0} ||r_h v_0 - v_0||_h = 0.$$

在 (2.32) 式中取 $v_h = r_h v_0$, 则有常数 m 和 M, 当 h 充分小时

$$||u_h||_h \leqslant a^{-1}(mM + ||f||_{0,\Omega})||u_h||_h + a^{-1}m||f||_{0,\Omega},$$

其中 a 为正定性常数.

由此可知存在与 h 无关的常数 C, 当 h 充分小时 $||u_h||_h \leqslant C$.

因 Ω 是凸的, 且 $u_h \in K_h$, 由引理 2.1 可从 $\{u_h\}_h$ 中选一个子列 (仍用 $\{u_h\}_h$ 表示), 使

$$u_h \to u^* \text{ 于 } L^2(\Omega), \text{ 且 } u^* \in K. \tag{2.33}$$

为了证明 $\lim_{h\to 0} ||u-u_h||_h = 0$, 我们在 (2.3) 式中取 $v_h = r_h u \in K_h$, 那么得

$$\begin{aligned} ||u-u_h||_h^2 &= \tilde{a}_h(u-u_h, u-u_h) = \tilde{a}_h(u-u_h, u-r_h u) + \tilde{a}_h(u-u_h, r_h u - u_h) \\ &\leqslant ||u-u_h||_h ||u-r_h u||_h + \tilde{a}_h(u, r_h u - u_h) - \int_\Omega f(r_h u - u_h)\mathrm{d}x. \end{aligned} \tag{2.34}$$

因 $u* \in K$, 故

$$\begin{aligned} &\tilde{a}_h(u, u-u^*) - \int_\Omega f(u-u^*)\mathrm{d}x \\ =& \tilde{a}(u, u-u^*) - \int_\Omega f(u-u^*)\mathrm{d}x \leqslant 0. \end{aligned} \tag{2.35}$$

设 g_h 是由 (2.7) 式所定义的映射中由 u_h 所确定的函数, 即

$$\int_\Omega \Delta g_h \Delta v \mathrm{d}x = \tilde{a}_h(u_h, v), \forall v \in H_0^2(\Omega), \tag{2.36}$$

其中 $g_h \in H_0^2(\Omega)$.

因 $u, u^* \in K \subset H_0^2(\Omega)$, 故

$$\tilde{a}(u, u-u^*) = a(u, u^*) - a(g_h, u). \tag{2.37}$$

类似于引理 2.1 的证明我们有

$$g_h \to u^* \text{ 于 } H_0^2(\Omega). \tag{2.38}$$

由式 (2.34)–(2.36) 得

$$\begin{aligned} ||u-u_h||_h^2 \leqslant & ||u-u_h||_h ||u-r_h u||_h + \tilde{a}_h(u, r_h u - u_h) + a(u, u^* - g_h) \\ & - \int_\Omega f(r_h u - u)\mathrm{d}x - \int_\Omega f(u^* - u_h)\mathrm{d}x \end{aligned} \tag{2.39}$$

由 (2.27)、(2.33)、(2.38)、(2.39) 式可知 (2.31) 式成立. 定理证毕. ■

3 非协调有限元离散问题迭代解

下面将讨论离散问题 (2.3) 的对偶迭代解, 为此引入问题 (2.3) 对应的 Lagrange 泛函 L_h.

$$V_h \times L_h \to R,$$

$$L(v_h,\mu_h)=\frac{1}{2}\tilde{a}_h(v_h,v_h)-\int_\Omega fv_h dx+\sum_{T\in\mathcal{T}_h}\Delta_T\mu_h\sum_{i=1}^{3}\frac{1}{3}(\varphi-v_h)(P_{iT}) \tag{3.1}$$

其中 Δ_T 为 $T\in\mathcal{T}_h$ 的面积, P_{iT} 为 T 的三个顶点 $(i=1,2,3)$.

令 $L_h=\{q_h\in L^2(\Omega);q_h|_T\in P_0,\forall T\in\mathcal{T}_h\}$

$$L_h^+=\{q_h\in L_h;q_h\geqslant 0\text{ 于 }T,\forall T\in\mathcal{T}_h\},$$

那么有下面的定理

定理3.1. (3.1) 式定义的 Lagrange 泛函 L 在 $V_h\times L_h^+$ 上有一个鞍点 $\{u_h,\lambda_h\}$, 进一步有, u_h 是问题 (2.3) 的解,

$$\sum_{i=1}^{3}(u_h-\varphi)(P_{iT})\lambda=0\text{ 于 }T,\quad \forall T\in\mathcal{T}_h \tag{3.2}$$

证明: V_h 和 L_h 是两个可分拓扑向量空间, V_h 和 L_h^+ 分别为 V_h 和 L_h 的紧凸子集. 同时 $\forall v\in V_h,\mu\to L(v,\mu)$ 是凹的和上半连续的; $V\mu\in L_h^+,v\to L(v,\mu)$ 是凸的和下半连续的, 因而由极小–极大定理 [8] 知鞍点 $\{u_h,\lambda_h\}$ 是存在的.

现设 $\{u_h,\lambda_h\}$ 是 L 在 $V_h\times L_h^+$ 上的一个鞍点, 那么,

$$\sum_{T\in\mathcal{T}_h}\Delta_T(\mu_h-\lambda_h)\frac{1}{3}\sum_{i=1}^{3}(\varphi-u_h)(P_{iT})\leqslant 0,\forall\mu_h\in L_h^+. \tag{3.3}$$

在 (3.3) 式中, 对 $\forall T\in\mathcal{T}_h$, 当 $x\in T$ 时, 分别取 $\mu_h=0$ 和 $\mu_h=2\lambda_h$, 在 $x\notin T$ 时, 取 $\mu_h=\lambda_h$, 可推得

$$\lambda_h\sum_{i=1}^{3}(\varphi-u_h)(P_{iT})=0\text{ 于}T.$$

从而 (3.2) 式成立. u_h 是问题 (2.3) 的解可由鞍点的定义及 (3.2) 得证. 定理证毕. ■

因 u_h 是离散问题的唯一解, 假设 $\{u_h,\lambda_h\}$ 和 $\{u_h,\lambda_h'\}$ 是 (3.1) 中定义的 L 在 $V_h\times L_h^+$ 上的两个鞍点, 令 $\delta_h=\lambda_h-\lambda_h'$, 则有

$$\tilde{a}_h(u_h,v_h)=\int_\Omega fv_h\mathrm{d}x+\sum_{T\in\mathcal{T}_h}\Delta_T\lambda_h\sum_{i=1}^{3}\frac{1}{3}v_h(P_{iT}),\forall v_h\in V_h,$$

$$\tilde{a}_h(u_h,v_h)=\int_\Omega fv_h\mathrm{d}x+\sum_{T\in\mathcal{T}_h}\Delta_T\lambda_h'\sum_{i=1}^{3}\frac{1}{3}v_h(P_{iT}),\forall v_h\in V_h,$$

由以上两式相减可得

$$\sum_{T\in\mathcal{T}_h}\Delta_T\delta_T\sum_{i=1}^{3}\frac{1}{3}v_h(P_{iT})=0,\forall v_h\in V_h,$$

从而 $\delta_h|_T=0$.

命题3.1. (3.1) 式中定义的泛函 L 在 $V_h \times L_h^+$ 上存在唯一鞍点 $\{u_h, \lambda_h\}$.

利用 Lagrange 泛函 L 可给出离散问题 (2.3) 的对偶迭代格式.

设 $\lambda_h^0 \in L_h^+$ 给定, 对 $n \geqslant 0, \lambda_n^0 \in L_h^+$ 已知, 我们从下面的条件来确定 u_h^n 和 λ_h^{n+1}.

$$\left\{\begin{array}{l}\left\{\begin{array}{l}u_h^n \in V_h,\\ \tilde{a}_h(u_h^n, v_h) = \displaystyle\int_\Omega f v_h \mathrm{d}x + \sum_{T\in\mathcal{T}_h} \Delta_T \lambda_h^n \sum_{i=1}^{3} \frac{1}{3} v_h(P_{iT}), \forall v_h \in V_h,\end{array}\right.\\ \lambda_h^{n+1} = \left[\lambda_h^n + \rho \dfrac{1}{3} \displaystyle\sum_{i=1}^{3} (\varphi - u_h)(P_{iT})\right]^+, \forall T \in \mathcal{T}_h,\end{array}\right. \tag{3.4}$$

其中 $\rho > 0$ 为参数.

关于上述算法有下面的收敛性结果.

定理3.2. 设 ρ 满足 $0 < \rho < \dfrac{2}{\sigma}$, 其中 σ 定义为

$$\sigma = \sup_{v\in V_h} \frac{\left|\displaystyle\sum_{T\in\mathcal{T}_h} \Delta_T \frac{1}{9} \left(\sum_{i=1}^{3} v_h(P_{iT})\right)^2\right|}{||v_h||_h^2}, \tag{3.5}$$

那么

$$\lim_{n\to\infty} u_h^n = u_h, \tag{3.6}$$

$$\lim_{n\to\infty} \lambda_h^n = \lambda_h, \tag{3.7}$$

其中 u_h 是离散问题 (2.3) 的解, λ_h 是使 $\{u_h, \lambda_h\}$ 为泛函 L 在 $V_h \times L_h^+$ 上的唯一鞍点.

证明: 设 $\{u_h, \lambda_h\}$ 是 L 在 $V_h \times L_h^+$ 上的鞍点, 则 u_h 是下面问题的解:

$$\left\{\begin{array}{l}u_h \in V_h,\\ \tilde{a}_h(u_h, v_h) = \displaystyle\int_\Omega f v_h \mathrm{d}x + \sum_{T\in\mathcal{T}_h} \Delta_T \lambda_h \sum_{i=1}^{3} \frac{1}{3} v_h(P_{iT}), \forall v_h \in V_h.\end{array}\right. \tag{3.8}$$

另一方面, 如果 $\rho > 0$, 由定理 3.1 知

$$\lambda_h|_T = \left[\lambda_h + \rho \frac{1}{3} \sum_{i=1}^{3} (\varphi - u_h)(P_{iT})\right]^+, \forall T \in \mathcal{T}_h.$$

令

$$\overline{u}_h^n = u_h^n - u_h, \overline{\lambda}_h^n = \lambda_h^n - \lambda_h. \tag{3.9}$$

从 $q_h \to q_h^+$ 是收缩映射得

$$|\overline{\lambda}_h^{n+1}|_{0,\Omega}^2 \leqslant |\overline{\lambda}_h^n|_{0,\Omega}^2 + \frac{\rho^2}{9} \sum_{T\in\mathcal{T}_h} \int_T \left(\sum_{i=1}^{3} \overline{u}_h^n(P_{iT})\right)^2 \mathrm{d}x - \frac{2\rho}{9} \sum_{T\in\mathcal{T}_h} \int_T \lambda_h^n \sum_{i=1}^{3} \frac{1}{3} \overline{u}_h^n(P_{iT}) \mathrm{d}x,$$

从而

$$|\overline{\lambda}_h^n|_{0,\Omega}^2 - |\overline{\lambda}_h^{n+1}|_{0,\Omega}^2 \geqslant 2\rho\frac{1}{9}\sum_{T\in\mathcal{T}_h}\int_T \lambda_h^n \sum_{i=1}^3 \frac{1}{3}\overline{u}_h^n(P_{iT})\mathrm{d}x - \frac{\rho^2}{9}\sum_{T\in\mathcal{T}_h}\int_T \left(\sum_{i=1}^3 \overline{u}_h^n(P_{iT})\right)^2 \mathrm{d}x.$$

$$L(v_h,\mu_h) = \frac{1}{2}\tilde{a}_h(v_h^n, v_h) - \int_\Omega f v_h \mathrm{d}x + \sum_{T\in\mathcal{T}_h}\Delta_T \mu_h \sum_{i=1}^3 \frac{1}{3}(\varphi - v_h)(P_{iT}) \tag{3.10}$$

从 (3.8) 和 (3.9) 式易知

$$\sum_{T\in\mathcal{T}_h}\Delta_T\lambda_h^n \sum_{i=1}^3 \overline{u}_h^n(P_{iT}) = ||\overline{u}_h^n||_h^2 \tag{3.11}$$

故由 (3.5)、(3.10) 和 (3.11) 式易推出

$$\begin{aligned}|\lambda_h^n|_{0,\Omega}^2 - |\lambda_h^{n+1}|_{0,\Omega}^2 &\geqslant 2\rho||\overline{u}_h^n||_h^2 - \rho^2\sum_{T\in\mathcal{T}_h}\int_T\left(\sum_{i=1}^3\overline{u}_h^n(P_{iT})\right)^2\mathrm{d}x.\\ &\geqslant 2\rho||\overline{u}_h^n||_h^2 - \rho^2\sigma||\overline{u}_h^n||_h^2 = \rho(2-\rho\sigma)||\overline{u}_h^n||_h^2\end{aligned}$$

亦即

$$|\lambda_h^n|_{0,\Omega}^2 - |\lambda_h^{n+1}|_{0,\Omega}^2 \geqslant \rho(2-\rho\sigma)||\overline{u}_h^n||_h^2 \tag{3.12}$$

故当 $0<\rho<\dfrac{2}{\sigma}$ 时, $||\overline{u}_h^n||_h^2 \to 0(n\to\infty)$, 从而 (3.6) 成立.

由 (3.12) 式当 $0<\rho<\dfrac{2}{\sigma}$ 时, 知 $\{\lambda_h^n\}$ 在 $L^2(\Omega)$ 中是有界的, 故必存在子列弱收敛于 λ_h^*, 其子列仍记为. 我们可以证明 $\{u_h,\lambda_h^*\}$ 是 L 在 $V_h\times L_h^+$ 上的鞍点. 由鞍点的唯一性知 $\lambda_h^*=\lambda_h$, 从而 (3.7) 式成立.

事实上由 (3.4) 式及 u_h^n 的收敛性有

$$\begin{cases} u_h\in V_h,\\ \tilde{a}_h(u_h,v_h) = \displaystyle\int_\Omega f v_h\mathrm{d}x + \sum_{T\in\mathcal{T}_h}\Delta_T\lambda_h^*\sum_{i=1}^3\frac{1}{3}v_h(P_{iT}), \quad \forall v_h\in V_h.\end{cases} \tag{3.13}$$

$$\sum_{T\in\mathcal{T}_h}\Delta_T(\mu_h-\lambda_h^{n+1})\left(\lambda_h^n + \rho\sum_{i=1}^3\frac{1}{3}(\varphi-u_h^n)(P_{iT}) - \lambda_h^{n+1}\right)\leqslant 0, \quad \forall\mu_h\in L_h^+. \tag{3.14}$$

类似于 (3.12) 式的证明得

$$\lim_{n\to\infty}(\lambda_h^{n+1}-\lambda_h^n) = 0 \text{ 于 } L^2(\Omega) \tag{3.15}$$

强收敛. 故由 (3.14) 令 $n\to\infty$ 并注意 (3.15) 及 u_h^n 的收敛性可得

$$\sum_{T\in\mathcal{T}_h}\Delta_T(\mu_h-\lambda_h^*)\rho\sum_{i=1}^3\frac{1}{3}(\varphi-u_h)(P_{iT})\leqslant 0, \quad \forall\mu_h\in L_h^+. \tag{3.16}$$

由 (3.15) 和 (3.16) 式我们知道 $\{u_h,\lambda_h^*\}$ 是 L 在 $V_h\times L_h^+$ 上的鞍点

以下只须证明 $\sigma < \infty$, 而 $\sigma > 0$ 是易见的.

由有限元空间 V_h 的定义知 $\forall v_h \in V_h, v_h|_T \in P_2, \forall T \in \mathcal{T}_h$, 故有

$$||v_h - v_h(P_{iT})||_{0,T} \leqslant Ch|v_h|_{1,T},$$

因此

$$\left(\int_T v_h(P_{iT})^2 \mathrm{d}x\right)^{\frac{1}{2}} \leqslant Ch|v_h|_{1,T} + ||v_h||_{0,T},$$

$$\int_T v_h(P_{iT})^2 \mathrm{d}x) \leqslant Ch^2|v_h|_{1,T} + ||v_h||_{0,T}^2 + Ch|v_h|_{1,T}||v_h||_{0,T},$$

$$\sum_{T\in\mathcal{T}_h}\sum_{i=1}^{3}\Delta_T v_h(P_{iT})^2 \leqslant Ch^2\sum_{T\in\mathcal{T}_h}|v_h|_{1,T} + C\sum_{T\in\mathcal{T}_h}||v_h||_{0,T}^2 + Ch\sum_{T\in\mathcal{T}_h}|v_h|_{1,T}||v_h||_{0,T}, \quad (3.17)$$

其中 C 为与 h 无关的常数, 在不同之处可不同. 再由离散的 Poincaré 不等式 [5] 知

$$\begin{cases} ||v_h||_{0,\Omega} \leqslant C||v_h||_h, \forall v_h \in V_h, \\ |v_h|_{1,\Omega} \leqslant C||v_h||_h, \forall v_h \in V_h, \end{cases} \quad (3.18)$$

由 (3.17) 和 (3.18) 可知

$$\sum_{T\in\mathcal{T}_h}\Delta_T\frac{1}{9}\left(\sum_{i=1}^{3}v_h(P_{iT})\right)^2 \leqslant C||v_h||_h^2.$$

证毕. ■

参考文献

[1] R. Glowinski and J. L. Lion, R. Tremolieres, *Approximation des inéquations de la mécanique et de la physique*, Dunod pais, 1976.

[2] R. Glowinski and L. D. Marini, M. Vidrascu, Finite-element approximations and iterative solutions of a fourth-order elliptic variational inequality, *IMA Journal of Numerical Analysis*, 4 (1984), 127-167.

[3] A. Fusciardi and F. Scarpini, A mixed finite element solution of some biharmonic unilateral problem, *Numerical Functional Analysis and Optimizzaiton*, 2 (1980), 397-420.

[4] A. Friedman, *Variational principles and free-boundary problems*, 1982, Newyork.

[5] V. Barros, Some results on the curved plate bending problem solved with non-conforming finite elements, *CALCOLO*, 15 (1978), 101-120.

[6] V.A. Kondratiev, Boundary-value problems for elliptic equations in domains with conical or angular point, *Trans. Moscow Math. Soc.* 16 (1967), 227-313.

[7] P. Lascaux and P. Lesaint, Some nonconforming finite elements for the plate bending problems, *RAIRO Modélisation Mathématique et Analyse Numérique*, 9 (1975), 9-53.

[8] J.Cèa 著, 胡毓达, 郑权译, 最优化理论与算法 (中译本), 高等教育出版社, 1982.

发表文章目录

1. Z.Q. Xie, P. Zhu, S.Z. Zhou. Uniform convergence of a coupled method for convection-diffusion problems in 2-D Shishkin mesh. *International Journal of Numerical Analysis & Modeling*, 10:4 (2013), 845-859.
2. 邹战勇, 周叔子. 求解 HJB 方程的拟变分不等式组的迭代法. 应用数学学报, 35:4 (2012), 747-755.
3. P. Zhu, Z.Q. Xie, S.Z. Zhou. A uniformly convergent continuous- discontinuous Galerkin method for singularly perturbed problems of convection-diffusion type. *Applied Mathematics and Computation*, 217:9 (2011), 4781-4790.
4. P. Zhu, Z.Q. Xie, S.Z. Zhou. A coupled continuous-discontinuous FEM approach for convection diffusion equations. *Acta Mathematica Scientia*, 31:2 (2011), 601-612.
5. P. Zhu, S.Z. Zhou. Relaxation Lax-Friedrichs sweeping for static Hamilton-Jacobi equations. *Numerical Algorithms*, 54:3 (2010), 325-342.
6. J.C. Zhang, S.Z. Zhou, X.Y. Hu. The (P, Q) generalized reflexive and anti-reflexive solutions of the matrix equation $AX = B$. *Applied Mathematics and Computation*, 209:2 (2009), 254-258.
7. S.Z. Zhou, Z.Y. Zou, G.H. Chen. Domain decomposition method for a system of quasivariational inequalities. *Acta Mathematicae Applicatae Sinica*, 25:1 (2009), 75-82.
8. S.Z. Zhou, Z.Y. Zou. A new iterative method for discrete HJB equations. *Numerische Mathematik*, 111:1 (2008), 159-167.
9. S.Z. Zhou, Z.Y. Zou. An iterative algorithm for a quasivariational inequality system related to HJB equation. *Journal of Computational and Applied Mathematics*, 219:1 (2008), 1-8.
10. S.Z. Zhou, M.R. Bai. On some feasibility conditions in MPEC. *Acta Mathematica Scientia*, 28:2 (2008), 289-294.
11. 周叔子, 史艳华. 抛物问题 Mortar 有限元的瀑布型多重网格法. 数值计算与计算机应用, 29:4 (2008), 241-250.
12. 周叔子, 李辉. 关于一类 HJ 方程的 Godunov 通量的一点注记. 应用数学, 21:1 (2008), 49-51.
13. D.X. Xie, S.Z. Zhou. A new minimization protocol for solving nonlinear Poisson-Boltzmann mortar finite element equation. *BIT*, 47:4 (2007), 853-871.
14. M.R. Bai, S.Z. Zhou, G.Y. Ni. On the generalized monotonicity of variational inequalities. *Computers & Mathematics with Applications*, 53:6 (2007), 910-917.
15. S.Z. Zhou, Z.Y. Zou. A relaxation scheme for Hamilton-Jacobi-Bellman equations. *Applied Mathematics and Computation*, 186:1 (2007), 806-813.

16. F. Long, Y.X. Wang, S.Z. Zhou. Existence and exponential stability of periodic solutions for a class of Cohen-Grossberg neural networks with bounded and unbounded delays. *Nonlinear Analysis Real World Applications*, 8:3 (2007), 797-810.
17. 周叔子, 孙佑兰. 关于 DC 函数最小化邻近点算法的注记. 应用数学学报, 30:2 (2007), 377-381.
18. 周叔子, 胡伯霞. 一类非对称变分不等式的非精确交替方向法. 湖南大学学报 (自然科学版), 34:4 (2007), 78-80.
19. M.R. Bai, S.Z. Zhou, G.Y. Ni. Variational-like inequalities with relaxed pseudomonotone mappings in Banach spaces. *Applied Mathematics Letters*, 19:6 (2006), 547-554.
20. 周叔子, 陈光华. 求解一类变分不等式离散问题的迭代法. 数值计算与计算机应用, 27:2 (2006), 81-85.
21. 王德华, 周叔子. 基于小波与多重网格方法的一类偏微分方程数值解. 21 (2006), 62-65.
22. X.J. Tong, S. Z. Zhou. A smoothing projected Newton-type method for semismooth equations with bound constraints. *Journal of Industrial and Management Optimization*, 1:2 (2005), 235-250.
23. S.Z. Zhou, D.H. Wang. Domain decomposition method for a parabolic variational inequality. *Applied Mathematics and Computation*, 166:1 (2005), 213-223.
24. 周叔子, 陈光华. 解离散 HJB 方程的一个单调迭代法. 应用数学, 18:4 (2005), 639-643.
25. 王奕宣, 周叔子. 对基于不同信度的两类奖惩系统的比较. 高等学校应用数学学报 B 辑 (英文版), 20:4 (2005), 407-415.
26. S.Z. Zhou, M.R. Bai. A new exceptional family of elements for a variational inequality problem on Hilbert space. *Applied Mathematics Letters*, 17:4 (2004), 423-428.
27. S.Z. Zhou, H.X. Hu. On the convergence of a cascadic multigrid method for semilinear elliptic problem. *Applied Mathematics and Computation*, 159:2 (2004), 407-417.
28. C.L. Li, J.P. Zeng, S.Z. Zhou. Convergence analysis of generalized Schwarz algorithms for solving obstacle problems with T-monotone operator. *Computers & Mathematics with Applications*, 48:3-4 (2004), 373-386.
29. X.J. Tong, S.Z. Zhou. A trust-region algorithm for nonlinear constrained optimization problem. *Journal of Mathematical Research and Exposition*, 24:3 (2004), 445-460.
30. 周叔子, 李荣军. 解半线性抛物问题的瀑布型多重网格法. 湖南大学学报 (自然科学版), 31:5 (2004), 104-105.
31. 白敏茹, 周叔子. 例外簇和变分不等式解的存在性. 湖南大学学报 (自然科学版), 31:2 (2004), 111-112.
32. 周叔子, 舒象改. 解抛物问题的一类新的瀑布型多重网格法. 应用数学, 17:3 (2004), 468-471.
33. S.Z. Zhou, W.P. Zhan. A new domain decomposition method for an HJB equation. *Journal of Computational and Applied Mathematics*, 159:1 (2003), 195-204.
34. X.J. Tong, S.Z. Zhou. Combining trust-region and line-search algorithms for minimization

subject to bounds. *Hokkaido Mathematical Journal*, 32:2 (2003), 355-369.

35. X.J. Tong, S.Z. Zhou. Combining trust region and line search methods for equality constrained optimization. *Numerical Functional Analysis and Optimization*, 24:1-2 (2003), 143-162.

36. X.J. Tong, S.Z. Zhou. A trust-region algorithm for nonlinear inequality constrained optimization. *Journal of Computational Mathematics*, 21:2 (2003), 207-220.

37. 童小娇, 周叔子. 等式与界约束非线性优化信赖域算法的全局收敛. 高等学校应用数学学报 B 辑 (英文版), 18:1 (2003), 83-94.

38. 万中, 周叔子. 非线性互补约束优化问题的可行性条件. 应用数学学报, 26:4 (2003), 646-651.

39. 周叔子, 曾金平, 单桂华. 一类非线性椭圆问题的 schwarz 算法. 计算数学, 25:2 (2003), 171-176.

40. G.Z. Gu, D.H. Li, L. Qi, S.Z. Zhou. Descent directions of quasi-Newton methods for symmetric nonlinear equations. *SIAM Journal on Numerical Analysis*, 40:5 (2002), 1763-1774.

41. J.P. Zeng, S.Z. Zhou. A domain decomposition method for a kind of optimization problems. *Journal of Computational and Applied Mathematics*, 146:1 (2002), 127-139.

42. J.P. Zeng, S.Z. Zhou. Block monotone iterative methods for elliptic variational inequalities. *Applied Mathematics and Computation*, 128:1 (2002), 109-127.

43. S.Z. Zhou, J.P. Zeng, W.P. Zhan. Monotonic iterative algorithms for an implicit two-sided obstacle problem. *Computers & Mathematics with Applications*, 43:1-2 (2002), 31-40.

44. 曾金平, 周叔子. 单障碍问题区域分解法的单调收敛性与收敛速度估计. 计算数学, 24:4 (2002), 395-404.

45. 童小娇, 周叔子. 等式与界约束非线性优化的信赖域增广 Lagrangian 算法. 计算数学, 24:1 (2002), 27-38.

46. 周叔子, 詹武平. 一类 Bellman 方程离散问题的解. 系统科学与数学, 22:4 (2002), 385-391.

47. 祝树金, 周叔子. 一类非线性椭圆问题的瀑布型多重网格法. 数学理论与应用, 22:1 (2002), 1-4.

48. 周叔子, 祝树金. 半线性问题的瀑布型多重网格法. 应用数学, 15:3 (2002), 136-139.

49. J.P. Zeng, S.Z. Zhou, L.H. Wang. Convergence rate analysis of multiplicative Schwarz algorithm for elliptic variational inequalities. *Journal of Systems Science and Complexity*, 14:3 (2001), 247-254.

50. S.Z. Zhou, W.P. Zhan, J.P. Zeng. Monotonic iterative algorithms for a quasicomplementarity problem. *Journal of Computational Mathematics*, 19:3 (2001), 293-298.

51. C.J. Zhang, S.Z. Zhou, X.X. Liao. D-convergence and GDN-stability of Runge-Kutta methods for a class of delay systems. *Applied Numerical Mathematics*, 37:1-2 (2001), 161-170.

52. X.J. Tong, S.Z. Zhou. Global convergence of a trust-region algorithm for inequality constrained optimization. *Hokkaido Mathematical Journal*, 30:1 (2001), 113-136.
53. 李郴良, 曾金平, 周叔子. 解含非线性源项的变分不等式问题的非重叠区域分解法. 计算数学, 23:1 (2001), 37-48.
54. 童小娇, 周叔子. 非线性等式与等式问题的信赖域算法. 数值计算与计算机应用, 22:1 (2001), 53-62.
55. 万中, 周叔子. MPEC 问题的带任意初值的一类算法. 湖南大学学报 (自然科学版), 28:2 (2001), 1-5.
56. Y.F. Yang, D.H. Li, S.Z. Zhou. A trust region method for a semismooth reformulation to variational inequality problems. *Optimization Methods and Software*, 14:1-2 (2000), 139-157.
57. 张诚坚, 廖晓昕, 周叔子. 变系数 MDDEs 系统的数值稳定性. 计算数学, 22:4 (2000), 409-416.
58. 童小娇, 周叔子. 非线性优化的非单调信赖域算法及全局收敛性. 高等学校应用数学学报 B 辑 (英文版), 15:2 (2000), 201-210.
59. 詹武平, 周叔子, 曾金平. 一类拟互补问题的迭代法. 应用数学学报, 23:4 (2000), 551-556.
60. 曾金平, 周叔子. 带非线性源项的变分不等式的区域分解法及其收敛速度分析. 应用数学学报, 23:2 (2000), 250-260.
61. 周叔子, 单桂华. 关于一类空间分解方法的收敛性. 湖南大学学报 (自然科学版), 27:2 (2000), 1-5.
62. 童小娇, 周叔子. 一类非线性规划问题的信赖域内点算法. 应用数学, 13:1 (2000), 70-74.
63. S.Z. Zhou, J.P. Zeng, X.M. Tang. Generalized Schwarz algorithm for obstacle problems. *Computers & Mathematics with Applications*, 38:7-8 (1999), 263-271.
64. C.J. Zhang, S.Z. Zhou. Stability analysis of LMMs for systems of neutral multidelay-differential equations. *Computers & Mathematics with Applications*, 38:3-4 (1999), 113-117.
65. 周叔子, 曾金平, 单桂华. On the convergence of a space cecomposition method for nonlinear pròblems. 数学进展, 28: 6 (1999), 541-542.
66. J.P. Zeng, S.Z. Zhou. On monotone and geometric convergence of Schwarz methods for two-sided obstacle problems. *SIAM Journal on Numerical Analysis*, 35:2 (1998), 600-616.
67. J.P. Zeng, S.Z. Zhou, Schwarz algorithm for the solution of variational inequalities with nonlinear source terms. *Applied Mathematics and Computation*, 97:1 (1998), 23-35.
68. D.H. Li, J.P. Zeng, S.Z. Zhou. Convergence of Broyden-like matrix. *Applied Mathematics Letters*, 11:5 (1998), 35-37.
69. C.J. Zhang, S.Z. Zhou. The asymptotic stability of theoretical and numerical solutions for systems of neutral multidelay-differential equations. *Science in China Series A: Mathematics*, 41:11 (1998), 1151-1157.
70. 张忠志, 周叔子. 利用离散正交性解一类二维抛物方程初值反问题. 数学杂志, 18:1 (1998),

63-70.

71. C.J. Zhang, S.Z. Zhou. Nonlinear stability and D-convergence of Runge-Kutta methods for delay differential equations. *Journal of Computational and Applied Mathematics*, 85:2 (1997), 225-237.
72. X.P. Yang, S.Z. Zhou, G.Y. Li. On an Axially Symmetric Elastic-Plastic Torsion Problem. *Applied Mathematics & Mechanics*, 18:7 (1997), 707-720.
73. 周叔子, 曾金平. 一类非线性算子障碍问题的 Schwarz 算法. 应用数学学报, 20:4 (1997), 521-530.
74. 周叔子, 丁立新. A parallel Schwarz algorithm for variational inequalities. *Chinese Science Bulletin*, 41:13 (1996), 1061-1064.
75. 杨余飞, 周叔子. 用 Uzawa 型算法解抛物方程右端反问题. 计算数学, 18:3 (1996), 269-278.
76. S. Larsson, V. Thomée, S.Z. Zhou. On multigrid methods for parabolic problems. *Journal of Computational Mathematics*, 13:3 (1995), 193-205.
77. S.Z. Zhou, D.H. Li, J.P. Zeng, A successive approximation quasi-Newton process for nonlinear complementarity problem. in: *Recent advances in nonsmooth optimization*, D-Z. Du, L. Qi, R. Womersley, eds., World Sci. Publishing, River Edge, NJ, 1995, 459-472.
78. 周叔子. 半线性抛物方程时间离散间断伽辽金方法的误差估计. 湖南大学学报, 22:1 (1995), 1-5,12.
79. J.P. Zeng, S.Z. Zhou. Two-side obstacle problem and its equivalent linear complementarity problem. *Chinese Science Bulletin*, 39 (1994), 1057-1061.
80. 周叔子, 文承标. 抛物问题非协调元多重网格法. 计算数学, 16:4 (1994), 372-381.
81. 周叔子, 陈永金. 求解一类 B 可微方程的阻尼牛顿法. 湖南大学学报, 21:2 (1994), 14-18, 27.
82. 周叔子, 冯钢. 重调和方程 Zienkiewicz 元逼近的多重网格法. 湖南大学学报, 20:2 (1993), 1-6.
83. S.Z. Zhou, Q.R. Yan. The Kantorovich theorem for nonlinear complementarity problems. *Chinese Science Bulletin*, 37:7 (1992), 529-533.
84. 周叔子, 冯钢. 变分不等式非协调元的收敛性. 湖南大学学报, 19:5（1992）, 6-14.
85. 周叔子, 李华夏. 非线性变分不等式的拟 Newton 法. 数值计算与计算机应用, 13:1 (1992), 51-57.
86. S.Z. Zhou. Perturbation for elliptic variational inequalities. *Science in China Ser A*, 34:6 (1991), 650-659.
87. 周叔子, 胡立辉. 双障碍问题的多重网格法. 数学理论与应用, 11:1-2 (1991), 20-30.
88. S.Z. Zhou. A direct method for the linear complementarity problem. *Journal of Computational Mathematics*, 8:2 (1990), 178-182.
89. 周叔子. 线性椭圆变分不等式的扰动. 湖南大学学报, 17:4 (1990), 10-14, 41.
90. 周叔子. 自由边界问题的样条有限元法. 计算数学, 11:2 (1989), 132-139.

91. 周叔子, 欧阳世芬. 奇异有限元的多重网格法. 高等学校计算数学学报, 11:4 (1989), 305-312.
92. S.Z. Zhou. On an axisymmetric free boundary problem. *Journal of Mathematical Analysis and Applications*, 121:2 (1987), 465-486.
93. 周叔子. 关于二步割线法的收敛阶. 湖南大学学报, 14:4 (1987), 99-103.
94. S.Z. Zhou. On the convergence of an algorithm of Uzawa's type for saddle-point problems. *Science Bulletin*, 31:20 (1986), 1375-1379.
95. 周叔子. 轴对称弹塑性扭转问题的数值解. 计算数学, 8:3 (1986), 242-250.
96. 周叔子. 关于鞍点问题 Uzawa 型算法的收敛性. 科学通报, 30:19 (1985), 1531-1534.
97. 周叔子. 一个退化的变分不等式的数值解. 高等学校计算数学学报, 6:1 (1984), 44-50.
98. S.Z. Zhou. The linear finite element method for a two-dimensional singular boundary value problem. *SIAM Journal on Numerical Analysis*, 20:5 (1983), 976-984.
99. C.W. Cryer, S.Z. Zhou. The solution of the free boundary problem for an axisymmetric partially penetrating well. *Annali di Matematica Pura ed Applicata*, 1: 135 (1983), 219-235.
100. 周叔子. 关于一个轴对称自由边界问题. 应用数学学报, 6:4 (1983), 420-432.
101. 周叔子, 齐东旭. 关于多结点 Hermite 样条. 湖南数学年刊, 1: 3 (1983),55-60.
102. 周叔子. 关于 $W^{m,p}$ 中一个稠密性定理的几点注记. 湖南大学学报, 9:3 (1982), 48-53.
103. 周叔子, 谭邦本. 关于有限单元半分析法的收敛性 (II). 湖南大学学报, 1 (1981), 51-59.
104. 周叔子, 谭邦本. 关于有限单元半分析法的收敛性 (I). 湖南大学学报, 2 (1979), 102-110.

后记

经过部分校友和学院全体同仁两年多的努力，周叔子教授的论文选顺利出版了. 在此，我们感谢为此付出不懈努力的各位校友和同事. 特别要感谢周叔子教授的学生们，他们提供了文选里所收集的论文原文并对相关文章进行了编辑、整理和校对.

编辑、整理数学文选是一件枯燥而烦琐的工作，需要花费大量的时间和精力. 文选里所收集的每一篇文章都需要逐字、逐句地输入原文，尤其是数学公式的输入，上标、下标多而繁杂，非常容易出错，需要极大的耐心和细致的工作；最后，必须对所有的原文逐篇进行仔细校对，尽最大的可能降低错漏. 所有这些困难都在全体同仁尤其是周叔子教授的学生们的不懈努力下得以克服.

特别感谢华南师范大学的李董辉教授和他的学生们、湖南大学白敏茹教授和她的学生们及湖北汽车工业学院的严钦容教授等，他们除了按期按质完成了自己参与论文的输入、编辑和校对工作外，还承担了周叔子教授单独署名的论文，以及周教授与海外学者合作论文的整理工作.

湖南大学的蒋月评教授、华南师范大学的李董辉教授在百忙之中抽出时间，全程负责论文的选取、收集、整理、编辑等一系列工作，在此，对他们的辛勤劳动和智慧结晶表示特别的敬意和由衷的谢意!

最后，感谢周叔子教授的夫人张碧纯老师及全家长期以来对论文选出版的关心和支持!

湖南大学数学学院